BASIC MATH & PRE-ALGEBRA

FOR

DUMMIES

A Wiley Brand

BY MARK ZEGARELLI

WILEY

U Can: Basic Math & Pre-Algebra For Dummies®

Published by:
John Wiley & Sons, Inc.,
111 River Street,
Hoboken, NJ 07030-5774,
www.wiley.com

For general information on our other products and services, please contact our Customer Care Department within the U.S. at 877-762-2974, outside the U.S. at 317-572-3993, or fax 317-572-4002. For technical support, please visit www.wiley.com/techsupport.

Wiley publishes in a variety of print and electronic formats and by print-on-demand. Some material included with standard print versions of this book may not be included in e-books or in print-on-demand. If this book refers to media such as a CD or DVD that is not included in the version you purchased, you may download this material at http://booksupport.wiley.com. For more information about Wiley products, visit www.wiley.com.

Library of Congress Control Number: 2014956254

ISBN 978-1-119-06796-2 (pbk); ISBN 978-1-119-06795-5 (ebk); ISBN 978-1-119-06794-8 (ebk)

Manufactured in the United States of America

10 9 8 7 6 5 4 3 2 1

CONTENTS AT A GLANCE

TABLE OF CONTENTS

INTRODUCTION

Does math really need to be so hard? Nope.

I say this speaking as a guy who has struggled with math as much as, if not more than, you have. Believe me. And a big part of the struggle often has more to do with the lack of clarity itself in how math is explained than with the math itself.

This is too bad, because the whole idea behind math is supposed to be *clarity*. In a world where so many things are unclear, 2 + 2 will always equal 4.

My second-greatest joy in teaching math is when a light breaks across a student's face as he or she suddenly understands something new. My greatest joy, though, is what often follows: a skepticism that it couldn't possibly be this easy.

When you approach math right, it's almost always easier than you think. And a lot of the stuff that hung you up when you first saw it probably isn't all that scary after all. Lots of students feel they got lost somewhere along the way on the road between learning to count to ten and their first day in an algebra class — and this may be true whether you're 14 or 104. If this is you, don't worry. You're not alone, and help is right here!

ABOUT THIS BOOK

This book brings together the three components you need to make sense of math:

- Clear explanations of each topic
- Example questions with step-by-step answers
- Plenty of practice problems (with more available online!)

Although you can certainly work through this book from beginning to end, you don't have to. Feel free to jump directly to whatever chapter has the type of problems you want to practice. When you've worked through enough problems in a section to your satisfaction, feel free to jump to a different section. If you find the problems in a section too difficult, flip back to an earlier section or chapter to practice the skills you need — just follow the cross-references.

FOOLISH ASSUMPTIONS

If you're planning to read this book, you likely fall into one of these categories:

- A student who wants a solid understanding of the basics of math for a class or test you're taking
- An adult who wants to improve skills in arithmetic, fractions, decimals, percentages, weights and measures, geometry, algebra, and so on for when you have to use math in the real world
- Someone who wants a refresher so you can help another person understand math

My only assumption about your skill level is that you can add, subtract, multiply, and divide. So, to find out whether you're ready for this book, take this simple test:

$$4 + 7 = \underline{\hspace{1cm}} \qquad 13 - 5 = \underline{\hspace{1cm}} \qquad 9 \times 3 = \underline{\hspace{1cm}} \qquad 35 \div 7 = \underline{\hspace{1cm}}$$

If you can answer these four questions correctly (the answers are 11, 8, 27, and 5), you're ready to begin.

BEYOND THE BOOK

In addition to the book you're reading right now, be sure to check out the free Cheat Sheet at www.dummies.com/cheatsheet/ucanbasicmathprealgebra for a set of quick reference notes on everything from the order of operations to basic algebra conventions to working with radicals and more.

In addition, at www.dummies.com/extras/ucanbasicmathprealgebra you can find free articles to extend your understanding of a variety of topics, such as probability, exponents, prime factorization, percent problems, repeating decimals, and more.

The online practice that comes free with this book offers 1,001 questions and answers that allow you to gain more practice with basic math and pre-algebra concepts. The beauty of the online questions is that you can customize your online practice to focus on the topic areas that give you the most trouble. Or if you're short on time but want to get a mixed bag of a limited number of questions, you can specify the number of questions you want to practice. Whether you practice a few hundred questions in one sitting or a couple dozen, and whether you focus on a few types of questions or practice every type, the online program keeps track of the questions you get right and wrong so you can monitor your progress and spend time studying exactly what you need.

To gain access to the online practice, all you have to do is register. Just follow these simple steps:

1. **Find your PIN code.**

   - **Print-book users:** If you purchased a hard copy of this book, turn to the front of this book to find your PIN.

   - **E-book users:** If you purchased this book as an e-book, you can get your PIN by registering your e-book at www.dummies.com/go/getaccess. Simply select your book from the drop-down menu, fill in your personal information, and then answer the security question to verify your purchase. You'll then receive an email with your PIN.

2. **Go to** http://onlinepractice.dummies.com.

3. **Enter your PIN.**

4. **Follow the instructions to create an account and establish your own login information.**

Now you're ready to go! You can come back to the online program as often as you want — simply log on with the username and password you created during your initial login. No need to enter the PIN a second time.

Tip: If you have trouble with your PIN or can't find it, contact Wiley Product Technical Support at 877-762-2974 or go to `http://wiley.custhelp.com`.

WHERE TO GO FROM HERE

You can use this book in a variety of ways. If you're reading without immediate time pressure from a test or homework assignment, start at the beginning and keep going, chapter by chapter, to the end. If you do this, you'll be surprised by how much of the math you may have been dreading will be almost easy. Additionally, setting up some solid groundwork is a great way to prepare for what follows later in the book.

If your time is limited — especially if you're taking a math course and you're looking for help with your homework or an upcoming test — skip directly to the topic you're studying. Wherever you open the book, you can find a clear explanation of the topic at hand, as well as a variety of hints and tricks. Read through the examples and try to do them yourself, or use them as templates to help you with assigned problems.

PART I

getting started
with

**BASIC MATH
AND PRE-ALGEBRA**

WEB EXTRA *Visit www.dummies.com for great Dummies content online.*

IN THIS PART . . .

- Identify four important sets of numbers: counting numbers, integers, rational numbers, and real numbers.

- Use place value to write numbers of any size.

- Round numbers to make calculating quicker.

1

PLAYING THE NUMBERS GAME

ne useful characteristic about numbers is that they're *conceptual,* which means that, in an important sense, they're all in your head. (This fact probably won't get you out of having to know about them, though — nice try!)

For example, you can picture three of anything: three cats, three baseballs, three cannibals, three planets. But just try to picture the concept of three all by itself, and you find it's impossible. Oh, sure, you can picture the numeral 3, but *threeness* itself — much like love or beauty or honor — is beyond direct understanding. But when you understand the *concept* of three (or four, or a million), you have access to an incredibly powerful system for understanding the world: mathematics.

In this chapter, I give you a brief history of how numbers came into being. I discuss a few common *number sequences* and show you how these connect with simple math *operations* like addition, subtraction, multiplication, and division.

After that, I describe how some of these ideas come together with a simple yet powerful tool: the *number line.* I discuss how numbers are arranged on the number line, and I also show you how to use the number line as a calculator for simple arithmetic. Finally, I describe how the *counting numbers* (1, 2, 3, . . .) sparked the invention of more unusual types of numbers, such as *negative numbers, fractions,* and *irrational numbers.* I also show you how these *sets of numbers* are *nested* — that is, how one set of numbers fits inside another, which fits inside another.

INVENTING NUMBERS

Historians believe that the first number systems came into being at the same time as agriculture and commerce. Before that, people in prehistoric, hunter-gatherer societies were pretty much content to identify bunches of things as "a lot" or "a little."

But as farming developed and trade between communities began, more precision was needed. So people began using stones, clay tokens, and similar objects to keep track of their goats, sheep, oil, grain, or whatever commodity they had. They exchanged these tokens for the objects they represented in a one-to-one exchange.

Eventually, traders realized that they could draw pictures instead of using tokens. Those pictures evolved into tally marks and, in time, into more complex systems. Whether they realized it or not, their attempts to keep track of commodities led these early humans to invent something entirely new: *numbers*.

Throughout the ages, the Babylonians, Egyptians, Greeks, Romans, Mayans, Arabs, and Chinese (to name just a few) all developed their own systems of writing numbers.

Although Roman numerals gained wide currency as the Roman Empire expanded throughout Europe and parts of Asia and Africa, the more advanced system that the Arabs invented turned out to be more useful. Our own number system, the Hindu–Arabic numbers (also called decimal numbers), is closely derived from these early Arabic numbers.

UNDERSTANDING NUMBER SEQUENCES

Although humans invented numbers for counting commodities, as I explain in the preceding section, they soon put them to use in a wide range of applications. Numbers were useful for measuring distances, counting money, amassing an army, levying taxes, building pyramids, and lots more.

But beyond their many uses for understanding the external world, numbers have an internal order all their own. So numbers are not only an *invention,* but equally a *discovery:* a landscape that seems to exist independently, with its own structure, mysteries, and even perils.

One path into this new and often strange world is the *number sequence:* an arrangement of numbers according to a rule. In the following sections, I introduce you to a variety of number sequences that are useful for making sense of numbers.

Evening the odds

One of the first facts you probably heard about numbers is that all of them are either even or odd. For example, you can split an even number of marbles *evenly* into two equal piles. But when you try to divide an odd number of marbles the same way, you always have one *odd,* leftover marble. Here are the first few even numbers:

$$2 \quad 4 \quad 6 \quad 8 \quad 10 \quad 12 \quad 14 \quad 16\ldots$$

You can easily keep the sequence of even numbers going as long as you like. Starting with the number 2, keep adding 2 to get the next number.

Similarly, here are the first few odd numbers:

$$1 \quad 3 \quad 5 \quad 7 \quad 9 \quad 11 \quad 13 \quad 15\ldots$$

The sequence of odd numbers is just as simple to generate. Starting with the number 1, keep adding 2 to get the next number.

Patterns of even or odd numbers are the simplest number patterns around, which is why kids often figure out the difference between even and odd numbers soon after learning to count.

Counting by threes, fours, fives, and so on

When you get used to the concept of counting by numbers greater than 1, you can run with it. For example, here's what counting by threes, fours, and fives looks like:

Threes:	3	6	9	12	15	18	21	24 …
Fours:	4	8	12	16	20	24	28	32 …
Fives:	5	10	15	20	25	30	35	40 …

Tip: Counting by a given number is a good way to begin learning the multiplication table for that number, especially for the numbers you're kind of sketchy on. (In general, people seem to have the most trouble multiplying by 7, but 8 and 9 are also unpopular.) In Chapter 3, I show you a few tricks for memorizing the multiplication table once and for all.

These types of sequences are also useful for understanding factors and multiples, which you get a look at in Chapter 7.

Getting square with square numbers

When you study math, sooner or later, you probably want to use visual aids to help you see what numbers are telling you. (Later in this book, I show you how one picture can be worth a thousand numbers when I discuss geometry in Chapter 15 and graphing in Chapter 16.)

The tastiest visual aids you'll ever find are those little square cheese-flavored crackers. (You probably have a box sitting somewhere in the pantry. If not, saltine crackers or any other square food works just as well.) Shake a bunch out of a box and place the little squares together to make bigger squares. Figure 1-1 shows the first few.

Figure 1-1:
Square
numbers.

© John Wiley & Sons, Inc.

Voilà! The square numbers:

$$1 \quad 4 \quad 9 \quad 16 \quad 25 \quad 36 \quad 49 \quad 64 \ldots$$

Tip: You get a *square number* by multiplying a number by itself, so knowing the square numbers is another handy way to remember part of the multiplication table. Although you probably remember without help that $2 \times 2 = 4$ you may be sketchy on some of the higher numbers, such as $7 \times 7 = 49$. Knowing the square numbers gives you another way to etch that multiplication table forever into your brain, as I show you in Chapter 3.

Square numbers are also a great first step on the way to understanding exponents, which I introduce later in this chapter and explain in more detail in Chapter 3.

Composing yourself with composite numbers

Some numbers can be placed in rectangular patterns. Mathematicians probably should call numbers like these "rectangular numbers," but instead they chose the term *composite numbers*. For example, 12 is a composite number because you can place 12 objects in rectangles of two different shapes, as in Figure 1-2.

Figure 1-2: The number 12 laid out in two rectangular patterns.

© John Wiley & Sons, Inc.

As with square numbers, arranging numbers in visual patterns like this tells you something about how multiplication works. In this case, by counting the sides of both rectangles, you find out the following:

$$3 \times 4 = 12$$
$$2 \times 6 = 12$$

Similarly, other numbers such as 8 and 15 can also be arranged in rectangles, as in Figure 1-3.

Figure 1-3: Composite numbers, such as 8 and 15, can form rectangles.

© John Wiley & Sons, Inc.

As you can see, both these numbers are quite happy being placed in boxes with at least two rows and two columns. And these visual patterns show this:

$$2 \times 4 = 8$$
$$3 \times 5 = 15$$

The word *composite* means that these numbers are *composed of* smaller numbers. For example, the number 15 is composed of 3 and 5 — that is, when you multiply these two smaller numbers, you get 15. Here are all the composite numbers from 1 to 16:

4   6   8   9   10   12   14   15   16

Notice that all the square numbers (see "Getting square with square numbers") also count as composite numbers because you can arrange them in boxes with at least two rows and two columns. Additionally, a lot of other non-square numbers are also composite numbers.

Stepping out of the box with prime numbers

Some numbers are stubborn. Like certain people you may know, these numbers — called *prime numbers* — resist being placed in any sort of a box. Look at how Figure 1-4 depicts the number 13, for example.

Figure 1-4: Unlucky 13, a prime example of a number that refuses to fit in a box.

© John Wiley & Sons, Inc.

Try as you may, you just can't make a rectangle out of 13 objects. (That fact may be one reason the number 13 got a bad reputation as unlucky.) Here are all the prime numbers less than 20:

2   3   5   7   11   13   17   19

As you can see, the list of prime numbers fills the gaps left by the composite numbers (see the preceding section). Therefore, every counting number is either prime or composite. The only exception is the number 1, which is neither prime nor composite. In Chapter 7, I give you a lot more information about prime numbers and show you how to *decompose* a number — that is, break down a composite number into its prime factors.

Multiplying quickly with exponents

Here's an old question whose answer may surprise you: Suppose you took a job that paid you just 1 penny the first day, 2 pennies the second day, 4 pennies the third day, and so on, doubling the amount every day, like this:

1   2   4   8   16   32   64   128   256   512 . . .

As you can see, in the first ten days of work, you would've earned a little more than $10 (actually, $10.23 — but who's counting?). How much would you earn in 30 days? Your answer may well be, "I wouldn't take a lousy job like that in the first place." At first glance, this looks like a good answer, but here's a glimpse at your second ten days' earnings:

$$\ldots 1{,}024 \quad 2{,}048 \quad 4{,}096 \quad 8{,}192 \quad 16{,}384 \quad 32{,}768 \quad 65{,}536 \quad 131{,}072 \quad 262{,}144 \quad 524{,}288 \ldots$$

By the end of the second 10 days, your total earnings would be over $10,000. And by the end of 30 days, your earnings would top out around $10,000,000! How does this happen? Through the magic of exponents (also called *powers*). Each new number in the sequence is obtained by multiplying the previous number by 2:

$$2^1 = 2$$
$$2^2 = 2 \times 2 = 4$$
$$2^3 = 2 \times 2 \times 2 = 8$$
$$2^4 = 2 \times 2 \times 2 \times 2 = 16$$

As you can see, the notation 2^4 means *multiply 2 by itself 4 times.*

You can use exponents on numbers other than 2. Here's another sequence you may be familiar with:

$$1 \quad 10 \quad 100 \quad 1{,}000 \quad 10{,}000 \quad 100{,}000 \quad 1{,}000{,}000 \ldots$$

In this sequence, every number is 10 times greater than the number before it. You can also generate these numbers using exponents:

$$10^1 = 10$$
$$10^2 = 10 \times 10 = 100$$
$$10^3 = 10 \times 10 \times 10 = 1{,}000$$
$$10^4 = 10 \times 10 \times 10 \times 10 = 10{,}000$$

This sequence is important for defining *place value,* the basis of the decimal number system, which I discuss in Chapter 2. It also shows up when I discuss decimals in Chapter 10 and scientific notation in Chapter 13. You find out more about exponents in Chapter 3.

FOUR IMPORTANT SETS OF NUMBERS

In the preceding section, you see how the number line grows in both the positive and negative directions and fills in with a lot of numbers in between. In this section, I provide a quick tour of how numbers fit together as a set of nested systems, one inside the other.

When I talk about a set of numbers, I'm really just talking about a group of numbers. You can use the number line to deal with four important sets of numbers:

- **Counting numbers (also called natural numbers):** The set of numbers beginning 1, 2, 3, 4 . . . and going on infinitely

- **Integers:** The set of counting numbers, zero, and negative counting numbers

- **Rational numbers:** The set of integers and fractions
- **Real numbers:** The set of rational and irrational numbers

The sets of counting numbers, integers, rational, and real numbers are nested, one inside another. This nesting of one set inside another is similar to the way that a city (for example, Boston) is inside a state (Massachusetts), which is inside a country (the United States), which is inside a continent (North America). The set of counting numbers is inside the set of integers, which is inside the set of rational numbers, which is inside the set of real numbers.

Counting on the counting numbers

The set of *counting numbers* is the set of numbers you first count with, starting with 1. Because they seem to arise naturally from observing the world, they're also called the *natural numbers:*

$$1 \quad 2 \quad 3 \quad 4 \quad 5 \quad 6 \quad 7 \quad 8 \quad 9 \ldots$$

The counting numbers are infinite, which means they go on forever.

When you add two counting numbers, the answer is always another counting number. Similarly, when you multiply two counting numbers, the answer is always a counting number. Another way of saying this is that the set of counting numbers is *closed* under both addition and multiplication.

Introducing integers

The set of *integers* arises when you try to subtract a larger number from a smaller one. For example, $4 - 6 = -2$. The set of integers includes the following:

- The counting numbers
- Zero
- The negative counting numbers

Here's a partial list of the integers:

$$\ldots -4 \quad -3 \quad -2 \quad -1 \quad 0 \quad 1 \quad 2 \quad 3 \quad 4 \ldots$$

Like the counting numbers, the integers are closed under addition and multiplication. Similarly, when you subtract one integer from another, the answer is always an integer. That is, the integers are also closed under subtraction.

Staying rational

Here's the set of *rational numbers:*

- Integers
  - Counting numbers
  - Zero
  - Negative counting numbers
- Fractions

Like the integers, the rational numbers are closed under addition, subtraction, and multiplication. Furthermore, when you divide one rational number by another, the answer is always a rational number. Another way to say this is that the rational numbers are closed under division.

Getting real

Even if you filled in all the rational numbers, you'd still have points left unlabeled on the number line. These points are the irrational numbers.

An *irrational number* is a number that's neither a whole number nor a fraction. In fact, an irrational number can only be approximated as a *non-repeating decimal.* In other words, no matter how many decimal places you write down, you can always write down more; furthermore, the digits in this decimal never become repetitive or fall into any pattern. (For more on repeating decimals, see Chapter 10.)

The most famous irrational number is π (you find out more about π when I discuss the geometry of circles in Chapter 15):

$$\pi = 3.14159265358979323846264338327950288419716939937510\ldots$$

Together, the rational and irrational numbers make up the *real numbers,* which comprise every point on the number line. In this book, I don't spend too much time on irrational numbers, but just remember that they're there for future reference.

2

IT'S ALL IN THE FINGERS: NUMBERS AND DIGITS

IN THIS CHAPTER

- Understanding how place value turns digits into numbers
- Distinguishing whether zeros are important placeholders or meaningless leading zeros
- Reading and writing long numbers
- Understanding how to round numbers and estimate values

When you're counting, ten seems to be a natural stopping point — a nice, round number. The fact that our ten fingers match up so nicely with numbers may seem like a happy accident. But of course, it's no accident at all. Fingers were the first calculator that humans possessed. Our number system — Hindu-Arabic numbers — is based on the number ten because humans have 10 fingers instead of 8 or 12. In fact, the very word *digit* has two meanings: numerical symbol and finger.

In this chapter, I show you how place value turns digits into numbers. I also show you when 0 is an important placeholder in a number and why leading zeros don't change the value of a number. And I show you how to read and write long numbers. After that, I discuss two important skills: rounding numbers and estimating values.

Telling the difference between numbers and digits

Sometimes people confuse numbers and digits. For the record, here's the difference:

- A digit is a single numerical symbol, from 0 to 9.
- A number is a string of one or more digits.

For example, 7 is both a digit and a number. In fact, it's a one-digit number. However, 15 is a string of two digits, so it's a number — a two-digit number. And 426 is a three-digit number. You get the idea.

In a sense, a digit is like a letter of the alphabet. By themselves, the uses of 26 letters, A through Z, are limited. (How much can you do with a single letter such as K or W?) Only when you begin using strings of letters as building blocks to spell words does the power of letters become apparent. Similarly, the ten digits, 0 through 9, have limited usefulness until you begin building strings of digits — that is, numbers.

KNOWING YOUR PLACE VALUE

The number system you're most familiar with — Hindu-Arabic numbers — has ten familiar digits:

0   1   2   3   4   5   6   7   8   9

Yet with only ten digits, you can express numbers as high as you care to go. In this section, I show you how it happens.

Counting to ten and beyond

The ten digits in our number system allow you to count from 0 to 9. All higher numbers are produced using place value. Place value assigns a digit a greater or lesser value, depending on where it appears in a number. Each place in a number is ten times greater than the place to its immediate right.

To understand how a whole number gets its value, suppose you write the number 45,019 all the way to the right in Table 2-1, one digit per cell, and add up the numbers you get.

Table 2-1   45,019 Displayed in a Place-Value Chart

Millions			Thousands			Ones		
Hundred Millions	Ten Millions	Millions	Hundred Thousands	Ten Thousands	Thousands	Hundreds	Tens	Ones
				4	5	0	1	9

You have 4 ten thousands, 5 thousands, 0 hundreds, 1 ten, and 9 ones. The chart shows you that the number breaks down as follows:

$$45,019 = 40,000 + 5,000 + 0 + 10 + 9$$

In this example, notice that the presence of the digit 0 in the hundreds place means that zero hundreds are added to the number.

Telling placeholders from leading zeros

Although the digit 0 adds no value to a number, it acts as a placeholder to keep the other digits in their proper places. For example, the number 5,001,000 breaks down into 5,000,000 + 1,000. Suppose, however, you decide to leave all the 0s out of the chart. Table 2-2 shows what you'd get.

Table 2-2   5,001,000 Displayed Incorrectly without Placeholding Zeros

Millions			Thousands			Ones		
Hundred Millions	Ten Millions	Millions	Hundred Thousands	Ten Thousands	Thousands	Hundreds	Tens	Ones
							5	1

The chart tells you that 5,001,000 = 50 + 1. Clearly, this answer is wrong!

Remember: As a rule, when a 0 appears to the right of at least one digit other than 0, it's a placeholder. Placeholding zeros are important — always include them when you write a number. However, when a 0 appears to the left of every digit other than 0, it's a leading zero. Leading zeros serve no purpose in a number, so dropping them is customary. For example, place the number 003,040,070 on the chart (see Table 2-3).

Table 2-3   3,040,070 Displayed with Two Leading Zeros

Millions			Thousands			Ones		
Hundred Millions	Ten Millions	Millions	Hundred Thousands	Ten Thousands	Thousands	Hundreds	Tens	Ones
0	0	3	0	4	0	0	7	0

The first two 0s in the number are leading zeros because they appear to the left of the 3. You can drop these 0s from the number, leaving you with 3,040,070. The remaining 0s are all to the right of the 3, so they're placeholders — be sure to write them in.

Reading long numbers

When you write a long number, you use commas to separate groups of three numbers. For example, here's about as long a number as you'll ever see:

$$234,845,021,349,230,467,304$$

Table 2-4 shows a larger version of the place-value chart.

Table 2-4   A Place-Value Chart Separated by Commas

Quintillions	Quadrillions	Trillions	Billions	Millions	Thousands	Ones
234	845	021	349	230	467	304

This version of the chart helps you read the number. Begin all the way to the left and read, "Two hundred thirty-four quintillion, eight hundred forty-five quadrillion, twenty-one trillion, three hundred forty-nine billion, two hundred thirty million, four hundred sixty-seven thousand, three hundred four."

Remember: When you read and write whole numbers, don't say the word and. In math, the word *and* means you have a decimal point. That's why, when you write a check, you save the word *and* for the number of cents, which is usually expressed as a decimal or sometimes as a fraction. (I discuss decimals in Chapter 11.)

CLOSE ENOUGH FOR ROCK 'N' ROLL: ROUNDING AND ESTIMATING

As numbers get longer, calculations become tedious, and you're more likely to make a mistake or just give up. When you're working with long numbers, simplifying your work by rounding numbers and estimating values is sometimes helpful.

When you round a number, you change some of its digits to placeholding zeros. And when you estimate a value, you work with rounded numbers to find an approximate answer to a problem. In this section, you build both skills.

Rounding numbers

Rounding numbers makes long numbers easier to work with. In this section, I show you how to round numbers to the nearest ten, hundred, thousand, and beyond.

Rounding numbers to the nearest ten

The simplest kind of rounding you can do is with two-digit numbers. When you round a two-digit number to the nearest ten, you simply bring it up or down to the nearest number that ends in 0. For example,

$$39 \rightarrow 40 \qquad 51 \rightarrow 50 \qquad 73 \rightarrow 70$$

Even though numbers ending in 5 are in the middle, always round them up to the next-highest number that ends in 0:

$$15 \rightarrow 20 \qquad 35 \rightarrow 40 \qquad 85 \rightarrow 90a$$

Numbers in the upper 90s get rounded up to 100:

$$99 \rightarrow 100 \qquad 95 \rightarrow 100 \qquad 94 \rightarrow 90$$

When you know how to round a two-digit number, you can round just about any number. For example, to round most longer numbers to the nearest ten, just focus on the ones and tens digits:

$$734 \rightarrow 730 \qquad 1,488 \rightarrow 1,490 \qquad 12,345 \rightarrow 12,350$$

Occasionally, a small change to the ones and tens digits affects the other digits. (This situation is a lot like when the odometer in your car rolls a bunch of 9s over to 0s.) For example:

$$899 \to 900 \qquad 1{,}097 \to 1{,}100 \qquad 9{,}995 \to 10{,}000$$

Rounding numbers to the nearest hundred and beyond

To round numbers to the nearest hundred, thousand, or beyond, focus only on two digits: the digit in the place you're rounding to and the digit to its immediate right. Change all other digits to the right of these two digits to 0s. For example, suppose you want to round 642 to the nearest hundred. Focus on the hundreds digit (6) and the digit to its immediate right (4):

$$642$$

I've underlined these two digits. Now just round these two digits as if you were rounding to the nearest ten, and change the digit to the right of them to a 0:

$$\underline{64}2 \to 600$$

Here are a few more examples of rounding numbers to the nearest hundred:

$$7{,}\underline{89}1 \to 7{,}900 \qquad 15{,}\underline{75}3 \to 15{,}800 \qquad 99{,}\underline{96}1 \to 100{,}000$$

When rounding numbers to the nearest thousand, underline the thousands digit and the digit to its immediate right. Round the number by focusing only on the two underlined digits and, when you're done, change all digits to the right of these to 0s:

$$\underline{4{,}9}84 \to 5{,}000 \qquad 78{,}\underline{5}21 \to 79{,}000 \qquad 1{,}099{,}\underline{3}04 \to 1{,}099{,}000$$

Even when rounding to the nearest million, the same rules apply:

$$1{,}\underline{2}34{,}567 \to 1{,}000{,}000 \qquad 78{,}\underline{8}83{,}958 \to 79{,}000{,}000$$

Estimating value to make problems easier

When you know how to round numbers, you can use this skill in estimating values. Estimating saves you time by allowing you to avoid complicated computations and still get an approximate answer to a problem.

Remember: When you get an approximate answer, you don't use an equals sign; instead, you use this wavy symbol, which means is *approximately equal to:* $\approx$.

Suppose you want to add these numbers: $722 + 506 + 383 + 1{,}279 + 91 + 811$. This computation is tedious, and you may make a mistake. But you can make the addition easier by first rounding all the numbers to the nearest hundred and then adding:

$$\approx 700 + 500 + 400 + 1{,}300 + 100 + 800 = 3{,}800$$

The approximate answer is 3,800. This answer isn't far off from the exact answer, which is 3,792.

PART II

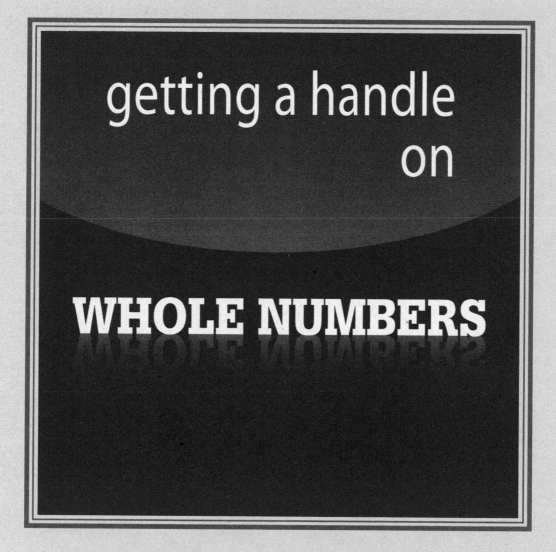

getting a handle on

WHOLE NUMBERS

WEB EXTRA *You can use prime factorization to find the greatest common factor (GCF) of difficult numbers. To find out how, go to www.dummies.com/extras/ucanbasicmathprealgebra.*

IN THIS PART . . .

- Add, subtract, multiply, and divide more complex calculations involving negative numbers, inequalities, exponents, square roots, and absolute value.

- Build equations and evaluate expressions.

- Understand arithmetic word problems.

- Employ a few quick tricks to determine whether one number is divisible by another.

- Find the factors and multiples of a number, and discover whether a number is prime or composite.

- Calculate the greatest common factor (GCF) and the least common multiple (LCM) of a set of numbers.

3
PUTTING THE BIG FOUR OPERATIONS TO WORK

IN THIS CHAPTER

- Working with the Big Four operations (addition, subtraction, multiplication, and division)
- Identifying which operations are inverses of each other
- Knowing the operations that are commutative, associative, and distributive
- Performing the Big Four operations on negative numbers
- Using four symbols for inequality
- Understanding exponents, roots, and absolute values

When you understand the Big Four operations — adding, subtracting, multiplying, and dividing — you can begin to look at math on a whole new level. In this chapter, you extend your understanding of the Big Four operations and move beyond them. I begin by focusing on four important properties of the Big Four operations: inverse operations, commutative operations, associative operations, and distribution. Then I show you how to perform the Big Four on negative numbers.

I continue by introducing you to some important symbols for inequality. Finally, you're ready to move beyond the Big Four by discovering three more advanced operations: exponents (also called *powers*), square roots (also called *radicals*), and absolute values.

THE BIG FOUR OPERATIONS

When most folks think of math, the first thing that comes to mind is four little (or not-so-little) words: addition, subtraction, multiplication, and division. I call these operations the *Big Four* all through the book.

In this chapter, I introduce you (or reintroduce you) to these little gems. Although I assume you're already familiar with the Big Four, this chapter reviews these operations, taking you from what you may have missed to what you need to succeed as you move onward and upward in math.

Adding things up: Addition

Addition is the first operation you find out about, and it's almost everybody's favorite. It's simple, friendly, and straightforward. No matter how much you worry about math, you've probably never lost a minute of sleep over addition. Addition is all about bringing things together, which is a positive goal. For example, suppose you and I are standing in line to buy tickets for a movie. I have $25 and you have only $5. I could lord it over you and make you feel crummy that I can go to the movies and you can't. Or instead, you and I can join forces, adding together my $25 and your $5 to make $30. Now, not only can we both see the movie, but we may even be able to buy some popcorn, too.

Addition uses only one sign — the plus sign (+): Your equation may read $2 + 3 = 5$, or $12 + 2 = 14$, or $27 + 44 = 71$, but the plus sign always means the same thing.

Remember: When you add two numbers together, those two numbers are called *addends*, and the result is called the *sum*. So in the first example, the addends are 2 and 3, and the sum is 5.

Take it away: Subtraction

Subtraction is usually the second operation you discover, and it's not much harder than addition. Still, there's something negative about subtraction — it's all about who has more and who has less. Suppose you and I have been running on treadmills at the gym. I'm happy because I ran 3 miles, but then you start bragging that you ran 10 miles. You subtract and tell me that I should be very impressed that you ran 7 miles farther than I did. (But with an attitude like that, don't be surprised if you come back from the showers to find your running shoes filled with liquid soap!)

As with addition, subtraction has only one sign: the minus sign (−). You end up with equations such as $4 − 1 = 3$, and $14 − 13 = 1$, and $93 − 74 = 19$.

Remember: When you subtract one number from another, the result is called the *difference*. This term makes sense when you think about it: When you subtract, you find the difference between a higher number and a lower one.

One of the first facts you probably heard about subtraction is that you can't take away more than you start with. In that case, the second number can't be larger than the first. And if the two numbers are the same, the result is always 0. For example, $3 − 3 = 0$; $11 − 11 = 0$; and $1,776 − 1,776 = 0$. Later someone breaks the news that you *can* take away more than you have. When you do, though, you need to place a minus sign in front of the difference to show that you have a negative number, a number below 0:

$$4 − 5 = −1$$
$$10 − 13 = −3$$
$$88 − 99 = −11$$

Tip: When subtracting a larger number from a smaller number, remember the words *switch* and *negate*: You *switch* the order of the two numbers and do the subtraction as you normally would, but at the end, you *negate* the result by attaching a minus sign. For example, to find $10 − 13$, you switch the order of these two numbers, giving you $13 − 10$, which equals 3; then you negate this result to get −3. That's why $10 − 13 = −3$.

Warning: The minus sign does double duty, so don't get confused. When you stick a minus sign between two numbers, it means the first number minus the second number. But when you attach it to the front of a number, it means that this number is a negative number.

A sign of the times: Multiplying

Multiplication is often described as a sort of shorthand for repeated addition. For example,

4×3 means add 4 to itself 3 times: $4 + 4 + 4 = 12$

9×6 means add 9 to itself 6 times: $9 + 9 + 9 + 9 + 9 + 9 = 54$

100×2 means add 100 to itself 2 times: $100 + 100 = 200$

Although multiplication isn't as warm and fuzzy as addition, it's a great timesaver. For example, suppose you coach a Little League baseball team, and you've just won a game against the toughest team in the league. As a reward, you promised to buy three hot dogs for each of the nine players on the team. To find out how many hot dogs you need, you can add 3 together 9 times. Or you can save time by multiplying 3 times 9, which gives you 27. Therefore, you need 27 hot dogs (plus a whole lot of mustard and sauerkraut).

Remember: When you multiply two numbers, the two numbers that you're multiplying are called *factors,* and the result is the *product.*

When you're first introduced to multiplication, you use the times sign ($\times$). As you move onward and upward on your math journey, you need to be aware of the conventions I discuss in the following sections.

Remember: The symbol $\cdot$ is sometimes used to replace the symbol $\times$. For example,

$4 \cdot 2 = 8$ means $4 \times 2 = 8$

$6 \cdot 7 = 42$ means $6 \times 7 = 42$

$53 \cdot 11 = 583$ means $53 \times 11 = 583$

In Parts I through IV of this book, I stick to the tried-and-true symbol $\times$ for multiplication. Just be aware that the symbol $\cdot$ exists so that you won't be stumped if your teacher or textbook uses it.

Remember: In math beyond arithmetic, using parentheses without another operator stands for multiplication. The parentheses can enclose the first number, the second number, or both numbers. For example,

$3(5) = 15$ means $3 \times 5 = 15$

$(8)7 = 56$ means $8 \times 7 = 56$

$(9)(10) = 90$ means $9 \times 10 = 90$

This switch makes sense when you stop to consider that the letter *x*, which is often used in algebra, looks a lot like the multiplication sign ×. So in this book, when I start using *x* in Part V, I also stop using × and begin using parentheses without another sign to indicate multiplication.

Doing division lickety-split

The last of the Big Four operations is division. Division literally means splitting things up. For example, suppose you're a parent on a picnic with your three children. You've brought along 12 pretzel sticks as snacks, and want to split them fairly so that each child gets the same number (don't want to cause a fight, right?).

Each child gets four pretzel sticks. This problem tells you that

$$12 \div 3 = 4$$

As with multiplication, division also has more than one sign: the division sign (÷) and the fraction slash (/) or fraction bar (—). So some other ways to write the same information are

$$12/3 = 4 \quad \text{and} \quad \frac{12}{3} = 4$$

Whichever way you write it, the idea is the same: When you divide 12 pretzel sticks equally among three people, each person gets 4 of them.

Remember: When you divide one number by another, the first number is called the *dividend,* the second is called the *divisor,* and the result is the *quotient.* For example, in the division from the earlier example, the dividend is 12, the divisor is 3, and the quotient is 4.

SWITCHING THINGS UP WITH INVERSE OPERATIONS AND THE COMMUTATIVE PROPERTY

The Big Four operations are actually two pairs of *inverse operations,* which means the operations can undo each other:

- **Addition and subtraction:** Subtraction undoes addition. For example, if you start with 3 and add 4, you get 7. Then, when you subtract 4, you undo the original addition and arrive back at 3:

$$3 + 4 = 7 \rightarrow 7 - 4 = 3$$

- This idea of inverse operations makes a lot of sense when you look at the number line. On a number line, 3 + 4 means *start at 3, up 4.* And 7 – 4 means *start at 7, down 4.* So when you add 4 and then subtract 4, you end up back where you started.

- **Multiplication and division:** Division undoes multiplication. For example, if you start with 6 and multiply by 2, you get 12. Then, when you divide by 2, you undo the original multiplication and arrive back at 6:

$$6 \times 2 = 12 \rightarrow 12 \div 2 = 6$$

Remember: The *commutative property of addition* tells you that you can change the order of the numbers in an addition problem without changing the result, and the *commutative property of multiplication* says you can change the order of the numbers in a multiplication problem without changing the result. For example,

$$2+5=7 \rightarrow 5+2=7$$
$$3 \times 4 = 12 \rightarrow 4 \times 3 = 12$$

Tip: In contrast, subtraction and division are *non-commutative* operations. When you switch the order of the numbers, the result changes.

Here's an example of how subtraction is non-commutative:

$$6-4=2 \quad \text{but} \quad 4-6=-2$$

Subtraction is non-commutative, so if you have \$6 and spend \$4, the result is *not* the same as if you have \$4 and spend \$6. In the first case, you still have \$2 left over. In the second case, you *owe* \$2. In other words, switching the numbers turns the result into a negative number. (I discuss negative numbers later in this chapter.)

And here's an example of how division is non-commutative:

$$5 \div 2 = 2r1 \quad \text{but} \quad 2 \div 5 = 0r2$$

For example, when you have five dog biscuits to divide between two dogs, each dog gets two biscuits and you have one biscuit left over. But when you switch the numbers and try to divide two biscuits among five dogs, you don't have enough biscuits to go around, so each dog gets none and you have two left over.

Through the commutative property and inverse operations, every equation has four alternative forms that contain the same information expressed in slightly different ways. For example, 2 + 3 = 5 and 3 + 2 = 5 are alternative forms of the same equation but tweaked using the commutative property. And 5 – 3 = 2 is the inverse of 2 + 3 = 5. Finally, 5 – 2 = 3 is the inverse of 3 + 2 = 5.

Tip: You can use alternative forms of equations to solve fill-in-the-blank problems. As long as you know two numbers in an equation, you can always find the remaining number. Just figure out a way to get the blank to the other side of the equal sign:

- When the *first* number is missing in any problem, use the inverse to turn the problem around:

$$\underline{\quad} + 6 = 10 \rightarrow 10 - 6 = \underline{\quad}$$

- When the *second* number is missing in an addition or multiplication problem, use the commutative property and then the inverse:

$$9 + \underline{\quad} = 17 \rightarrow \underline{\quad} + 9 = 17 \rightarrow 17 - 9 = \underline{\quad}$$

- When the *second* number is missing in a subtraction or multiplication problem, just switch around the two values that are next to the equal sign (that is, the blank and the equal sign):

$$15 - \underline{\quad} = 8 \rightarrow 15 - 8 = \underline{\quad}$$

Examples

Q. What's the inverse equation to $16 - 9 = 7$?

A. **7 + 9 = 16.** In the equation $16 - 9 = 7$, you start at 16 and subtract 9, which brings you to 7. The inverse equation undoes this process, so you start at 7 and add 9, which brings you back to 16:

$$16 - 9 = 7 \rightarrow 7 + 9 = 16$$

Q. What's the inverse equation to $6 \times 7 = 42$?

A. **42 ÷ 7 = 6.** In the equation $6 \times 7 = 42$, you start at 6 and multiply by 7, which brings you to 42. The inverse equation undoes this process, so you start at 42 and divide by 7, which brings you back to 6:

$$6 \times 7 = 42 \rightarrow 42 \div 7 = 6$$

Q. Use inverse operations and the commutative property to find three alternative forms of the equation $7 - 2 = 5$.

A. **5 + 2 = 7, 2 + 5 = 7, and 7 – 5 = 2.** First, use inverse operations to change subtraction to addition:

$$7 - 2 = 5 \rightarrow 5 + 2 = 7$$

Now use the commutative property to change the order of this addition:

$$5 + 2 = 7 \rightarrow 2 + 5 = 7$$

Finally, use inverse operations to change addition to subtraction:

$$2 + 5 = 7 \rightarrow 7 - 5 = 2$$

Q. Fill in the blank: _____ ÷ 3 = 13.

A. **39.** Use inverse operations to turn the problem from division to multiplication:

$$\underline{\quad} \div 3 = 13 \rightarrow 13 \times 3 = \underline{\quad}$$

Now you can solve the problem by multiplying $13 \times 3 = 39$.

Q. Solve this problem by filling in the blank: $16 +$ _____ $= 47$.

A. **31.** First, use the commutative property to reverse the addition:

$$16 + \underline{\quad} = 47 \rightarrow \underline{\quad} + 16 = 47$$

Now use inverse operations to change the problem from addition to subtraction:

$$\underline{\quad} + 16 = 47 \rightarrow 47 - 16 = \underline{\quad}$$

At this point, you can solve the problem by subtracting $47 - 16 = 31$.

Q. Fill in the blank: $64 - \underline{\quad} = 15$.

A. **49.** Switch around the last two numbers in the problem:

$$64 - \underline{\quad} = 15 \rightarrow 64 - 15 = \underline{\quad}$$

Now you can solve the problem by subtracting $64 - 15 = 49$.

Practice Questions

1. Using inverse operations, write down an alternative form of each equation:

    a. $8 + 9 = 17$

    b. $23 - 13 = 10$

    c. $15 \times 5 = 75$

    d. $132 \div 11 = 12$

2. Use the commutative property to write down an alternative form of each equation:

    a. $19 + 35 = 54$

    b. $175 + 88 = 263$

    c. $22 \times 8 = 176$

    d. $101 \times 99 = 9{,}999$

3. Use inverse operations and the commutative property to find all three alternative forms for each equation:

    a. $7 + 3 = 10$

    b. $12 - 4 = 8$

    c. $6 \times 5 = 30$

    d. $18 \div 2 = 9$

4. Fill in the blanks in each question:

   a. ____ − 74 = 36

   b. ____ × 7 = 105

   c. 45 + ____ = 132

   d. 273 − ____ = 70

   e. 8 × ____ = 648

   f. 180 ÷ ____ = 9

Practice Answers

1. a. 17 − 9 = 8

   b. 10 + 13 = 23

   c. 75 ÷ 5 = 15

   d. 12 × 11 = 132

2. a. 35 + 19 = 54

   b. 88 + 175 = 263

   c. 8 × 22 = 176

   d. 99 × 101 = 9,999

3. a. 10 − 3 = 7, 3 + 7 = 10, and 10 − 7 = 3

   b. 8 + 4 = 12, 4 + 8 = 12, and 12 − 8 = 4

   c. 30 ÷ 5 = 6, 5 × 6 = 30, and 30 ÷ 6 = 5

   d. 9 × 2 = 18, 2 × 9 = 18, 18 ÷ 9 = 2

4. **a.** **110.** Rewrite ____ – 74 = 36 as its inverse: 36 + 74 = ____. Therefore, 36 + 74 = 110.

  b. **15.** Rewrite ____ × 7 = 105 as its inverse: 105 ÷ 7 = ____. So, 105 ÷ 7 = 15.

  c. **87.** Rewrite 45 + ____ = 132 using the commutative property: ____ + 45 = 132.
  Now rewrite this equation as its inverse: 132 – 45 = ____. Therefore, 132 – 45 = 87.

  d. **203.** Rewrite 273 – ____ = 70 by switching around the two numbers next to the
  equal sign: 273 – 70 = ____. So, 273 – 70 = 203.

  e. **81.** Rewrite 8 × ____ = 648 using the commutative property: ____ × 8 = 648.
  Now rewrite this equation as its inverse: 648 ÷ 8 = ____. So, 648 ÷ 8 = 81.

  f. **20.** Rewrite 180 ÷ ____ = 9 by switching around the two numbers next to the
  equal sign: 180 ÷ 9 = ____. So, 180 ÷ 9 = 20.

GETTING WITH THE IN-GROUP: PARENTHESES AND THE ASSOCIATIVE PROPERTY

Remember: Parentheses group operations together, telling you to do any operations inside a set of parentheses *before* you do operations outside of it. Parentheses can make a big difference in the result you get when solving a problem, especially in a problem with mixed operations. In two important cases, however, moving parentheses doesn't change the answer to a problem:

- The *associative property of addition* says that when every operation is addition, you can group numbers however you like and choose which pair of numbers to add first; you can move parentheses without changing the answer.

- The *associative property of multiplication* says you can choose which pair of numbers to multiply first, so when every operation is multiplication, you can move parentheses without changing the answer.

In contrast, subtraction and division are non-associative operations. This means that grouping them in different ways changes the result.

Warning: Don't confuse the commutative property with the associative property. The commutative property tells you that it's okay *to switch* two numbers that you're adding or multiplying. The associative property tells you that it's okay to *regroup* three (or more) numbers using parentheses.

Tip: Taken together, the associative property and the commutative property (which I discuss in the preceding section) allow you to completely rearrange all the numbers in any problem that's either all addition or all multiplication.

Examples

Q. What's $(21 - 6) \div 3$? What's $21 - (6 \div 3)$?

A. **5 and 19.** To calculate $(21 - 6) \div 3$, first do the operation inside the parentheses — that is, $21 - 6 = 15$:

$$(21-6) \div 3 = 15 \div 3$$

Now finish the problem by dividing: $15 \div 3 = 5$.

To solve $21 - (6 \div 3)$, first do the operation inside the parentheses — that is, $6 \div 3 = 2$:

$$21-(6 \div 3) = 21 - 2$$

Finish up by subtracting $21 - 2 = 19$. Notice that the placement of the parentheses changes the answer.

Q. Solve $1 + (9 + 2)$ and $(1 + 9) + 2$.

A. **12 and 12.** To solve $1 + (9 + 2)$, first do the operation inside the parentheses — that is, $9 + 2 = 11$:

$$1+(9+2) = 1 + 11$$

Finish up by adding $1 + 11 = 12$.

To solve $(1 + 9) + 2$, first do the operation inside the parentheses — that is, $1 + 9 = 10$:

$$(1+9)+2 = 10 + 2$$

Finish up by adding $10 + 2 = 12$. Notice that the only difference between the two problems is the placement of the parentheses, but because both operations are addition, moving the parentheses doesn't change the answer.

Q. Solve $2 \times (4 \times 3)$ and $(2 \times 4) \times 3$.

A. **24 and 24.** To solve $2 \times (4 \times 3)$, first do the operation inside the parentheses — that is, $4 \times 3 = 12$:

$$2 \times (4 \times 3) = 2 \times 12$$

Finish by multiplying $2 \times 12 = 24$.

To solve $(2 \times 4) \times 3$, first do the operation inside the parentheses — that is, $2 \times 4 = 8$:

$$(2 \times 4) \times 3 = 8 \times 3$$

Finish by multiplying $8 \times 3 = 24$. No matter how you group the multiplication, the answer is the same.

Q.   Solve 41 × 5 × 2.

A.   **410.** The last two numbers are small, so place parentheses around these numbers:

$$41 \times 5 \times 2 = 41 \times (5 \times 2)$$

First, do the multiplication inside the parentheses:

$$41 \times (5 \times 2) = 41 \times 10$$

Now you can easily multiply 41 × 10 = 410.

Practice Questions

1.   Find the value of (8 × 6) + 10.

2.   Find the value of 123 ÷ (145 − 144).

3.   Solve the following two problems:

   a.   (40 ÷ 2) + 6 = ?

   b.   40 ÷ (2 + 6) = ?

   Do the parentheses make a difference in the answers?

4.   Solve the following two problems:

   a.   (16 + 24) + 19

   b.   16 + (24 + 19)

   Do the parentheses make a difference in the answers?

5.   Solve the following two problems:

   a.   (18 × 25) × 4

   b.   18 × (25 × 4)

   Do the parentheses make a difference in the answers?

6.   Find the value of 93,769 × 2 × 5. (**Hint:** Use the associative property for multiplication to make the problem easier.)

Practice Answers

1. **58.** First, do the multiplication inside the parentheses:

$$(8 \times 6) + 10 = 48 + 10$$

   Now add: 48 + 10 = 58.

2. **123.** First, do the subtraction inside the parentheses:

$$123 \div (145 - 144) = 123 \div 1$$

   Now simply divide 123 ÷ 1 = 123.

3. **a.** **26.** (40 ÷ 2) + 6 = 20 + 6 = 26.

    b. **5.** 40 ÷ (2 + 6) = 40 ÷ 8 = 5.

    Yes, the placement of parentheses changes the result.

4. **a.** **59.** (16 + 24) + 19 = 40 + 19 = 59.

    b. **59.** 16 + (24 + 19) = 16 + 43 = 59.

    No, because of the associative property of addition, the placement of parentheses doesn't change the result.

5. **a.** **1,800.** (18 × 25) × 4 = 450 × 4 = 1,800.

    b. **1,800.** 18 × (25 × 4) = 18 × 100 = 1,800.

    No, because of the associative property of multiplication, the placement of parentheses doesn't change the result.

6. **937,690.** The problem is easiest to solve by placing parentheses around 2 × 5:

$$93,769 \times (2 \times 5) = 93,769 \times 10 = 937,690$$

Distribution to lighten the load

If you've ever tried to carry a heavy bag of groceries, you may have found that distributing the contents into two smaller bags is helpful. This same concept also works for multiplication.

In math, *distribution* (also called the *distributive property of multiplication over addition*) allows you to split a large multiplication problem into two smaller ones and add the results to get the answer.

For example, suppose you want to multiply these two numbers:

$$17 \times 101$$

You can go ahead and just multiply them, but distribution provides a different way to think about the problem that you may find easier. Because 101 = 100 + 1, you can split this problem into two easier problems, as follows:

$$= 17 \times (100 + 1)$$
$$= (17 \times 100) + (17 \times 1)$$

You take the number outside the parentheses, multiply it by each number inside the parentheses one at a time, and then add the products. At this point, you may be able to calculate the two multiplications in your head and then add them up easily:

$$= 1,700 + 17 = 1,717$$

Distribution becomes even more useful when you get to algebra in Part V.

Example

Q.   Multiply 82 × 4 using distribution:

$$82 \times 4 = (80 \times 4) + (2 \times 4) = \underline{\quad} + \underline{\quad} = \underline{\quad}$$

A.   **328.** 82 × 4 = (80 × 4) + (2 × 4) = 320 + 8 = 328

Practice Questions

1.   Multiply 53 × 3 using distribution:

$$53 \times 3 = (50 \times 3) + (3 \times 3) = \underline{\quad} + \underline{\quad} = \underline{\quad}$$

2.   Multiply 71 × 6 using distribution:

$$71 \times 6 = (70 \times 6) + (1 \times 6) = \underline{\quad} + \underline{\quad} = \underline{\quad}$$

3.   Multiply 95 × 7 using distribution:

$$95 \times 7 = (90 \times 7) + (5 \times 7) = \underline{\quad} + \underline{\quad} = \underline{\quad}$$

4.   Multiply 116 × 9 using distribution:

$$116 \times 9 = (110 \times 9) + (6 \times 9) = \underline{\quad} + \underline{\quad} = \underline{\quad}$$

Practice Answers

1. **159.** $53 \times 3 = (50 \times 3) + (3 \times 3) = 150 + 9 = 159$

2. **426.** $71 \times 6 = (70 \times 6) + (1 \times 6) = 420 + 6 = 426$

3. **665.** $95 \times 7 = (90 \times 7) + (5 \times 7) = 630 + 35 = 665$

4. **1,044.** $116 \times 9 = (110 \times 9) + (6 \times 9) = 990 + 54 = 1,044$

UNDERSTANDING WHERE NEGATIVE NUMBERS COME FROM

When you first discovered subtraction, you were probably told that you can't take a small number minus a greater number. For example, if you start with four marbles, you can't subtract six because you can't take away more marbles than you have. This rule is true for marbles, but in other situations, you *can* subtract a big number from a small one.

In real-world applications, negative numbers represent debt. For example, if you have only five chairs to sell but a customer pays for eight of them, you owe her three more chairs. Even though you may have trouble picturing –3 chairs, you still need to account for this debt, and negative numbers are the right tool for the job.

As another example, if you have $4 and you buy something that costs $6, you end up with less than $0 dollars — that is, –$2, which means a debt of $2.

A number with a minus sign in front of it, like –2, is called a *negative number*. You call the number –2 either *negative two* or *minus two*. Negative numbers appear on the number line to the left of 0, as shown in Figure 3-1.

Figure 3-1:
Negative numbers on the number line.

Remember: When you don't have a number line to work with, here's a simple rule for subtracting a small number minus a big number: Switch the two numbers around and take the big number minus the small one; then attach a negative sign to the result.

Examples

Q. Use the number line to subtract 5 – 8.

A. **–3.** On the number line, 5 – 8 means

Q. What is 11 – 19?

A. **–8.** Because 11 is less than 19, subtract 19 – 11, which equals 8, and attach a minus sign to the result. Therefore, 11 – 19 = –8.

Practice Questions

1. Using the number line, subtract the following numbers:

   a. 1 – 4 = _____

   b. 3 – 7 = _____

   c. 6 – 8 = _____

   d. 7 – 14 = _____

2. Find the answers to the following subtraction problems:

   a. 15 – 22 = _____

   b. 27 – 41 = _____

   c. 89 – 133 = _____

   d. 1,000 – 1,234 = _____

Practice Answers

1. **a.** **–3.** Start at 1, down 4.

   b. **–4.** Start at 3, down 7.

   c. **–2.** Start at 6, down 8.

   d. **–7.** Start at 7, down 14.

2. **a.** **–7.** Fifteen is less than 22, so subtract 22 – 15 = 7 and attach a minus sign to the result: –7.

   b. **–14.** Twenty-seven is less than 41, so subtract 41 – 27 = 14 and attach a minus sign to the result: –14.

   c. **–44.** Eighty-nine is less than 133, so subtract 133 – 89 = 44 and attach a minus sign to the result: –44.

   d. **–234.** One thousand is less than 1,234, so subtract 1,234 – 1,000 = 234 and attach a minus sign to the result: –234.

SIGN-SWITCHING: UNDERSTANDING NEGATION AND ABSOLUTE VALUE

Remember: When you attach a minus sign to any number, you *negate* that number. Negating a number means changing its sign to the opposite sign, so

- Attaching a minus sign to a positive number makes it negative.

- Attaching a minus sign to a negative number makes it positive. The two *adjacent* (side-by-side) minus signs cancel each other out.

- Attaching a minus sign to 0 doesn't change its value, so –0 = 0.

In contrast to negation, placing two bars around a number gives you the absolute value of that number. *Absolute value* is the number's distance from 0 on the number line — that is, it's the positive value of a number, regardless of whether you started out with a negative or positive number:

- The absolute value of a positive number is the same number.

- The absolute value of a negative number makes it a positive number.

- Placing absolute value bars around 0 doesn't change its value, so $|0| = 0$.

- Placing a minus sign outside absolute value bars gives you a negative result — for example, $-|6| = -6$, and $-|-6| = -6$.

Examples

Q. Negate the number 7.

A. **–7.** Negate 7 by attaching a negative sign to it: –7.

Q. Find the negation of –3.

A. **3.** The negation of –3 is – (–3). The two adjacent minus signs cancel out, which gives you 3.

Q. What's the negation of 7 – 12?

A. **5.** First do the subtraction, which tells you 7 – 12 = –5. To find the negation of –5, attach a minus sign to the answer: –(–5). The two *adjacent* minus signs cancel out, which gives you 5.

Q. What number does |9| equal?

A. **9.** The number 9 is already positive, so the absolute value of 9 is also 9.

Q. What number does |–17| equal?

A. **17.** Because –17 is negative, the absolute value of –17 is 17.

Q. Solve this absolute value problem: –|9 – 13| = ?

A. **–4.** Do the subtraction first: 9 – 13 = –4, which is negative, so the absolute value of –4 is 4. But the minus sign on the left (outside the absolute value bars in the original expression) negates this result, so the answer is –4.

Practice Questions

1. Negate each of the following numbers and expressions by attaching a minus sign and then canceling out minus signs when possible:

    a. 6

    b. –29

    c. 0

    d. 10 + 4

    e. 15 – 7

    f. 9 – 10

2. Solve the following absolute value problems:

a. $|7| = ?$

b. $|-11| = ?$

c. $|3 + 15| = ?$

d. $-|10 - 1| = ?$

e. $|1 - 10| = ?$

f. $|0| = ?$

Practice Answers

1. **a.** **–6.** To negate 6, attach a minus sign: –6.

b. **29.** To negate –29, attach a minus sign: –(–29). Now cancel adjacent minus signs: –(–29) = 29.

c. **0.** Zero is its own negation.

d. **–14.** Add first: 10 + 4 = 14, and the negation of 14 is –14.

e. **–8.** Subtract first: 15 – 7 = 8, and the negation of 8 is –8.

f. **1.** Begin by subtracting: 9 – 10 = –1, and the negation of –1 is 1.

2. **a.** **7.** Seven is already positive, so the absolute value of 7 is also 7.

b. **11.** The number –11 is negative, so the absolute value of –11 is 11.

c. **18.** First, do the addition inside the absolute value bars: 3 + 15 = 18, which is positive. The absolute value of 18 is 18.

d. **–9.** First do the subtraction: 10 – 1 = 9, which is positive. The absolute value of 9 is 9. You have a negative sign outside the absolute value bars, so negate your answer to get –9.

e. **9.** Begin by subtracting: 1 – 10 = –9, which is negative. The absolute value of –9 is 9.

f. **0.** The absolute value of 0 is 0.

ADDITION AND SUBTRACTION WITH NEGATIVE NUMBERS

The great secret to adding and subtracting negative numbers is to turn every problem into a series of ups and downs on the number line. When you know how to do this, you find that all these problems are quite simple.

So in this section, I explain how to add and subtract negative numbers on the number line. Don't worry about memorizing every little bit of this procedure. Instead, just follow along so you get a sense of how negative numbers fit onto the number line. (If you need a quick refresher on how the number line works, see Chapter 1.)

Starting with a negative number

When you're adding and subtracting on the number line, starting with a negative number isn't much different from starting with a positive number. For example, suppose you want to calculate –3 + 4. Using the up and down rules, you start at –3 and go up 4:

So –3 + 4 = 1.

Similarly, suppose you want to calculate –2 – 5. Again, the up and down rules help you out. You're subtracting, so move to the left: start at –2, down 5:

So –2 – 5 = –7.

Adding a negative number

Suppose you want to calculate –2 + (–4). You already know to start at –2, but where do you go from there? Here's the up and down rule for adding a negative number:

Remember: Adding a negative number is the same as subtracting a positive number — go *down* on the number line.

By this rule, –2 + (–4) is the same as –2 – 4, so start at –2, down 4:

So –2 + (–4) = –6.

Tip: If you rewrite a subtraction problem as an addition problem — for instance, rewriting 3 – 7 as 3 + (–7) — you can use the commutative and associative properties of addition, which I discuss earlier in this chapter. Just remember to keep the negative sign attached to the number when you rearrange: (–7) + 3.

Subtracting a negative number

The last rule you need to know is how to subtract a negative number. For example, suppose you want to calculate 2 – (–3). Here's the up and down rule:

Remember: Subtracting a negative number is the same as adding a positive number — go *up* on the number line.

This rule tells you that 2 – (–3) is the same as 2 + 3, so start at 2, up 3:

So 2 – (–3) = 5.

Tip: When subtracting negative numbers, you can think of the two minus signs canceling each other out to create a positive.

Examples

Q. Use the number line to add –3 + 5.

A. **2.** On the number line, –3 + 5 means *start at –3, up 5,* which brings you to 2:

Q. Use the number line to add 6 + (–2).

A. **4.** On the number line, 6 + (–2) means *start at 6, down 2,* which brings you to 4:

Q. Use the number line to add –3 + (–4).

A. **–7.** On the number line, –3 + (–4) means *start at –3, down 4,* which brings you to –7:

Q. Add –23 + 39.

A. **16.** Switch around the two numbers with the signs attached:

$$-23 + 39 = +39 - 23$$

Now you can drop the plus sign and use the minus sign for subtraction:

$$39 - 23 = 16$$

Q. Use the number line to subtract –1 – 4.

A. **–5.** On the number line, –1 – 4 means start at –1, down 4, which brings you to –5:

Q. Use the number line to subtract –5 – (–3).

A. **–2.** The expression –5 – (–3) means start at –5, up 3 (because the two minus signs cancel each other out), which brings you to –2:

Practice Questions

1. Use the number line to solve the following addition problems:

   a. $-5 + 6$

   b. $-1 + (-7)$

   c. $4 + (-6)$

   d. $-3 + 9$

   e. $2 + (-1)$

   f. $-4 + (-4)$

2. Solve the following addition problems without using the number line:

   a. $-17 + 35$

   b. $29 + (-38)$

   c. $-61 + (-18)$

   d. $70 + (-63)$

   e. $-112 + 84$

   f. $-215 + (-322)$

3. Use the number line to solve the following subtraction problems:

   a. $-3 - 4$

   b. $5 - (-3)$

   c. $-1 - (-8)$

   d. $-2 - 4$

   e. $-4 - 2$

   f. $-6 - (-10)$

4. Solve the following subtraction problems without using the number line:

   a. $17 - (-26)$

   b. $-21 - 45$

   c. $-42 - (-88)$

   d. $-67 - 91$

   e. $75 - (-49)$

   f. $-150 - (-79)$

A

Practice Answers

1. **a.** **1.** Start at -5, go up 6.

   b. **-8.** Start at -1, go down 7.

   c. **-2.** Start at 4, go down 6.

   d. **6.** Start at -3, go up 9.

   e. **1.** Start at 2, go down 1.

   f. **-8.** Start at -4, go down 4.

2. **a.** **18.** Switch around the numbers (with their signs) to turn the problem into subtraction:

$$-17 + 35 = 35 - 17 = 18$$

   b. **-9.** Drop the plus sign to turn the problem into subtraction:

$$29 + (-38) = 29 - 38 = -9$$

   c. **-79.** Drop the signs, add the numbers, and negate the result:

$$61 + 18 = 79, \text{ so } -61 + (-18) = -79$$

   d. **7.** Turn the problem into subtraction:

$$70 + -63 = 70 - 63 = 7$$

   e. **-28.** Turn the problem into subtraction:

$$-112 + 84 = 84 - 112 = -28$$

   f. **-537.** Drop the signs, add the numbers, and negate the result:

$$215 + 322 = 537, \text{ so } -215 + -322 = -537$$

3. **a.** **–7.** Start at –3, down 4.

   b. **8.** Start at 5, up 3.

   c. **7.** Start at –1, up 8.

   d. **–6.** Start at –2, down 4.

   e. **–6.** Start at –4, down 2.

   f. **4.** Start at –6, up 10.

4. **a.** **43.** Cancel the adjacent minus signs to turn the problem into addition:

$$17 - (-26) = 17 + 26 = 43$$

   b. **–66.** Drop the signs, add the numbers, and negate the result:

$$21 + 45 = 66, \text{ so } -21 - 45 = -66$$

   c. **46.** Cancel the adjacent minus signs to turn the problem into addition:

$$-42 - (-88) = -42 + 88$$

Now switch around the numbers (with their signs) to turn the problem back into subtraction:

$$88 - 42 = 46$$

   d. **–158.** Drop the signs, add the numbers, and negate the result:

$$67 + 91 = 158, \text{ so } -67 - 91 = -158$$

   e. **124.** Cancel the adjacent minus signs to turn the problem into addition:

$$75 - (-49) = 75 + 49 = 124$$

   f. **–71.** Cancel the adjacent minus signs to turn the problem into addition:

$$-150 - (-79) = -150 + 79$$

Now switch around the numbers (with their signs) to turn the problem back into subtraction:

$$79 - 150 = -71$$

KNOWING SIGNS OF THE TIMES (AND DIVISION) FOR NEGATIVE NUMBERS

Multiplication and division with negative numbers is virtually the same as with positive numbers. The presence of one or more minus signs (–) doesn't change the numerical part of the answer. The only question is whether the sign is positive or negative:

Remember: Just remember that when you multiply or divide two numbers,

- If the numbers have the *same sign,* the result is always positive.
- If the numbers have *opposite signs,* the result is always negative.

For example,

$$2 \times 3 = 6 \quad 2 \times -3 = -6$$
$$-2 \times -3 = 6 \quad -2 \times 3 = -6$$

As you can see, the numerical portion of the answer is always 6. The only question is whether the complete answer is 6 or –6. That's where the rule of same or opposite signs comes in.

Tip: Another way of thinking of this rule is that the two negatives cancel each other out to make a positive.

Similarly, look at these four division equations:

$$10 \div 2 = 5 \quad 10 \div -2 = -5$$
$$-10 \div -2 = 5 \quad -10 \div 2 = -5$$

In this case, the numerical portion of the answer is always 5. When the signs are the same, the result is positive, and when the signs are different, the result is negative.

Examples

Q.   Solve the following four multiplication problems:

a. $5 \times 6 = \underline{\quad}$

b. $-5 \times 6 = \underline{\quad}$

c. $5 \times (-6) = \underline{\quad}$

d. $-5 \times (-6) = \underline{\quad}$

A.   As you can see from this example, the numerical part of the answer (30) doesn't change. Only the sign of the answer changes, depending on the signs of the two numbers in the problem.

a. $5 \times 6 = 30$

b. $-5 \times 6 = (-30)$

c. $5 \times -6 = (-30)$

d. $-5 \times -6 = 30$

Q. Solve the following four division problems:

    a. $18 \div 3 =$ _____

    b. $-18 \div 3 =$ _____

    c. $18 \div (-3) =$ _____

    d. $-18 \div (-3) =$ _____

A. The numerical part of the answer (6) doesn't change. Only the sign of the answer changes, depending on the signs of the two numbers in the problem.

    a. $18 \div 3 = 6$

    b. $-18 \div 3 = -6$

    c. $18 \div (-3) = -6$

    d. $-18 \div (-3) = 6$

Q. What is -84×21?

A. **$-1,764$.** First, drop the signs and multiply:

$$84 \times 21 = 1,764$$

The numbers -84 and 21 have different signs, so the answer is negative: $-1,764$.

Q. What is $-580 \div (-20)$?

A. **29.** Drop the signs and divide:

$$580 \div 20 = 29$$

The numbers -580 and -20 have the same signs, so the answer is positive: 29.

Practice Questions

    1. Solve the following multiplication problems:

        a. $7 \times 11 =$ _____

        b. $-7 \times 11 =$ _____

        c. $7 \times (-11) =$ _____

        d. $-7 \times (-11) =$ _____

2. Solve the following division problems:

    a. $32 \div (-8) =$ _____

    b. $-32 \div (-8) =$ _____

    c. $-32 \div 8 =$ _____

    d. $32 \div 8 =$ _____

3. What is -65×23?

4. Find $-143 \times (-77)$.

5. Calculate $216 \div (-9)$.

6. What is $-3,375 \div (-25)$?

Practice Answers

1.   **a.** $7 \times 11 = \mathbf{77}$

    b. $-7 \times 11 = \mathbf{-77}$

    c. $7 \times (-11) = \mathbf{-77}$

    d. $-7 \times (-11) = \mathbf{77}$

2.   **a.** $32 \div (-8) = \mathbf{-4}$

    b. $-32 \div (-8) = \mathbf{4}$

    c. $-32 \div 8 = \mathbf{-4}$

    d. $32 \div 8 = \mathbf{4}$

3. **–1,495.** First, drop the signs and multiply:

$$65 \times 23 = 1,495$$

The numbers –65 and 23 have different signs, so the answer is negative: –1,495.

4. **11,011.** Drop the signs and multiply:

$$143 \times 77 = 11,011$$

The numbers –143 and –77 have the same sign, so the answer is positive: 11,011.

5.   **–24.** Drop the signs and divide (use long division, as I show you in Chapter 1):

$$216 \div 9 = 24$$

The numbers 216 and –9 have different signs, so the answer is negative: –24.

6.   **135.** First, drop the signs and divide:

$$3,375 \div 25 = 135$$

The numbers –3,375 and –25 have the same sign, so the answer is positive: 135.

UNDERSTANDING INEQUALITIES

Sometimes you want to talk about when two quantities are different. These statements are called *inequalities*. In this section, I discuss six types of inequalities: ≠ (doesn't equal), < (less than), > (greater than), ≤ (less than or equal to), ≥ (greater than or equal to), and ≈ (approximately equals).

Doesn't equal (≠)

The simplest inequality is ≠, which you use when two quantities are not equal. For example,

$$2 + 2 \neq 5$$
$$3 \times 4 \neq 34$$
$$999,9000 \neq 1,000,000$$

You can read ≠ as "doesn't equal" or "is not equal to." Therefore, read $2 + 2 \neq 5$ as "two plus two doesn't equal five."

Less than (<) and greater than (>)

The symbol < means *less than*. For example, the following statements are true:

$$4 < 5$$
$$100 < 1,000$$
$$2 + 2 < 5$$

Similarly, the symbol > means *greater than*. For example,

$$5 > 4$$
$$100 > 99$$
$$2 + 2 > 3$$

Tip: The two symbols < and > are similar and easily confused. Here are two simple ways to remember which is which:

- Notice that the < looks sort of like an *L*. This *L* should remind you that it means *less than*.

- Remember that, in any true statement, the *large* open mouth of the symbol is on the side of the *greater* amount, and the *small* point is on the side of the *lesser* amount.

Less than or equal to (≤) and greater than or equal to (≥)

The symbol ≤ means *less than or equal to*. For example, the following statements are true:

$$100 \leq 1,000$$
$$2+2 \leq 5$$
$$2+2 \leq 4$$

Similarly, the symbol ≥ means *greater than or equal to*. For example,

$$100 \geq 99$$
$$2+2 \geq 3$$
$$2+2 \geq 4$$

Tip: The symbols ≤ and ≥ are called *inclusive inequalities* because they *include* (allow) the possibility that both sides are equal. In contrast, the symbols < and > are called *exclusive inequalities* because they *exclude* (don't allow) this possibility.

Approximately equals (≈)

In Chapter 2, I show you how rounding numbers makes large numbers easier to work with. In that chapter, I also introduce ≈, which means *approximately equals*.

For example,

$$49 \approx 50$$
$$1,024 \approx 1,000$$
$$999,999 \approx 1,000,000$$

You can also use ≈ when you estimate the answer to a problem:

$$1,000,487 + 2,001,932 + 5,000,032$$
$$\approx 1,000,000 + 2,000,000 + 5,000,000$$
$$= 8,000,000$$

Examples

Q. Place the correct symbol (=, >, or <) in the blank: 2 + 2 ____ 5.

A. <. Because 2 + 2 = 4 and 4 is less than 5, use the symbol that means *is less than*.

Q. Place the correct symbol (=, >, or <) in the blank: 42 – 19 ____ 5 × 4.

A. >. Because 42 – 19 = 23 and 5 × 4 = 20, and 23 is greater than 20, use the symbol that means *is greater than*.

Q. Sam worked 7 hours for his parents at $8 an hour, and his parents paid him with a $50 bill. Use the symbol ≠ to point out why Sam was upset.

A. **$50 ≠ $56.** Sam worked 7 hours for $8 an hour, so here's how much he earned:

$$7 \times \$8 = \$56$$

He was upset because his parents didn't pay him the correct amount: $50 ≠ $56.

Q. Find an approximate solution to 2,000,398 + 6,001,756.

A. **8,000,000.** The two numbers are both in the millions, so you can use ≈ to round them to the nearest million:

$$2,000,398 + 6,001,756 \approx 2,000,000 + 6,000,000$$

Now it's easy to add 2,000,000 + 6,000,000 = 8,000,000.

Practice Questions

1. Place the correct symbol (=, >, or <) in the blanks:

  a. 4 + 6 ____ 13

  b. 9 × 7 ____ 62

  c. 33 – 16 ____ 60 ÷ 3

  d. 100 ÷ 5 ____ 83 – 63

2. Change the ≠ signs to either > or <:

  a. 17 + 14 ≠ 33

  b. 144 – 90 ≠ 66

  c. 11 × 14 ≠ 98

  d. 150 ÷ 6 ≠ 20

3. Tim's boss paid him for 40 hours of work last week. Tim accounted for his time by saying that he spent 19 hours with clients, 11 hours driving, and 7 hours doing paperwork. Use ≠ to show why Tim's boss was unhappy with Tim's work.

4. Find an approximate solution to 10,002 – 6,007.

Practice Answers

1. **a.** 4 + 6 = 10, and 10 < 13

  b. 9 × 7 = 63, and 63 > 62

  c. 33 – 16 = 17 and 60 ÷ 3 = 20, so 17 < 20.

  d. 100 ÷ 5 = 20 and 83 – 63 = 20, so 20 = 20.

2. **a.** 17 + 14 = 31, and 31 < 33

  b. 144 – 90 = 54, and 54 < 66

  c. 11 × 14 = 154, and 154 > 98

  d. 150 ÷ 6 = 25, and 25 > 20

3. 37 ≠ 40

4. 4,000

MOVING BEYOND THE BIG FOUR: EXPONENTS AND SQUARE ROOTS

In this section, I introduce you to three new operations that you need as you move on with math: exponents, square roots, and absolute value. As with the Big Four operations, these three operations tweak numbers in various ways.

To tell the truth, these three operations have fewer everyday applications than the Big Four. But you'll be seeing a lot more of them as you progress in your study of math. Fortunately, they aren't difficult, so this is a good time to become familiar with them.

Understanding exponents

Exponents (also called *powers*) are shorthand for repeated multiplication. For example, 2^3 means to multiply 2 by itself three times. To do that, use the following notation:

$$2^3 = 2 \times 2 \times 2 = 8$$

In this example, 2 is the *base number* and 3 is the *exponent*. You can read 2^3 as "2 to the third power" or "2 to the power of 3" (or even "2 cubed," which has to do with the formula for finding the value of a cube — see Chapter 15 for details).

Here's another example:

10^5 means to multiply 10 by itself five times

That works out like this:

$$10^5 = 10 \times 10 \times 10 \times 10 \times 10 = 100,000$$

This time, 10 is the base number and 5 is the exponent. Read 10^5 as "10 to the fifth power" or "10 to the power of 5."

Tip: When the base number is 10, figuring out any exponent is easy. Just write down a 1 and that many 0s after it:

1 with two 0s	1 with seven 0s	1 with twenty 0s
$10^2 = 100$	$10^7 = 10,000,000$	$10^{20} = 100,000,000,000,000,000,000$

Exponents with a base number of 10 are important in scientific notation, which I cover in Chapter 13.

The most common exponent is the number 2. When you take any whole number to the power of 2, the result is a square number. (For more information on square numbers, see Chapter 1.) For this reason, taking a number to the power of 2 is called *squaring* that number. You can read 3^2 as "three squared," 4^2 as "four squared," and so forth. Here are some squared numbers:

$$3^2 = 3 \times 3 = 9$$
$$4^2 = 4 \times 4 = 16$$
$$5^2 = 5 \times 5 = 25$$

Remember: Any number (except 0) raised to the 0 power equals 1. So 1^0, 37^0, and $999,999^0$ are equivalent, or equal, because they all equal 1.

Discovering your roots

Earlier in this chapter, in "Knowing Properties of the Big Four Operations," I show you how addition and subtraction are inverse operations. I also show you how multiplication and division are inverse operations. In a similar way, roots are the inverse operation of exponents.

The most common root is the square root. A *square root* undoes an exponent of 2. For example,

$$3^2 = 3 \times 3 = 9, \text{ so } \sqrt{9} = 3$$
$$4^2 = 4 \times 4 = 16, \text{ so } \sqrt{16} = 4$$
$$5^2 = 5 \times 5 = 25, \text{ so } \sqrt{25} = 5$$

You can read the symbol $\sqrt{}$ either as "the square root of" or as "radical." So read $\sqrt{9}$ as either "the square root of 9" or "radical 9."

As you can see, when you take the square root of any square number, the result is the number that you multiplied by itself to get that square number in the first place. For example, to find $\sqrt{100}$, you ask the question, "What number when multiplied by itself equals 100?" The answer here is 10 because

$$10^2 = 10 \times 10 = 100, \text{ so } \sqrt{100} = 10$$

You probably won't use square roots much until you get to algebra, but at that point, they become handy.

Examples

Q. What is 3^4?

A. **81.** The expression 3^4 tells you to multiply 3 by itself 4 times:

$$3 \times 3 \times 3 \times 3 = 81$$

Q. What is 10^6?

A. **1,000,000.** Using the power of ten rule, 10^6 is 1 followed by six 0s, so $10^6 = 1,000,000$.

Q. What is $\sqrt{36}$?

A. **6.** To find $\sqrt{36}$, you want to find a number that, when multiplied by itself, equals 36. You know that $6 \times 6 = 36$, so $\sqrt{36} = 6$.

Q. What is $\sqrt{256}$?

A. **16.** To find $\sqrt{256}$, you want to find a number that, when multiplied by itself, equals 256. Try guessing to narrow down the possibilities. Start by guessing 10:

$$10 \times 10 = 100$$

256 > 100, so $\sqrt{256}$ is greater than 10. Guess 20:

$$20 \times 20 = 400$$

256 < 400, so $\sqrt{256}$ is between 10 and 20. Guess 15:

$$15 \times 15 = 225$$

256 > 225, so $\sqrt{256}$ is between 15 and 20. Guess 16:

$$16 \times 16 = 256$$

This is correct, so $\sqrt{256} = 16$.

Practice Questions

1. a. 6^2

    b. 3^5

    c. 2^7

    d. 2^8 (*Hint:* You can make your work easier by using the answer to *c.*)

2. a. 10^4

    b. 10^{10}

    c. 10^{15}

    d. 10^1

A Practice Answers

1.   a.   **36.** $6^2 = 6 \times 6 = 36$.

     b.   **243.** $3^5 = 3 \times 3 \times 3 \times 3 \times 3 = 243$.

     c.   **128.** $2^7 = 2 \times 2 \times 2 \times 2 \times 2 \times 2 \times 2 = 128$.

     d.   **256.** $2^8 = 2 \times 2 \times 2 \times 2 \times 2 \times 2 \times 2 \times 2 = 256$. You already know from part c that $2^7 = 128$, so multiply this number by 2 to get your answer: $128 \times 2 = 256$.

2.   a.   **10,000.** Write 1 followed by four 0s.

     b.   **10,000,000,000.** Write 1 followed by ten 0s.

     c.   **1,000,000,000,000,000.** Write 1 followed by fifteen 0s.

     d.   **10.** Any number raised to the power of 1 is that number.

4

A QUESTION OF VALUES: EVALUATING ARITHMETIC EXPRESSIONS

IN THIS CHAPTER

- Understanding the Three E's of math — equations, expressions, and evaluation
- Using order of operations to evaluate expressions containing the Big Four operations
- Working with expressions that contain exponents
- Evaluating expressions with parentheses

In this chapter, I introduce you to what I call the Three E's of math: equations, expressions, and evaluation. You'll likely find the Three E's of math familiar because, whether you realize it or not, you've been using them for a long time. Whenever you add up the cost of several items at the store, balance your checkbook, or figure out the area of your room, you're evaluating expressions and setting up equations. In this section, I shed light on this stuff and give you a new way to look at it.

You probably already know that an *equation* is a mathematical statement that has an equals sign (=) — for example, 1 + 1 = 2. An *expression* is a string of mathematical symbols that can be placed on one side of an equation — for example, 1 + 1. And *evaluation* is finding out the *value* of an expression as a number — for example, finding out that the expression 1 + 1 is equal to the number 2.

Throughout the rest of the chapter, I show you how to turn expressions into numbers using a set of rules called the *order of operations* (or *order of precedence*). These rules look complicated, but I break them down so you can see for yourself what to do next in any situation.

SEEKING EQUALITY FOR ALL: EQUATIONS

An *equation* is a mathematical statement that tells you that two things have the same value — in other words, it's a statement with an equals sign. The equation is one of the most important concepts in mathematics because it allows you to boil down a bunch of complicated information into a single number.

Mathematical equations come in a lot of varieties: arithmetic equations, algebraic equations, differential equations, partial differential equations, Diophantine equations, and many more. In this book, I look at only two types: arithmetic equations and algebraic equations.

In this chapter, I discuss only *arithmetic equations,* which are equations involving numbers, the Big Four operations, and the other basic operations I introduce in Chapter 3 (absolute values, exponents, and roots). In Part V, I introduce you to algebraic equations. Here are a few examples of simple arithmetic equations:

$$2 + 2 = 4$$
$$3 \times 4 = 12$$
$$20 \div 2 = 10$$

And here are a few examples of more-complicated arithmetic equations:

$$1,000 - 1 - 1 - 1 = 997$$
$$(3 + 5) \div (9 - 7) = 4$$
$$4^2 - \sqrt{256} = (791 - 842) \times 0$$

Hey, it's just an expression

An *expression* is any string of mathematical symbols that can be placed on one side of an equation. Mathematical expressions, just like equations, come in a lot of varieties. In this chapter, I focus only on *arithmetic expressions,* which are expressions that contain numbers, the Big Four operations, and a few other basic operations (see Chapter 4). In Part V, I introduce you to algebraic expressions. Here are a few examples of simple expressions:

$$2 + 2$$
$$-17 + (-1)$$
$$14 \div 7$$

And here are a few examples of more-complicated expressions:

$$(88 - 23) \div 13$$
$$100 + 2 - 3 \times 17$$
$$\sqrt{441} + |-2^3|$$

Evaluating the situation

At the root of the word *evaluation* is the word *value.* In other words, when you evaluate something, you find its value. Evaluating an arithmetic expression is also referred to as *simplifying, solving,* or *finding the value of an expression.* The words may change, but the idea is the same — boiling down a string of numbers and math symbols to a single number.

When you evaluate an arithmetic expression, you simplify it to a single numerical value — in other words, you find the number that it's equal to. For example, evaluate the following arithmetic expression:

$$7 \times 5$$

How? Simplify it to a single number:

Putting the Three E's together

I'm sure you're dying to know how the Three E's — equations, expressions, and evaluation — are all connected. *Evaluation* allows you to take an *expression* containing more than one number and reduce it to a single number. Then you can make an *equation,* using an equals sign, to connect the expression and the number. For example, here's an *expression* containing four numbers:

$$1+2+3+4$$

When you *evaluate* it, you reduce it to a single number:

$$10$$

And now you can make an *equation* by connecting the expression and the number with an equals sign:

$$1+2+3+4=10$$

INTRODUCING ORDER OF OPERATIONS

When you were a kid, did you ever try putting on your shoes first and then your socks? If you did, you probably discovered this simple rule:

1. **Put on socks.**

2. **Put on shoes.**

Thus, you have an order of operations: The socks have to go on your feet before your shoes. So in the act of putting on your shoes and socks, your socks have precedence over your shoes. A simple rule to follow, right?

In this section, I outline a similar set of rules for evaluating expressions, called the *order of operations* (sometimes called *order of precedence*). Don't let the long name throw you. Order of operations is just a set of rules to make sure you get your socks and shoes on in the right order, mathematically speaking, so you always get the right answer.

Note: Through most of this book, I introduce overarching themes at the beginning of each section and then explain them later in the chapter instead of building them and finally revealing the result. But order of operations is a bit too confusing to present that way. Instead, I start with a list of four rules and go into more detail about them later in the chapter. Don't let the complexity of these rules scare you off before you work through them!

Remember: Evaluate arithmetic expressions from left to right according to the following order of operations:

1. **Parentheses**

2. **Exponents (also called powers)**

3. **Multiplication and division**

4. **Addition and subtraction**

Don't worry about memorizing this list right now. I break it to you slowly in the remaining sections of this chapter, starting from the bottom and working toward the top.

Generally speaking, the Big Four expressions come in the three types in the following table.

Expression	*Example*	*Rule*
Contains only addition and subtraction	$12 + 7 - 6 - 3 + 8$	Evaluate left to right.
Contains only multiplication and division	$18 \div 3 \times 7 \div 14$	Evaluate left to right.
Mixed-operator expression: contains a combination of addition/subtraction and multiplication/division	$9 + 6 \div 3$	1. Evaluate multiplication and division left to right. 2. Evaluate addition and subtraction left to right.

In this section, I show you how to identify and evaluate all three types of expressions.

Expressions with only addition and subtraction

Some expressions contain only addition and subtraction. When this is the case, the rule for evaluating the expression is simple.

Remember: When an expression contains only addition and subtraction, evaluate it step by step from left to right.

E

Example

Q. Find $7 + -3 - 6 - (-10)$.

A. 8. Start at the left with the first two numbers, $7 + -3 = 4$:

$$7 + \underline{-3} - 6 - (-10) = \underline{4} - 6 - (-10)$$

Continue with the next two numbers, $4 - 6 = -2$:

$$\underline{4 - 6} - (-10) = -\underline{2} - (-10)$$

Finish up with the last two numbers, remembering that subtracting a negative number is the same thing as adding a positive number:

$$-2 - (-10) = -2 + 10 = 8$$

Practice Questions

1. What's 9 – 3 + 8 – 7?

2. Evaluate 11 – 5 – 2 + 6 – 12.

3. Find 17 – 11 – (–4) + –10 – 8.

4. –7 + –3 – (–11) + 8 – 10 + –20 = ?

Practice Answers

1. **7.** Add and subtract from left to right:

$$9 - 3 + 8 - 7$$
$$= 6 + 8 - 7$$
$$= 14 - 7 = 7$$

2. **–2.**

$$11 - 5 - 2 + 6 - 12$$
$$= 6 - 2 + 6 - 12$$
$$= 4 + 6 - 12$$
$$= 10 - 12 = -2$$

3. **–8.**

$$17 - 11 - (-4) + -10 - 8$$
$$= 6 - (-4) + -10 - 8$$
$$= 10 + -10 - 8$$
$$= 0 - 8 = -8$$

4. **–21.**

$$-7 + -3 - (-11) + 8 - 10 + -20$$
$$= -10 - (-11) + 8 - 10 + -20$$
$$= 1 + 8 - 10 + -20$$
$$= 9 - 10 + -20$$
$$= -1 + -20 = -21$$

Expressions with only multiplication and division

Some expressions contain only multiplication and division. Unless you've been using your free time to practice your Persian, Arabic, or Hebrew reading skills, you're probably used to reading from left to right. Luckily, that's the direction of choice for multiplication and division problems, too. When an expression has *only* multiplication and division, in any combination, you should have no trouble evaluating it: Just start with the first two numbers and continue from left to right.

Examples

Q. Evaluate this expression:

$$9 \times 2 \div 6 \div 3 \times 2$$

A. **2.** The expression contains only multiplication and division, so you can move from left to right, starting with 9×2:

$$= 18 \div 6 \div 3 \times 2$$
$$= 3 \div 3 \times 2$$
$$= 1 \times 2$$
$$= 2$$

Notice that the expression shrinks one number at a time until all that's left is 2. So $9 \times 2 \div 6 \div 3 \times 2 = 2$.

Q. Evaluate $-10 \times 2 \times -3 \div -4$.

A. **−15.** The same procedure applies when you have negative numbers (just make sure you use the correct rule for multiplying or dividing by negative numbers, as I explain in Chapter 3). Start at the left with the first two numbers, $-10 \times 2 = -20$:

$$-10 \times 2 \times -3 \div -4 = \underline{-20} \times -3 \div -4$$

Continue with the next two numbers, $-20 \times -3 = 60$:

$$-20 \times -3 \div -4 = \underline{60} \div -4$$

Finish up with the last two numbers:

$$60 \div -4 = -15$$

Practice Questions

1. Find $18 \div 6 \times 10 \div 6$.

2. Evaluate $20 \div 4 \times 8 \div 5 \div -2$.

3. $12 \div -3 \times -9 \div 6 \times -7 = ?$

4. Solve $-90 \div 9 \times -8 \div -10 \div 4 \times -15$.

Practice Answers

1. **5.** Divide and multiply from left to right:

$$18 \div 6 \times 10 \div 6$$
$$= 3 \times 10 \div 6$$
$$= 30 \div 6 = 5$$

2. **−4.**

$$20 \div 4 \times 8 \div 5 \div -2$$
$$= 5 \times 8 \div 5 \div -2$$
$$= 40 \div 5 \div -2$$
$$= 8 \div -2 = -4$$

3. **−42.**

$$12 \div -3 \times -9 \div 6 \times -7$$
$$= -4 \times -9 \div 6 \times -7$$
$$= 36 \div 6 \times -7$$
$$= 6 \times -7 = -42$$

4. **30.**

$$-90 \div 9 \times -8 \div -10 \div 4 \times -15$$
$$= -10 \times -8 \div -10 \div 4 \times -15$$
$$= 80 \div -10 \div 4 \times -15$$
$$= -8 \div 4 \times -15$$
$$= -2 \times -15$$
$$= 30$$

Mixed-operator expressions

Often an expression contains

- At least one addition or subtraction operator
- At least one multiplication or division operator

I call these *mixed-operator expressions.* To evaluate them, you need some stronger medicine.

Remember: Evaluate mixed-operator expressions as follows:

1. **Evaluate the multiplication and division from left to right.**
2. **Evaluate the addition and subtraction from left to right.**

Example

Q. What's $-15 \times 3 \div -5 - (-3) \times -4$?

A. **–3.** Start by underlining all the multiplication and division in the problem; then evaluate all multiplication and division from left to right:

$$\underline{-15 \times 3 \div -5} - \underline{(-3) \times -4}$$
$$= -45 \div -5 - (-3) \times -4$$
$$= 9 - (-3) \times -4$$
$$= 9 - 12$$

Finish up by evaluating the addition and subtraction from left to right:

$$= -3$$

Practice Questions

1. Evaluate $8 - 3 \times 4 \div 6 + 1$.

2. Find $10 \times 5 - (-3) \times 8 \div -2$.

3. $-19 - 7 \times 3 + -20 \div 4 - 8 = ?$

4. What's $60 \div -10 - (-2) + 11 \times 8 \div 2$?

Practice Answers

1.   **7.** Start by underlining and evaluating all multiplication and division from left to right:

$$8 - \underline{3 \times 4} \div 6 + 1$$
$$= 8 - \underline{12 \div 6} + 1$$
$$= 8 - \underline{2} + 1$$

Now evaluate all addition and subtraction from left to right:

$$= 6 + 1 = 7$$

2.   **38.** Start by underlining and evaluating all multiplication and division from left to right:

$$\underline{10 \times 5} - \underline{(-3) \times 8} \div -2$$
$$= 50 - \underline{(-3) \times 8} \div -2$$
$$= 50 - \underline{(-24) \div -2}$$
$$= 50 - 12$$

Now evaluate the subtraction:

$$= 38$$

3.   **−53.** Start by underlining and evaluating all multiplication and division from left to right:

$$-19 - \underline{7 \times 3} + \underline{-20 \div 4} - 8$$
$$= -19 - 21 + \underline{-20 \div 4} - 8$$
$$= -19 - 21 + -5 - 8$$

Then evaluate all addition and subtraction from left to right:

$$= -40 + -5 - 8 = -45 - 8 = -53$$

4.   **40.** Start by underlining and evaluating all multiplication and division from left to right:

$$\underline{60 \div -10} - (-2) + \underline{11 \times 8} \div 2$$
$$= -6 - (-2) + \underline{11 \times 8} \div 2$$
$$= -6 - (-2) + \underline{88 \div 2}$$
$$= -6 - (-2) + 44$$

Now evaluate all addition and subtraction from left to right:

$$= -4 + 44 = 40$$

HANDLING POWERS RESPONSIBLY

You may have heard that power corrupts, but rest assured that when mathematicians deal with powers, the order of operations usually keeps them in line. When an expression contains one or more powers, evaluate all powers from left to right before moving on to the Big Four operators. Here's the breakdown:

1. **Evaluate all powers from left to right.**

   In Chapter 3, I show you that raising a number to a power simply means multiplying the number by itself that many times. For example, $2^3 = 2 \times 2 \times 2 = 8$. Remember that anything raised to the 0 power equals 1.

2. **Evaluate all multiplication and division from left to right.**

3. **Evaluate addition and subtraction from left to right.**

If you compare this numbered list with the one in the preceding section, you'll notice the only difference is that I've now inserted a new rule at the top.

Example

Q. Evaluate $7 - 4^2 \div 2^4 + 9 \times 2^3$.

A. **78.** Evaluate all powers from left to right, starting with $4^2 = 4 \times 4 = 16$:

$$= 7 - 16 \div 2^4 + 9 \times 2^3$$

Move on to evaluate the remaining two powers:

$$= 7 - 16 \div 16 + 9 \times 8$$

Next, evaluate all multiplication and division from left to right:

$$= 7 - 1 + 9 \times 8$$
$$= 7 - 1 + 72$$

Finish up by evaluating the addition and subtraction from left to right:

$$= 6 + 72$$
$$= 78$$

Practice Questions

1. Evaluate $3^2 - 2^3 \div 2^2$.

2. Find $5^2 - 4^2 - (-7) \times 2^2$.

3. $70^1 - 3^4 \div -9 \times -7 + 123^0 = ?$

4. What's $11^2 - 2^7 + 3^5 \div 3^3$?

Practice Answers

1. **7.** First, evaluate all powers:

    $$3^2 - 2^3 \div 2^2 = 9 - 8 \div 4$$

    Next, evaluate the division:

    $$= 9 - 2$$

    Finally, evaluate the subtraction:

    $$9 - 2 = 7$$

2. **37.** Evaluate all powers:

    $$5^2 - 4^2 - (-7) \times 2^2 = 25 - 16 - (-7) \times 4$$

    Evaluate the multiplication:

    $$= 25 - 16 - (-28)$$

    Finally, evaluate the subtraction from left to right:

    $$= 9 - (-28) = 37$$

3. **8.** Evaluate all powers:

    $$70^1 - 3^4 \div -9 \times -7 + 123^0 = 70 - 81 \div -9 \times -7 + 1$$

    Next, evaluate the multiplication and division from left to right:

    $$= 70 - (-9) \times -7 + 1 = 70 - 63 + 1$$

    Evaluate the addition and subtraction from left to right:

    $$= 7 + 1 = 8$$

4. 2. First, evaluate all powers:

$$11^2 - 2^7 + 3^5 \div 3^3 = 121 - 128 + 243 \div 27$$

Evaluate the division:

$$= 121 - 128 + 9$$

Finally, evaluate the addition and subtraction from left to right:

$$= -7 + 9 = 2$$

Prioritizing parentheses

Did you ever go to the post office and send a package high-priority so that it'd arrive as soon as possible? Parentheses work just like that. Parentheses — ( ) — allow you to indicate that a piece of an expression is high-priority — that is, it has to be evaluated before the rest of the expression.

Remember: When an expression includes parentheses with only Big Four operators, just do the following:

1. **Evaluate the contents of parentheses, from left to right, removing parentheses as you go.**

2. **Evaluate the rest of the expression.**

Example

Q. Evaluate (6 – 2) + (10 – 15 ÷ 3).

A. 9. Start by evaluating the contents of the first set of parentheses:

$$(6 - 2) + (10 - 15 \div 3) = 4 + (10 - 15 \div 3)$$

Move on to the next set of parentheses. This contains a mixed-operator expression, so evaluate it in two steps:

$$= 4 + (10 - 5)$$
$$= 4 + 5$$

To finish up, evaluate the addition:

$$= 9$$

Practice Questions

1. Evaluate $4 \times (3 + 4) - (16 \div 2)$.

2. What's $(5 + -8 \div 2) + (3 \times 6)$?

3. Find $(4 + 12 \div 6 \times 7) - (3 + 8)$.

4. $(2 \times -5) - (10 - 7) \times (13 + -8) = ?$

Practice Answers

1. **20.** Start by evaluating what's inside the first set of parentheses:

$$4 \times (3+4) - (16 \div 2)$$
$$= 4 \times 7 - (16 \div 2)$$

Next, evaluate the contents of the second set of parentheses:

$$4 \times 7 - 8$$

Evaluate the multiplication and then the subtraction:

$$= 28 - 8 = 20$$

2. **19.** Inside the first set of parentheses, evaluate the division first and then the addition:

$$(5 + -8 \div 2) + (3 \times 6)$$
$$= (5 + -4) + (3 \times 6)$$
$$= 1 + (3 \times 6)$$

Next, evaluate the contents of the second set of parentheses:

$$= 1 + 18$$

Finish up by evaluating the addition:

$$1 + 18 = 19$$

3. **7.** Begin by focusing on the first set of parentheses, handling all multiplication and division from left to right:

$$(4 + 12 \div 6 \times 7) - (3 + 8)$$
$$= (4 + 2 \times 7) - (3 + 8)$$
$$= (4 + 14) - (3 + 8)$$

Now do the addition inside the first set of parentheses:

$$= 18 - (3 + 8)$$

Next, evaluate the contents of the second set of parentheses:

$$= 18 - 11$$

Finish up by evaluating the subtraction:

$$18 - 11 = 7$$

4. **–25.** Evaluate the first set of parentheses, then the second, and then the third:

$$(2 \times -5) - (10 - 7) \times (13 + -8)$$
$$= -10 - (10 - 7) \times (13 + -8)$$
$$= -10 - 3 \times (13 + -8)$$
$$= -10 - 3 \times 5$$

Next, do multiplication and then finish up with the subtraction:

$$= -10 - 15 = -25$$

Pulling apart parentheses and powers

When an expression has parentheses and powers, evaluate it in the following order:

1. **Contents of parentheses**
2. **Powers from left to right**

After you've evaluated all parentheses and powers, the expression will be left with only the Big Four operations. Finish up by evaluating these four operations as I discuss earlier in "Mixed-operator expressions."

Examples

Q. Evaluate $(8 + 6^2) \div (2^3 - 4)$.

A. **11.** Begin by evaluating the contents of the first set of parentheses. Inside this set, evaluate the power first and do the addition next:

$$(8 + 6^2) \div (2^3 - 4)$$
$$= (8 + 36) \div (2^3 - 4)$$
$$= 44 \div (2^3 - 4)$$

Move to the next set of parentheses, evaluating the power first and then the subtraction:

$$= 44 \div (8-4) = 44 \div 4$$

Finish up by evaluating the division: $44 \div 4 = 11$.

Q. Find the value of $-1 + (-20 + 3^3)^{8-6}$.

A. **48.** When the entire contents of a set of parentheses is raised to a power, evaluate what's inside the parentheses before evaluating the power. *Inside* this set, evaluate the power first and the addition next:

$$-1 + (-20 + 3^3)^{8-6} = -1 + (-20 + 27)^{8-6} = -1 + 7^{8-6}$$

Next, evaluate the power. Begin by evaluating the expression in the exponent, $8 - 6$, and then evaluate the power itself:

$$7^{8-6} = 7^2 = 7 \times 7 = 49 :$$
$$= -1 + 49$$

Finish up by evaluating the addition: $-1 + 49 = 48$

Practice Questions

1. Find $(6^2 - 12) \div (16 \div 2^3)$.

2. Evaluate $-10 - (2 + 3^2 \times -4)$.

3. $7^2 - (3 + 3^2 \div -9)^5 = ?$

4. What is $(10 - 1^{14} \times 8)^{4 \div 4 + 5}$

Practice Answers

1. **12.** Focusing on the contents of the first set of parentheses, evaluate the power and then the subtraction:

$$(6^2 - 12) \div (16 \div 2^3)$$
$$= (36 - 12) \div (16 \div 2^3)$$
$$= 24 \div (16 \div 2^3)$$

Next, work inside the second set of parentheses, evaluating the power first and then the division:

$$= 24 \div (16 \div 8) = 24 \div 2$$

Finish by evaluating the division:

$$= 24 \div 2 = 12$$

2. **24.** Focusing on the contents of the parentheses, evaluate the power first, then the multiplication, and then the addition:

$$-10 - (2 + 3^2 \times -4) = -10 - (2 + 9 \times -4) = -10 - (2 + -36) = -10 - (-34)$$

Finish by evaluating the subtraction:

$$-10 - (-34) = 24$$

3. **17.** Focusing *inside* the parentheses, evaluate the power first, then the division, and then the addition:

$$7^2 - (3 + 3^2 \div -9)^5$$
$$= 7^2 - (3 + 9 \div -9)^5$$
$$= 7^2 - (3 + -1)^5$$
$$= 7^2 - 2^5$$

Next, evaluate both powers in order:

$$= 49 - 2^5 = 49 - 32$$

To finish, evaluate the subtraction:

$$49 - 32 = 17$$

4. **64.** Focusing inside the first set of parentheses, evaluate the power first, then the multiplication, and then the subtraction:

$$(10 - 1^{14} \times 8)^{4 \div 4 + 5} = (10 - 1 \times 8)^{4 \div 4 + 5} = (10 - 8)^{4 \div 4 + 5} = 2^{4 \div 4 + 5}$$

Next, handle the expression in the exponent, evaluating the division first and then the addition:

$$2^{1+5} = 2^6$$

To finish, evaluate the power:

$$2^6 = 64$$

Figuring out nested parentheses

Have you ever seen nested wooden dolls? (They originated in Russia, where their ultra-cool name is *matryoshka*.) This curiosity appears to be a single carved piece of wood in the shape of a doll. But when you open it up, you find a smaller doll nested inside it. And when you open up the smaller doll, you find an even smaller one hidden inside that one — and so on.

Like these Russian dolls, some arithmetic expressions contain sets of *nested* parentheses — one set of parentheses inside another set. To evaluate a set of nested parentheses, start by evaluating the inner set of parentheses and work your way outward.

Tip: Parentheses — ( ) — come in a number of styles, including brackets — [ ] — and braces — { }. These different styles help you keep track of where a statement in parentheses begins and ends. No matter what they look like, to the mathematician these different styles are all parentheses, so they all get treated the same.

Example

Q. Find the value of $\{3 \times [10 \div (6 - 4)]\} + 2$.

A. **17.** Begin by evaluating what's inside the innermost set of parentheses: $6 - 4 = 2$:

$$\{3 \times [10 \div (6 - 4)]\} + 2 = \{3 \times [10 \div 2]\} + 2$$

The result is an expression with one set of parentheses inside another set, so evaluate what's inside the inner set: $10 \div 2 = 5$:

$$= \{3 \times 5\} + 2$$

Now, evaluate what's inside the final set of parentheses:

$$= 15 + 2$$

Finish up by evaluating the addition: $15 + 2 = 17$.

Practice Questions

1. Evaluate $7 + \{[(10 - 6) \times 5] + 13\}$.

2. Find the value of $[(2 + 3) - (30 \div 6)] + (-1 + 7 \times 6)$.

3. $-4 + \{[-9 \times (5 - 8)] \div 3\} = ?$

4. Evaluate $\{(4 - 6) \times [18 \div (12 - 3 \times 2)]\} - (-5)$.

A Practice Answers

1. **40.** First evaluate the inner set of parentheses:

$$7 + \{[(10-6) \times 5] + 13\} = 7 + \{[4 \times 5] + 13\}$$

Move outward to the next set of parentheses:

$$= 7 + \{20 + 13\}$$

Next, handle the remaining set of parentheses:

$$= 7 + 33$$

To finish, evaluate the addition:

$$7 + 33 = 40$$

2. **41.** Start by focusing on the first set of parentheses. This set contains two inner sets of parentheses, so evaluate these two sets from left to right:

$$[(2+3) - (30 \div 6)] + (-1 + 7 \times 6)$$
$$= [(5) - (30 \div 6)] + (-1 + 7 \times 6)$$
$$= [5 - 5] + (-1 + 7 \times 6)$$

Now, the expression has two separate sets of parentheses, so evaluate the first set:

$$= 0 + (-1 + 7 \times 6)$$

Handle the remaining set of parentheses, evaluating the multiplication first and then the addition:

$$= 0 + (-1 + 42) = 0 + 41$$

To finish, evaluate the addition:

$$0 + 41 = 41$$

3. **5.** Start by evaluating the inner set of parentheses:

$$-4 + \{[-9 \times (5-8)] \div 3\} = -4 + \{[-9 \times -3)] \div 3\}$$

Move outward to the next set of parentheses:

$$= -4 + [27 \div 3]$$

Next, handle the remaining set of parentheses:

$$= -4 + 9$$

Finally, evaluate the addition:

$$-4 + 9 = 5$$

4.   **−1.** Focus on the inner set of parentheses, $(12 - 3 \times 2)$. Evaluate the multiplication first and then the subtraction:

$$\{(4-6) \times [18 \div (12 - 3 \times 2)]\} - (-5)$$
$$= \{(4-6) \times [18 \div (12-6)]\} - (-5)$$
$$= \{(4-6) \times [18 \div 6]\} - (-5)$$

Now the expression is an outer set of parentheses with two inner sets. Evaluate these two inner sets of parentheses from left to right:

$$= \{-2 \times [18 \div 6]\} - (-5) = \{-2 \times 3\} - (-5)$$

Next, evaluate the final set of parentheses:

$$= -6 - (-5)$$

Finish by evaluating the subtraction:

$$-6 - (-5) = -1$$

BRINGING IT ALL TOGETHER: THE ORDER OF OPERATIONS

Remember: Throughout this chapter, you work with a variety of rules for deciding how to evaluate arithmetic expressions. These rules all give you a way to decide the order in which an expression gets evaluated. All together, this set of rules is called the *order of operations* (or sometimes, the *order of precedence*). Here's the complete order of operations for arithmetic:

1. **Contents of parentheses from the inside out**

2. **Powers from left to right**

3. **Multiplication and division from left to right**

4. **Addition and subtraction from left to right**

The only difference between this list and the one in "Pulling Apart Parentheses and Powers" is that I've now added a few words to the end of Step 1 to cover nested parentheses (which I discuss in the preceding section).

Example

Q. Evaluate $[(8 \times 4 + 2^3) \div 10]^{7-5}$.

A. 16. Start by focusing on the inner set of parentheses, evaluating the power, then the multiplication, and then the addition:

$$[(8 \times 4 + 2^3) \div 10]^{7-5}$$
$$= [(8 \times 4 + 8) \div 10]^{7-5}$$
$$= [(32 + 8) \div 10]^{7-5}$$
$$= [40 \div 10]^{7-5}$$

Next, evaluate what's inside the parentheses and the expression that makes up the exponent:

$$= 4^{7-5} = 4^2$$

Finish by evaluating the remaining power: $4^2 = 16$.

Practice Questions

1. Evaluate $1 + [(2^3 - 4) + (10 \div 2)^2]$.

2. $(-7 \times -2 + 6^2 \div 4)^{9 \times 2 - 17}$

3. What is $\{6^2 - [12 \div (-13 + 14)^2] \times 2\}^2$?

4. Find the value of $[(123 - 11^2)^4 - (6^2 \div 2^{20-3 \times 6})]^2$.

Practice Answers

1. **30.** Start by focusing on the first of the two inner sets of parentheses, $(2^3 - 4)$. Evaluate the power first and then the subtraction:

$$1 + [(2^3 - 4) + (10 \div 2)^2] = 1 + [(8 - 4) + (10 \div 2)^2] = 1 + [4 + (10 \div 2)^2]$$

Continue by focusing on the remaining inner set of parentheses:

$$= 1 + [4 + 5^2]$$

Next, evaluate what's inside the last set of parentheses, evaluating the power first and then the addition:

$$= 1 + [4 + 25] = 1 + 29$$

Finish by adding the remaining numbers:

$$1 + 29 = 30$$

2. **23.** Start with the first set of parentheses. Evaluate the power first, then the multiplication and division from left to right, and then the addition:

$$(-7 \times -2 + 6^2 \div 4)^{9 \times 2 - 17}$$
$$= (-7 \times -2 + 36 \div 4)^{9 \times 2 - 17}$$
$$= (14 + 36 \div 4)^{9 \times 2 - 17}$$
$$= (14 + 9)^{9 \times 2 - 17}$$
$$= 23^{9 \times 2 - 17}$$

Next, work on the exponent, evaluating the multiplication first and then the subtraction:

$$= 23^{18 - 17} = 23^1$$

Finish by evaluating the power:

$$23^1 = 23$$

3. **144.** Start by evaluating the inner set of parentheses $(-13 + 14)$:

$$\{6^2 - [12 \div (-13 + 14)^2] \times 2\}^2$$
$$= \{6^2 - [12 \div 1^2] \times 2\}^2$$

Move outward to the next set of parentheses, $[12 \div 1^2]$, evaluating the power and then the division:

$$= \{6^2 - [12 \div 1] \times 2\}^2$$
$$= \{6^2 - 12 \times 2\}^2$$

Next, work on the remaining set of parentheses, evaluating the power, then the multiplication, and then the subtraction:

$$= \{36 - 12 \times 2\}^2$$
$$= \{36 - 24\}^2$$
$$= 12^2$$

Finish by evaluating the power:

$$12^2 = 144$$

4. **49.** Start by working on the exponent, $20 - 3 \times 6$, evaluating the multiplication and then the subtraction:

$$[(123 - 11)^2)^4 - (6^2 \div 2^{20 - 3 \times 6})]^2$$
$$= [(123 - 11)^2)^4 - (6^2 \div 2^{20 - 18})]^2$$
$$= [(123 - 11)^2)^4 - (6^2 \div 2^2)]^2$$

The result is an expression with two inner sets of parentheses. Focus on the first of these two sets, evaluating the power and then the subtraction:

$$= [(123 - 121)^4 - (6^2 \div 2^2)]^2$$

Work on the remaining inner set of parentheses, evaluating the two powers from left to right and then the division:

$$= [2^4 - (36 \div 2^2)]^2$$
$$= [2^4 - (36 \div 4)]^2$$
$$= [2^4 - 9]^2$$

Now evaluate what's left inside the parentheses, evaluating the power and then the subtraction:

$$= [16 - 9]^2$$
$$= 7^2$$

Finish by evaluating the power: $7^2 = 49$.

5

SAY WHAT? TURNING WORDS INTO NUMBERS

IN THIS CHAPTER

- Dispelling myths about word problems
- Knowing the four steps to solving a word problem
- Jotting down simple word equations that condense the important information
- Writing more-complex word equations
- Plugging numbers into the word equations to solve the problem
- Attacking more-complex word problems with confidence

The very mention of word problems — or story problems, as they're sometimes called — is enough to send a cold shiver of terror into the bones of the average math student. Many would rather swim across a moat full of hungry crocodiles than "figure out how many bushels of corn Farmer Brown picked" or "help Aunt Sylvia decide how many cookies to bake." But word problems help you understand the logic behind setting up equations in real-life situations, making math actually useful — even if the scenarios in the word problems you practice on are pretty far-fetched.

In this chapter, I dispel a few myths about word problems. Then I show you how to solve a word problem in four simple steps. After you understand the basics, I show you how to solve more-complex problems. Some of these problems have longer numbers to calculate, and others may have more complicated stories. In either case, you can see how to work through them step by step.

DISPELLING TWO MYTHS ABOUT WORD PROBLEMS

Here are two common myths about word problems:

- Word problems are always hard.
- Word problems are only for school — after that, you don't need them.

Both of these ideas are untrue. But they're so common that I want to address them head-on.

Word problems aren't always hard

Word problems don't have to be hard. For example, here's a word problem that you may have run into in first grade:

Adam had 4 apples. Then Brenda gave him 5 more apples. How many apples does Adam have now?

You can probably do the math in your head, but when you were starting out in math, you may have written it down:

$$4 + 5 = 9$$

Finally, if you had one of those teachers who made you write out your answer in complete sentences, you wrote "Adam has 9 apples." (Of course, if you were the class clown, you probably wrote, "Adam doesn't have any apples because he ate them all.")

Word problems seem hard when they get too complex to solve in your head and you don't have a system for solving them. In this chapter, I give you a system and show you how to apply it to problems of increasing difficulty. And in Chapters 5 and 12, I give you further practice solving more difficult word problems.

Word problems are useful

In the real world, math rarely comes in the form of equations. It comes in the form of situations that are very similar to word problems.

Whenever you paint a room, prepare a budget, bake a double batch of oatmeal cookies, estimate the cost of a vacation, buy wood to build a shelf, do your taxes, or weigh the pros and cons of buying a car versus leasing one, you need math. And the math skill you need most is understanding how to turn the *situation* you're facing into numbers that you calculate.

Word problems give you practice turning situations — or stories — into numbers.

SOLVING BASIC WORD PROBLEMS

Generally, solving a word problem involves four steps:

1. **Read through the problem and set up a *word equation*— that is, an equation that contains words as well as numbers.**

2. **Plug in numbers in place of words wherever possible to set up a regular math equation.**

3. **Use math to solve the equation.**

4. **Answer the question the problem asks.**

Most of this book is about Step 3. This chapter and Chapters 5 and 12 are all about Steps 1 and 2. I show you how to break down a word problem sentence by sentence, jot down the information you need to solve the problem, and then substitute numbers for words to set up an equation.

When you know how to turn a word problem into an equation, the hard part is done. Then you can use the rest of what you find in this book to figure out how to do Step 3 — solve the equation. From there, Step 4 is usually pretty easy, though at the end of each example, I make sure you understand how to do it.

Turning word problems into word equations

The first step to solving a word problem is reading it and putting the information you find into a useful form. In this section, I show you how to squeeze the juice out of a word problem and leave the pits behind!

Jotting down information as word equations

Most word problems give you information about numbers, telling you exactly how much, how many, how fast, how big, and so forth. Here are some examples:

Nunu is spinning 17 plates.

The width of the house is 80 feet.

If the local train is going 25 miles per hour . . .

You need this information to solve the problem. And paper is cheap, so don't be afraid to use it. (If you're concerned about trees, write on the back of all that junk mail you get.) Have a piece of scrap paper handy and jot down a few notes as you read through a word problem.

For example, here's how you can jot down "Nunu is spinning 17 plates":

$$\text{Nunu} = 17$$

Here's how to note that "the width of the house is 80 feet":

$$\text{width} = 80$$

The third example tells you, "If the local train is going 25 miles per hour. . . ." So you can jot down the following:

$$\text{local} = 25$$

Remember: Don't let the word *if* confuse you. When a problem says "If so-and-so were true . . ." and then asks you a question, assume that it *is* true and use this information to answer the question.

When you jot down information this way, you're really turning words into a more useful form called a *word equation*. A word equation has an equals sign like a math equation, but it contains both words and numbers.

Example

Q. Write a word equation for the statement "James bought 36 bagels."

A. James = 36.

Practice Questions

1. Write a word equation for the statement "Ralphie took 45 minutes to clean his room."

2. Write a word equation for the statement "Mavis has saved $2,000 for a new car."

3. Write a word equation for the statement "Ms. Archer lives 80 miles from her workplace."

4. Write a word equation for the statement "Pinewood Drive is 700 feet in length."

Practice Answers

1. **Ralphie = 45.**

2. **Mavis = 2,000.**

3. **Ms. Archer = 80.**

4. **Pinewood Drive = 700.**

Writing relationships: Turning more-complex statements into word equations

When you start doing word problems, you notice that certain words and phrases show up over and over again. For example,

Bobo is spinning five fewer plates than Nunu.

The height of a house is half as long as its width.

The express train is moving three times faster than the local train.

You've probably seen statements such as these in word problems since you were first doing math. Statements like these look like English, but they're really math, so spotting them is important. You can represent each of these types of statements as word equations that also use Big Four operations. Look again at the first example:

Bobo is spinning five fewer plates than Nunu.

You don't know the number of plates that either Bobo or Nunu is spinning. But you know that these two numbers are related.

You can express this relationship like this:

$$\text{Bobo} + 5 = \text{Nunu}$$

This word equation is shorter than the statement it came from. And as you see in the next section, word equations are easy to turn into the math you need to solve the problem.

Here's another example:

> The height of a house is half as long as its width.

You don't know the width or height of the house, but you know that these numbers are connected.

You can express this relationship between the width and height of the house as the following word equation:

$$\text{height} = \text{width} \div 2$$

With the same type of thinking, you can express "The express train is moving three times faster than the local train" as this word equation:

$$\text{express} = 3 \times \text{local}$$

Remember: As you can see, each of the examples allows you to set up a word equation using one of the Big Four operations — adding, subtracting, multiplying, and dividing.

Examples

Q. Given the statement "Abigail has six fewer dresses than Gwendolyn," which girl has the greater number of dresses?

A. **Gwendolyn has more dresses than Abigail.**

Q. Given the word equation Myra = Samuel ÷ 2, does Myra or Samuel have the greater number of objects?

A. **Samuel has more objects than Myra.**

Q. Write a word equation for the statement "The obelisk is 51 feet shorter than the radio tower."

A. **Obelisk = Radio Tower – 51; Obelisk + 51 = Radio Tower; Radio Tower – Obelisk = 51**

Practice Questions

1. In each of the following statements, which person has the greater number of objects?

   a. David received five more balloons than Jessica.

   b. Fred has ten fewer action figures than Wally.

   c. Larry owns half as many antique coins as Mei Ling.

   d. Ronaldo has collected four times as many butterflies as Patricia.

2. In each of the following word equations, which person has the greater number of objects?

   a. Scott = Violet + 3

   b. Carly − 15 = Noelle

   c. Mr. Grace × 4 = Ms. Pardee

   d. Aunt Eleanor = Aunt Maddie ÷ 6

3. Consider the statement "Kathleen owns 22 more comic books than Ulysses." Which of the following word equations is a correct translation of this statement. (More than one answer may be correct.)

   a. Kathleen + 22 = Ulysses

   b. Kathleen = 22 + Ulysses

   c. Kathleen − 22 = Ulysses

   d. Kathleen − Ulysses = 22

4. Write a word equation for each of the following statements:

   a. Kyle owns 9 more goldfish than Drew.

   b. Geoff bought half as many bags of seeds as Shari.

   c. The hare runs five times as fast as the tortoise.

   d. Peggy Sue saved $150 less than Lisa Marie.

Practice Answers

1. a. **David has more balloons than Jessica.**

   b. **Wally has more action figures than Fred.**

   c. **Mei Ling has more antique coins than Larry.**

   d. **Ronaldo has more butterflies than Patricia.**

2. **a.** Scott has more items than Violet.

    b. Carly has more items than Noelle.

    c. Ms. Pardee has more items than Mr. Grace.

    d. Aunt Maddie has more items than Aunt Eleanor.

3. **a.** Wrong

    b. Right

    c. Right

    d. Right

4. **a.** Kyle = Drew + 9; Kyle − 9 = Drew; Kyle − Drew = 9

    b. Geoff = Shari ÷ 2; Geoff × 2 = Shari; Shari ÷ Geoff = 2

    c. Hare = Tortoise × 5; Hare ÷ 5 = Tortoise; Hare ÷ Tortoise = 5

    d. Peggy Sue = Lisa Marie − 150; Peggy Sue + 150 = Lisa Marie; Lisa Marie − Peggy Sue = 150

Figuring out what the problem's asking

The end of a word problem usually contains the question you need to answer to solve the problem. You can use word equations to clarify this question so you know right from the start what you're looking for.

For example, you can write the question, "All together, how many plates are Bobo and Nunu spinning?" as

$$Bobo + Nunu = ?$$

You can write the question "How tall is the house" as:

$$height = ?$$

Finally, you can rephrase the question "What's the difference in speed between the express train and the local train?" in this way:

$$express - local = ?$$

Example

Q. For a word problem about the cost of groceries, write a word equation for the question "What was the total cost of the chips, the dip, and the salsa?"

A. Chips + Dip + Salsa = ?

Q **Practice Questions**

1. For a word problem about three children baking cookies for a bake sale, write a word equation for the question "How many cookies did Greg bake?"

2. For a word problem about runners preparing for a marathon, write a word equation for the question "How many more miles did Sarah run as compared with Vicki?"

3. For a word problem about saving money, write a word equation for the question "How much money do Melissa and Tyrone have altogether?"

4. For a word problem about the cost of a new car, write a word equation for the question "If the salesman gave Yolanda a $700 discount off the sticker price, how much did she actually pay for the car?"

A **Practice Answers**

1. **Greg = ?**

2. **Sarah – Vicki = ?**

3. **Melissa + Tyrone = ?**

4. **Sticker price – 700 = ?**

Plugging in numbers for words

After you've written out a bunch of word equations, you have the facts you need in a form you can use. You can often solve the problem by plugging numbers from one word equation into another. In this section, I show you how to use the word equations you built in the last section to solve three problems.

Send in the clowns

Some problems involve simple addition or subtraction. Here's an example:

Bobo is spinning five fewer plates than Nunu. (Bobo dropped a few.) Nunu is spinning 17 plates. All together, how many plates are Bobo and Nunu spinning?

Here's what you have already, just from reading the problem:

$$\text{Nunu} = 17$$
$$\text{Bobo} + 5 = \text{Nunu}$$

Plugging in the information gives you the following:

$$\text{Bobo} + 5 = 17$$

If you see how many plates Bobo is spinning, feel free to jump ahead. If not, here's how you rewrite the addition equation as a subtraction equation (see Chapter 4 for details):

$$Bobo = 17 - 5 = 12$$

The problem wants you to find out how many plates the two clowns are spinning together. So you need to find out the following:

$$Bobo + Nunu = ?$$

Just plug in the numbers, substituting 12 for Bobo and 17 for Nunu:

$$12 + 17 = 29$$

So Bobo and Nunu are spinning 29 plates.

Our house in the middle of our street

At times, a problem notes relationships that require you to use multiplication or division. Here's an example:

> The height of a house is half as long as its width, and the width of the house is 80 feet. How tall is the house?

You already have a head start from what you determined earlier:

$$width = 80$$
$$height = width \div 2$$

You can plug in information as follows, substituting 80 for the word *width*:

$$height = 80 \div 2 = 40$$

So you know that the height of the house is 40 feet.

I hear the train a-comin'

Pay careful attention to what the question is asking. You may have to set up more than one equation. Here's an example:

> The express train is moving three times faster than the local train. If the local train is going 25 miles per hour, what's the difference in speed between the express train and the local train?

Here's what you have so far:

$$local = 25$$
$$express = 3 \times local$$

Plug in the information you need:

$$express = 3 \times 25 = 75$$

In this problem, the question at the end asks you to find the difference in speed between the express train and the local train. Finding the difference between two numbers is subtraction, so here's what you want to find:

$$\text{express} - \text{local} = ?$$

You can get what you need to know by plugging in the information you've already found:

$$75 - 25 = 50$$

Therefore, the difference in speed between the express train and the local train is 50 miles per hour.

Examples

Q. The Panthers have won 4 fewer baseball games this season than the Cobras, and the Cobras have won twice as many games as the Fireballs. If the Cobras have won 28 games, how many more games did the Panthers win than the Fireballs?

   a. Write a word equation for the statement "The Panthers have won 4 fewer baseball games this season than the Cobras."

   b. Write a word equation for the statement "The Cobras have won twice as many games as the Fireballs."

   c. Write a word equation for the statement "The cobras have won 28 games."

   d. Plug in the information from c into the word equations in a and b.

   e. Solve these equations.

   f. Write a word equation for the question "How many more games did the Panthers win than the Fireballs?"

   g. Solve this equation.

   h. Write the answer to the question in a complete sentence.

A.   **a.**   **Panthers = Cobras − 4**

    b.   **Cobras = Fireballs × 2**

    c.   **Cobras = 28**

    d.   **Panthers = 28 − 4; 28 = Fireballs × 2**

    e.   **Panthers = 24; Fireballs = 14**

    f.   **Panthers − Fireballs = ?**

    g.   **24 − 14 = 10**

    h.   **The Panthers have won 10 more games than the Fireballs.**

Q. Together, Andre and Toby sold 35 boxes of candy to support their school play. Toby sold twice as many boxes as Samantha. If Samantha sold 12 boxes, how many boxes did Andre and Samantha sell together?

    a. How many boxes did Samantha sell?

    b. How many boxes did Toby sell?

    c. How many boxes did Andre sell?

    d. What is the problem asking? Write the answer in a complete sentence.

A.   **a.**   **Samantha sold 12 boxes.**

    b.   **Toby sold 24 boxes. (12 × 2 = 24)**

    c.   **Andre sold 11 boxes. (35 – 24 = 11)**

    d.   **Together, Andre and Samantha sold 23 boxes. (11 + 12 = 23)**

Practice Questions

    1. Mason Street has five times as many houses as London Avenue. And London Avenue has 40 fewer houses than River Road. How many more houses are on Mason Street than London Avenue, assuming that River Road has 56 houses?

        a. Write a word equation for the statement "London Avenue has 40 fewer houses than River Road."

        b. Write a word equation for the statement "River Road has 56 houses."

        c. Plug the information from b into the word equation in a and solve. How many houses are on London Avenue?

        d. Write a word equation for the statement "Mason Street has five times as many houses as London Avenue."

        e. Plug the information from c into the word equation in d and solve. How many houses are on Mason Street?

        f. What is the problem asking? Write the answer in a complete sentence.

    2. Last week, Dominic worked twice as many hours as Raymond, and Carter worked 6 fewer hours than Dominic. If Carter worked 36 hours, how many hours did the three men work altogether?

        a. How many hours did Carter work?

        b. How many hours did Dominic work?

        c. How many hours did Raymond work?

        d. What is the question asking? Write the answer in a complete sentence.

3. At rush hour, the bus from Center City to Hackettstown takes three times longer to reach its destination than the train, which takes 35 minutes less than a car. If a car takes 55 minutes, how much longer is the trip by bus than by car?

   a. How long does the trip by car take?

   b. How long does the trip by train take?

   c. How long does the trip by bus take?

   d. What is the problem asking? Write the answer in a complete sentence.

Practice Answers

1. **a.** **London Avenue = River Road – 40.**

   b. **River Road = 56.**

   c. **London Avenue = 56 – 40 = 16.**

   d. **Mason Street = London Avenue × 5.**

   e. **Mason Street = 16 × 5 = 80.**

   f. **Mason Street has 64 more houses than London Avenue** (Mason Street – London Avenue = 80 – 16 = 64).

2. **a.** **Carter worked 36 hours.**

   b. **Dominic worked 42 hours.** (36 + 6 = 42)

   c. **Raymond worked 21 hours.** (42 ÷ 2 = 21)

   d. **The three men worked a total of 99 hours.** (36 + 42 + 21 = 99)

3. **a.** **The trip by car takes 55 minutes.**

   b. **The trip by train takes 20 minutes.** (55 – 35 = 20)

   c. **The trip by bus takes 60 minutes.** (20 × 3 = 60)

   d. **The trip by bus takes 5 minutes longer than the trip by car.** (60 – 55 = 5)

SOLVING MORE-COMPLEX WORD PROBLEMS

The skills I show you previously in "Solving Basic Word Problems" are important for solving any word problem because they streamline the process and make it simpler. What's more, you can use those same skills to find your way through more complex problems. Problems become more complex when

- The calculations become harder. (For example, instead of a dress costing $30, it costs $29.95.)

- The amount of information in the problem increases. (For example, instead of two clowns, you have five.)

Don't let problems like these scare you. In this section, I show you how to use your new problem-solving skills to solve more-difficult word problems.

When numbers get serious

A lot of problems that look tough aren't much more difficult than the problems I show you in the previous sections. For example, consider this problem:

> Aunt Effie has $732.84 hidden in her pillowcase, and Aunt Jezebel has $234.19 less than Aunt Effie has. How much money do the two women have all together?

One question you may have is how these women ever get any sleep with all that change clinking around under their heads. But moving on to the math, even though the numbers are larger, the principle is still the same as in problems in the earlier sections. Start reading from the beginning: "Aunt Effie has $732.84 . . . ." This text is just information to jot down as a simple word equation:

$$Effie = \$732.84$$

Continuing, you read, "Aunt Jezebel has *$234.19 less than* Aunt Effie has." It's another statement you can write as a word equation:

$$Jezebel = Effie - \$234.19$$

Now you can plug in the number $732.84 where you see Aunt Effie's name in the equation:

$$Jezebel = \$732.84 - \$234.19$$

So far, the big numbers haven't been any trouble. At this point, though, you probably need to stop to do the subtraction:

$$\begin{array}{r} \$732.84 \\ - \$234.19 \\ \hline \$498.65 \end{array}$$

Now you can jot this information down, as always:

$$Jezebel = \$498.65$$

The question at the end of the problem asks you to find out how much money the two women have all together. Here's how to represent this question as an equation:

$$Effie + Jezebel = ?$$

You can plug information into this equation:

$$\$732.84 + \$498.65 = ?$$

Again, because the numbers are large, you probably have to stop to do the math:

$$\begin{array}{r} \$732.84 \\ +\ \$498.65 \\ \hline \$1231.49 \end{array}$$

So all together, Aunt Effie and Aunt Jezebel have $1,231.49.

As you can see, the procedure for solving this problem is basically the same as for the simpler problems in the earlier sections. The only difference is that you have to stop to do some addition and subtraction.

Example

Q. Daisy Mae is a circus elephant that weighs 12,587 pounds. Together, she and her twin sister, Luci Jo, weigh 25,322 pounds. How much more does Luci Jo weigh than her sister?

   a. How much does Daisy Mae weigh?

   b. How much does Luci Jo weigh?

   c. What is the problem asking? Write the answer in a complete sentence.

A. **a.** **Daisy Mae weighs 12,587 lbs.**

   b. **Luci Jo weighs 12,735 lbs.** (25,322 – 12,587 = 12,735)

   c. **Luci Jo weighs 148 lbs more than Daisy Mae.** (12,735 – 12,587 = 148)

Practice Questions

  1. Amelia bought lunch for herself and her two daughters at a restaurant. When the check arrived, it said that she owed $35.65. She asked the server to add on a $7.50 tip, and gave him a $50.00 bill. How much money should the server return to Amelia?

    a. How much was the check for?

    b. How much did Amelia want to leave as a tip?

    c. How much money did Amelia want to pay, including the check and the tip?

    d. What is the problem asking? Write the answer in a complete sentence.

2. The washing machine that Todd wanted to buy costs $58.75 dollars more than the dryer. But he found that buying both items together would give him a $75 discount. If the dryer costs $392.50, how much did Todd end up spending for both items?

    a. What is the price of the dryer?

    b. What is the price of the washing machine?

    c. What would be the combined price of both items if bought separately?

    d. What is the problem asking? Write the answer in a complete sentence.

3. Last month, Josh spent exactly 5 times more on rent than he spent on electricity and his Internet connection combined. If his Internet bill was $86 and he paid half that amount for electricity, how much did he spend on all three combined?

    a. How much did Josh spend on the Internet?

    b. How much did he spend on electricity?

    c. How much did he spend on Internet and electricity combined?

    d. How much did he spend on rent?

    e. What is the problem asking? Write the answer in a complete sentence.

4. Three brothers are all keeping track of how many pushups they can do in one week. Yesterday, Tyler had completed 405 more than Stephen, and Stephen had completed exactly half as many as Andrew. But today, only Stephen had time to do more pushups, so he did 475, bringing his total up to 1,140. How many pushups have the three boys completed all together?

    a. How many pushups has Stephen completed?

    b. How many pushups had Stephen completed yesterday?

    c. How many pushups has Tyler completed?

    d. How many pushups has Andrew completed?

    e. What is the problem asking? Write the answer in a complete sentence.

Practice Answers

1. a. **$35.65.**

    b. **$7.50.**

    c. **$43.15.** ($35.65 + $7.50 = $43.15)

    d. **$6.85.** ($50.00 - $43.15 = $6.85) The server should return $6.85 to Amelia.

2. a. **$392.50.**

    b. **$451.25.** ($392.50 + $58.75 = $451.25)

    c. **$843.75.** ($392.50 + $451.25 = $843.75)

    d. **$768.75.** ($843.75 – $75.00 = $768.75) Todd spent $768.75 for both items.

3. **a.** **$86.**

   b. **$43.** ($86 ÷ 2 = $43)

   c. **$129.** ($86 + $43 = $129)

   d. **$645.** ($129 × 5 = $645)

   e. **Josh spent a total of $774 for rent, electricity, and the Internet.** ($645 + $86 + 43 = $774)

4. **a.** **1,140**

   b. **665** (1,140 − 475 = 665)

   c. **1,070** (665 + 405 = 1,070)

   d. **1,330** (665 × 2 = 1,330)

   e. **All together, the three boys have completed 3,540 pushups.** (1,140 + 1,070 + 1,330 = 3,540)

Too much information

When the going gets tough, knowing the system for writing word equations really becomes helpful. Here's a word problem that's designed to scare you off — but with your new skills, you're ready for it:

> Four women collected money to save the endangered Salt Creek tiger beetle. Keisha collected $160, Brie collected $50 more than Keisha, Amy collected twice as much as Brie, and together Amy and Sophia collected $700. How much money did the four women collect all together?

If you try to do this problem all in your head, you'll probably get confused. Instead, take it line by line and just jot down word equations as I discuss earlier in this chapter.

First, "Keisha collected $160." So jot down the following:

$$\text{Keisha} = 160$$

Next, "Brie collected $50 dollars more than Keisha," so write

$$\text{Brie} = \text{Keisha} + 50$$

After that, "Amy collected twice as much as Brie":

$$\text{Amy} = \text{Brie} \times 2$$

Finally, "together, Amy and Sophia collected $700":

$$\text{Amy} + \text{Sophia} = 700$$

That's all the information the problem gives you, so now you can start working with it. Keisha collected $160, so you can plug in 160 anywhere you find Keisha's name:

$$\text{Brie} = 160 + 50 = 210$$

Now you know how much Brie collected, so you can plug this information into the next equation:

$$\text{Amy} = 210 \times 2 = 420$$

This equation tells you how much Amy collected, so you can plug this number into the last equation:

$$420 + \text{Sophia} = 700$$

To solve this problem, change it from addition to subtraction using inverse operations, as I show you in Chapter 4:

$$\text{Sophia} = 700 - 420 = 280$$

Now that you know how much money each woman collected, you can answer the question at the end of the problem:

$$\text{Keisha} + \text{Brie} + \text{Amy} + \text{Sophia} = ?$$

You can plug in this information easily:

$$160 + 210 + 420 + 280 = 1{,}070$$

So you can conclude that the four women collected $1,070 all together.

Here's one final example putting together everything from this chapter:

> On a recent shopping trip, Travis bought six shirts for $19.95 each and two pairs of pants for $34.60 each. He then bought a jacket that cost $37.08 less than he paid for both pairs of pants. If he paid the cashier with three $100 bills, how much change did he receive?

On the first read-through, you may wonder how Travis found a store that prices jackets that way. Believe me — it was quite a challenge. Anyway, back to the problem. You can jot down the following word equations:

$$\text{shirts} = \$19.95 \times 6$$
$$\text{pants} = \$34.60 \times 2$$
$$\text{jacket} = \text{pants} - \$37.08$$

The numbers in this problem are probably longer than you can solve in your head, so they require some attention:

$$\begin{array}{r} \$19.95 \\ \times \quad 6 \\ \hline \$119.70 \end{array} \qquad \begin{array}{r} \$34.60 \\ \times \quad 2 \\ \hline \$69.20 \end{array}$$

With this done, you can fill in some more information:

$$\text{shirts} = \$119.70$$
$$\text{pants} = \$69.20$$
$$\text{jacket} = \text{pants} - \$37.08$$

Now you can plug in $69.20 for *pants:*

$$\text{jacket} = \$69.20 - \$37.08$$

Again, because the numbers are long, you need to solve this equation separately:

$$\begin{array}{r} \$69.20 \\ -\$37.08 \\ \hline \$32.12 \end{array}$$

This equation gives you the price of the jacket:

$$\text{jacket} = \$32.12$$

Now that you have the price of the shirts, pants, and jacket, you can find out how much Travis spent:

$$\text{amount Travis spent} = \$119.70 + \$69.20 + \$32.12$$

Again, you have another equation to solve:

$$\begin{array}{r} \$119.70 \\ \$69.20 \\ +\$32.12 \\ \hline \$221.02 \end{array}$$

So you can jot down the following:

$$\text{amount Travis spent} = \$221.02$$

The problem is asking you to find out how much change Travis received from $300, so jot this down:

$$\text{change} = \$300 - \text{amount Travis spent}$$

You can plug in the amount that Travis spent:

$$change = \$300 - \$221.02$$

And do just one more equation:

$$
\begin{array}{r}
\$300.00 \\
-\$221.02 \\
\hline
\$78.98
\end{array}
$$

So you can jot down the answer:

$$change = \$78.98$$

Therefore, Travis received $78.98 in change.

E

Example

Q. Emily loves counting money from her parents' change bowl. Yesterday, she counted 3 times as many pennies as nickels, 46 more nickels than dimes, and 7 times more dimes than quarters. If she counted 285 pennies, how many more pennies were in the bowl than quarters?

    a. How many pennies were in the bowl?

    b. How many nickels?

    c. How many dimes?

    d. How many quarters?

    e. What is the problem asking? Write the answer in a complete sentence.

A.   **a.** **285 pennies.**

    b. **95 nickels.** $(285 \div 3 = 95)$

    c. **49 dimes.** $(95 - 46 = 49)$

    d. **7 quarters.** $(49 \div 7 = 7)$

    e. **There were 278 more pennies in the bowl than quarters.** $(285 - 7 = 278)$

Practice Questions

1. At Seaborne High School, each of the 355 students cast a vote for one of four candidates for the president of student council. Marjorie received exactly one-fifth of the vote, which was 24 fewer votes than Leo, and Leo received 38 more votes than Glen. If Britney won the election by capturing all the rest of the votes, how many votes did she receive?

    a. How many votes were cast in total?

    b. How many votes did Marjorie receive?

    c. How many votes did Leo receive?

    d. How many votes did Glen receive?

    e. How many votes did Marjorie, Leo, and Glen receive together?

    f. What is the problem asking? Write the answer in a complete sentence.

2. If the Amaranth Building were three times taller, it would be exactly 24 feet taller than the Quinoa Hotel. And if the Quinoa Hotel were half as tall, it would be just 19 feet taller than the Teff Tower. Assuming that the Teff Tower is 74 feet, how much taller is the Quinoa Hotel than the Amaranth Building.

    a. How tall is the Teff Tower?

    b. How tall is the Quinoa Hotel?

    c. How tall is the Amaranth Building?

    d. What is the problem asking? Write the answer in a complete sentence.

3. Evan baked four types of cookies for a bake sale. He made twice as many chocolate chip cookies as oatmeal cookies, and a total of 120 mint julep cookies and peanut butter cookies combined. He made 23 more oatmeal cookies than mint julep cookies, and 66 peanut butter cookies. How many cookies did he make all together?

    a. How many peanut butter cookies did Evan make?

    b. How many mint julep cookies did he make?

    c. How many oatmeal cookies did he make?

    d. How many chocolate chip cookies did he make?

    e. What is the problem asking? Write the answer in a complete sentence.

4. A school has 16 times as many students as teachers, and 5 more teachers than administrators. The number of male students is 7 times the number of non-teaching staff workers, and there are 10 fewer male students than female students. If there are exactly 101 female students in the school, what is the total number of teacher, staff workers, and administrators combined?

    a. How many female students are there?

    b. How many male students are there?

    c.   What is the total number of students?

    d.   How many staff workers are there?

    e.   How many teachers are there?

    f.   How many administrators are there?

    g.   What is the problem asking? Write the answer in a complete sentence.

5.   Five members of an extended family all have birthdays on the same day. On their most recent birthday, Brad's age was three times Sara's, Sara was 16 years younger than Allen. When you add Allen's and Brad's ages together, the result was Eve's age. And Virginia is exactly four times older than Allen. If Virginia had just turned 84 years old, how old is Eve?

6.   The express train starting at Watmore and ending in Yarborough passes through Cramerville, Middle City, and Danbury, in that order. The leg from Watmore to Cramerville takes half as many minutes as the leg from Cramerville to Middle City. And the trip from Watmore to Middle City takes three times longer than the trip from Danbury to Yarborough. If the trip from Middle City to Danbury takes 18 minutes and the trip from Middle City to Yarborough takes 33 minutes, how long does the trip from Cramerville to Danbury take?

7.   If five donuts cost $3.25, two donuts and two small cups of coffee cost $3.00, and a cup of coffee and a toasted bagel with cream cheese costs $2.50, what is the cost of one donut, two toasted bagels with cream cheese, and three cups of coffee?

Ⓐ Practice Answers

1.   **a.** **A total of 355 were cast.**

    b. **Marjorie received 71 votes.** ($355 \div 5 = 71$)

    c. **Leo received 95 votes.** ($71 + 24 = 95$)

    d. **Glen received 57 votes.** ($95 - 38 = 57$)

    e. **Together, Marjorie, Leo, and Glen received 223 votes.** ($71 + 95 + 57 = 223$)

    f. **Britney won the election with 132 votes.** ($355 - 223 = 132$)

2.   **a.** **74 feet**

    b. **186 feet** $(74 + 19) \times 2 = 186$

    c. **70 feet** $(186 + 24) \div 3 = 70$

    d. **The Quinoa Hotel is 116 feet taller than the Amaranth Building.** ($186 - 70 = 116$)

3. a. **66**

    b. **54** $(120 - 66 = 54)$

    c. **77** $(54 + 23 = 77)$

    d. **154** $(77 \times 2 = 154)$

    e. **Evan made a total of 351 cookies.** $(66 + 54 + 77 + 154 = 351)$

4. a. **101**

    b. **91** $(101 - 10 = 91)$

    c. **192** $(101 + 91 = 192)$

    d. **13** $(91 \div 7 = 13)$

    e. **12** $(192 \div 16 = 12)$

    f. **7** $(12 - 5 = 7)$

    g. **There are a total of 32 teachers, staff workers, and administrators.** $(13 + 12 + 7 = 32)$

5. **36 years old.** Virginia was 84, so Allen was 21 $(84 \div 4 = 21)$. Sara was five $(21 - 16 = 5)$, so Brad was 15 $(5 \times 3 = 15)$. Therefore, Eve was 36 $(21 + 15 = 36)$.

6. **48 minutes.** The trip from Middle City to Danbury takes 18 minutes and the trip from Middle City to Yarborough takes 33 minutes, so the trip from Danbury to Yarborough takes 15 minutes $(33 - 18 = 15)$. So the trip from Watmore to Middle City takes 45 minutes $(15 \times 3 = 45)$. These 45 minutes consist of two legs, the first of which (from Watmore to Cramerville) takes half as long as the trip second (from Cramerville to Middle City). So, the trip from Watmore to Cramerville takes 15 minutes and the trip from Cramerville to Middle City takes 30 minutes. Therefore, the trip from Cramerville to Danbury takes 48 minutes.

7. **$6.50.** Five donuts cost $3.25, so each donut costs $.65 $(\$3.25 \div 5 = 65)$. So two donuts cost $1.30 $(\$.65 \times 2 = \$1.30)$. The combined cost of two donuts and two coffees $3.00, so two coffees cost $1.70 $(\$3.00 - \$1.30 = \$1.70)$; therefore, one coffee costs $.85 $(\$1.70 \div 2 = \$.85)$. The combined cost of a coffee and a bagel is $2.50, so a bagel costs $1.65 $(\$2.50 - \$.85 = \$1.65)$. Therefore, here's how you calculate the cost of a donut, two bagels, and three coffees:

$$\$.65 + (\$1.65 \times 2) + (\$.85 \times 3) = \$.65 + \$3.30 + \$2.55 = \$6.50$$

6

DIVISIBILITY AND PRIME NUMBERS

IN THIS CHAPTER

- Checking final digits to find out whether a number is divisible by 2, 5, 10, 100, or 1,000
- Adding up digits to discover when a number is divisible by 3, 9, or 11
- Subtracting to know whether a number is divisible by 4 or 8
- Combining rules to test for divisibility by 6, 12, 15, or 18
- Seeing the difference between prime numbers and composite numbers

hen one number is *divisible* by another, you can divide the first number by the second number without getting a remainder (see Chapter 3 for details on division). In this chapter, I explore divisibility from a variety of angles.

To start, I show you a bunch of handy tricks for discovering whether one number is divisible by another without actually doing the division. (In fact, you don't find long division anywhere in this chapter!) After that, I talk about prime numbers and composite numbers (which I introduce briefly in Chapter 1).

This discussion, plus what follows in Chapter 8, can help make your encounter with fractions in Part III a lot friendlier.

KNOWING THE DIVISIBILITY TRICKS

As you begin to work with fractions in Part III, the question of whether one number is divisible by another comes up a lot. In this section, I give you a bunch of time-saving tricks for finding out whether one number is divisible by another without actually making you do the division.

Counting everyone in: Numbers you can divide everything by

Every number is divisible by 1. As you can see, when you divide any number by 1, the answer is the number itself, with no remainder:

$$2 \div 1 = 2$$
$$17 \div 1 = 17$$
$$431 \div 1 = 431$$

Similarly, every number (except 0) is divisible by itself. Clearly, when you divide any number by itself, the answer is 1:

$$5 \div 5 = 1$$
$$28 \div 28 = 1$$
$$873 \div 873 = 1$$

Warning: You can't divide any number by 0. Mathematicians say that dividing by 0 is *undefined*.

In the end: Looking at the final digits

You can tell whether a number is divisible by 2, 5, 10, 100, or 1,000 simply by looking at how the number ends — no calculations required.

Divisible by 2

Every even number — that is, every number that ends in 2, 4, 6, 8, or 0 — is divisible by 2. For example, the following bolded numbers are divisible by 2:

$$\mathbf{6} \div 2 = 3$$
$$\mathbf{22} \div 2 = 11$$
$$\mathbf{538} \div 2 = 269$$
$$\mathbf{6,790} \div 2 = 3,395$$
$$\mathbf{77,144} \div 2 = 38,572$$
$$\mathbf{212,116} \div 2 = 106,058$$

Divisible by 5

Every number that ends in either 5 or 0 is divisible by 5. The following bolded numbers are divisible by 5:

$$\mathbf{15} \div 5 = 3$$
$$\mathbf{625} \div 5 = 125$$
$$\mathbf{6,970} \div 5 = 1,394$$
$$\mathbf{44,440} \div 5 = 8,888$$
$$\mathbf{511,725} \div 5 = 102,345$$
$$\mathbf{9,876,630} \div 5 = 1,975,326$$

Divisible by 10, 100, or 1,000

Every number that ends in 0 is divisible by 10. The following bolded numbers are divisible by 10:

$$\mathbf{20} \div 10 = 2$$
$$\mathbf{170} \div 10 = 17$$
$$\mathbf{56{,}720} \div 10 = 5{,}672$$

Every number that ends in 00 is divisible by 100:

$$\mathbf{300} \div 100 = 3$$
$$\mathbf{8{,}300} \div 100 = 83$$
$$\mathbf{634{,}900} \div 100 = 6{,}349$$

And every number that ends in 000 is divisible by 1,000:

$$\mathbf{6{,}000} \div 1{,}000 = 6$$
$$\mathbf{99{,}000} \div 1{,}000 = 99$$
$$\mathbf{1{,}234{,}000} \div 1{,}000 = 1{,}234$$

In general, every number that ends with a string of 0s is divisible by the number you get when you write 1 followed by that many 0s. For example,

900,000 is divisible by 100,000.

235,000,000 is divisible by 1,000,000.

820,000,000,000 is divisible by 10,000,000,000.

Examples

Q. Is 348 divisible by 2?

A. **Yes,** because it's an even number (check: 348 ÷ 2 = 174).

Q. Is 551 divisible by 5?

A. **No,** because it ends in 1, not 5 or 0 (check: 551 ÷ 5 = 110 r 1).

Q. Is 1,620 divisible by 10?

A. **Yes,** because it ends in 0 (check: 1,620 ÷ 10 = 162).

Q. Is 3,050 divisible by 100?

A. **No,** because it ends in 50, not 00 (check: 3,050 ÷ 100 = 30 r 50).

Practice Questions

1. Which of the following numbers are divisible by 2?

   a. 37

   b. 82

   c. 111

   d. 75,316

2. Which of the following numbers are divisible by 5?

   a. 75

   b. 103

   c. 230

   d. 9,995

3. Which of the following numbers are divisible by 10?

   a. 40

   b. 105

   c. 200

   d. 60,001

4. Which of the following numbers are divisible by 100?

   a. 660

   b. 900

   c. 10,200

   d. 500,080

Practice Answers

1.   **a.** **No,** because it's odd (check: $37 \div 2 = 18$ r 1).

     b. **Yes,** because it's even (check: $82 \div 2 = 41$).

     c. **No,** because it's odd (check: $111 \div 2 = 55$ r 1).

     d. **Yes,** because it's even (check: $75,316 \div 2 = 37,658$).

2. **a.** **Yes,** because it ends in 5 (check: 75 ÷ 5 = 25).

   b. **No,** because it ends in 3, not 0 or 5 (check: 103 ÷ 5 = 20 r 3).

   c. **Yes,** because it ends in 0 (check: 230 ÷ 5 = 46).

   d. **Yes,** because it ends in 5 (check: 9,995 ÷ 5 = 1,999).

3. **a.** **Yes,** because it ends in 0 (check: 40 ÷ 10 = 4).

   b. **No,** because it ends in 5, not 0 (check: 105 ÷ 10 = 10 r 5).

   c. **Yes,** because it ends in 0 (check: 200 ÷ 10 = 20).

   d. **No,** because it ends in 5, not 0 (check: 60,001 ÷ 10 = 6,000 r 1).

4. **a.** **No,** because it ends in 60, not 00 (check: 660 ÷ 100 = 6 r 60).

   b. **Yes,** because it ends in 00 (check: 900 ÷ 100 = 9).

   c. **Yes,** because it ends in 00 (check: 10,200 ÷ 100 = 102).

   d. **No,** because it ends in 80, not 00 (check: 500,080 ÷ 100 = 5,000 r 80).

Count it up: Checking divisibility by adding and subtracting digits

Sometimes you can check divisibility by adding up all or some of the digits in a number. The sum of a number's digits is called its *digital root*. Finding the digital root of a number is easy, and it's handy to know.

Remember: To find the digital root of a number, just add up the digits and repeat this process until you get a one-digit number. Here are some examples:

The digital root of 24 is 6 because 2 + 4 = 6.

The digital root of 143 is 8 because 1 + 4 + 3 = 8.

The digital root of 51,111 is 9 because 5 + 1 + 1 + 1 + 1 = 9.

Sometimes you need to do this process more than once. Here's how to find the digital root of the number 87,482. You have to repeat the process three times, but eventually you find that the digital root of 87,482 is 2:

$$8 + 7 + 4 + 8 + 2 = 29$$
$$2 + 9 = 11$$
$$1 + 1 = 2$$

Read on to find out how sums of digits can help you check for divisibility by 3, 9, or 11.

Divisible by 3

Remember: Every number whose digital root is 3, 6, or 9 is divisible by 3.

First, find the digital root of a number by adding its digits until you get a single-digit number. Here are the digital roots of 18, 51, and 975:

18:	1 + 8 = 9
51:	5 + 1 = 6
975:	9 + 7 + 5 = 21; 2 + 1 = 3

With the numbers 18 and 51, adding the digits leads immediately to digital roots 9 and 6, respectively. With 975, when you add up the digits, you first get 21, so you then add up the digits in 21 to get the digital root 3. Thus, these three numbers are all divisible by 3. If you do the actual division, you find that 18 ÷ 3 = 6, 51 ÷ 3 = 17, and 975 ÷ 3 = 325, so the method checks out.

However, when the digital root of a number is anything other than 3, 6, or 9, the number *isn't* divisible by 3:

1,037:	1 + 0 + 3 + 7 = 11; 1 + 1 = 2

Because the digital root of 1,037 is 2, 1,037 *isn't* divisible by 3. If you try to divide by 3, you end up with 345 r 2.

Divisible by 9

Remember: Every number whose digital root is 9 is divisible by 9.

To test whether a number is divisible by 9, find its digital root by adding up its digits until you get a one-digit number. Here are some examples:

36:	3 + 6 = 9
243:	2 + 4 + 3 = 9
7,587:	7 + 5 + 8 + 7 = 27; 2 + 7 = 9

With the numbers 36 and 243, adding the digits leads immediately to digital roots of 9 in both cases. With 7,587, however, when you add up the digits, you get 27, so you then add up the digits in 27 to get the digital root 9. Thus, all three of these numbers are divisible by 9. You can verify this by doing the division:

$$36 \div 9 = 4 \quad 243 \div 9 = 27 \quad 7,857 \div 9 = 873$$

However, when the digital root of a number is anything other than 9, the number isn't divisible by 9. Here's an example:

706:	$7 + 0 + 6 = 13; 1 + 3 = 4$

Because the digital root of 706 is 4, 706 *isn't* divisible by 9. If you try to divide 706 by 9, you get 78 r 4.

Divisible by 11

Two-digit numbers that are divisible by 11 are hard to miss because they simply repeat the same digit twice. Here are all the numbers less than 100 that are divisible by 11:

$$11 \quad 22 \quad 33 \quad 44 \quad 55 \quad 66 \quad 77 \quad 88 \quad 99$$

Tip: For numbers between 100 and 200, use this rule: Every three-digit number whose first and third digits add up to its second digit is divisible by 11. For example, suppose you want to decide whether the number 154 is divisible by 11. Just add the first and third digits:

$$1 + 4 = 5$$

Because these two numbers add up to the second digit, 5, the number 154 is divisible by 11. If you divide, you get $154 \div 11 = 14$, a whole number.

Now suppose you want to figure out whether 136 is divisible by 11. Add the first and third digits:

$$1 + 6 = 7$$

Because the first and third digits add up to 7 instead of 3, the number 136 isn't divisible by 11. You can find that $136 \div 11 = 12$ r 4.

Tip: For numbers of any length, the rule is slightly more complicated, but it's still often easier than doing long division. To find out when a number is divisible by 11, place plus and minus signs alternatively in front of every digit, then calculate the result. If this result is divisible by 11 (including 0), the number is divisible by 11; otherwise, the number isn't divisible by 11.

For example, suppose you want to discover whether the number 15,983 is divisible by 11. To start out, place plus and minus signs in front of alternate digits (every other digit):

$$+1 - 5 + 9 - 8 + 3 = 0$$

Because the result is 0, the number 15,983 is divisible by 11. If you check the division, $15,983 \div 11 = 1,453$.

Now suppose you want to find out whether 9,181,909 is divisible by 11. Again, place plus and minus signs in front of alternate digits and calculate the result:

$$+9-1+8-1+9-0+9=33$$

Because 33 is divisible by 11, the number 9,181,909 is also divisible by 11. The actual answer is

$$9,181,909 \div 11 = 834,719$$

Examples

Q. Which of the following numbers are divisible by 3?

   a. 31

   b. 54

   c. 768

   d. 2,809

A. Add the digits to determine each number's digital root; if the digital root is 3, 6, or 9, the number is divisible by 3:

   a. **No,** because 3 + 1 = 4 (check: 31 ÷ 3 = 10 r 1).

   b. **Yes,** because 5 + 4 = 9 (check: 54 ÷ 3 = 18).

   c. **Yes,** because 7 + 6 + 8 = 21 and 2 + 1 = 3 (check: 768 ÷ 3 = 256).

   d. **No,** because 2 + 8 + 0 + 9 = 19, 1 + 9 = 10, and 1 + 0 = 1 (check: 2,809 ÷ 3 = 936 r 1).

Q. Which of the following numbers are divisible by 11?

   a. 71

   b. 154

   c. 528

   d. 28,094

A. Place + and - signs between the numbers and determine whether the result is 0 or a multiple of 11:

   a. **No,** because +7 - 1 = 6 (check: 71 ÷ 11 = 6 r 5).

   b. **Yes,** because +1 - 5 + 4 = 0 (check: 154 ÷ 11 = 14).

   c. **Yes,** because +5 - 2 + 8 = 11 (check: 528 ÷ 11 = 48).

   d. **Yes,** because +2 - 8 + 0 - 9 + 4 = -11 (check: 28,094 ÷ 11 = 2,554).

Practice Questions

1. Which of the following numbers are divisible by 3?

    a. 81

    b. 304

    c. 986

    d. 1,027

2. Which of the following numbers are divisible by 3?

    a. 20,103

    b. 541,836

    c. 2,345,678

    d. 4,444,444

3. Which of the following numbers are divisible by 9?

    a. 107

    b. 765

    c. 9,876

    d. 1,111,248

4. Which of the following numbers are divisible by 11?

    a. 42

    b. 187

    c. 726

    d. 1,969

Practice Answers

1. Which of the following numbers are divisible by 3? *Note:* The digital root is 3, 6, or 9 for numbers divisible by 3:

    a. **Yes,** because 8 + 1 = 9 (check: 81 ÷ 3 = 27).

    b. **No,** because 3 + 0 + 4 = 7 (check: 304 ÷ 3 = 101 r 1).

    c. **No,** because 9 + 8 + 6 = 23 and 2 + 3 = 5 (check: 986 ÷ 3 = 328 r 2).

    d. **No,** because 1 + 2 + 7 = 10 and 1 + 0 = 1 (check: 1,027 ÷ 3 = 342 r 1).

2.   Which of the following numbers are divisible by 3? *Note:* The digital root is 3, 6, or 9 for numbers divisible by 3:

   a.   **Yes,** because $2 + 0 + 1 + 0 + 3 = 6$ (check: $20{,}103 \div 3 = 6{,}701$).

   b.   **Yes,** because $5 + 4 + 1 + 8 + 3 + 6 = 27$ and $2 + 7 = 9$ (check: $541{,}836 \div 3 = 180{,}612$).

   c.   **No,** because $2 + 3 + 4 + 5 + 6 + 7 + 8 = 35$ and $3 + 5 = 8$ (check: $2{,}345{,}678 \div 3 = 78{,}189$ r 2).

   d.   **No,** because $4 + 4 + 4 + 4 + 4 + 4 + 4 = 28$, $2 + 8 = 10$, and $1 + 0 = 1$ (check: $4{,}444{,}444 \div 3 = 1{,}481{,}481$ r 1).

3.   Which of the following numbers are divisible by 9? *Note:* The digital root is 9 for numbers divisible by 9:

   a.   **No,** because $1 + 0 + 7 = 8$ (check: $107 \div 9 = 11$ r 8).

   b.   **Yes,** because $7 + 6 + 5 = 18$ and $1 + 8 = 9$ (check: $765 \div 9 = 85$).

   c.   **No,** because $9 + 8 + 7 + 6 = 30$ and $3 + 0 = 3$ (check: $9{,}876 \div 9 = 1{,}097$ r 3).

   d.   **Yes,** because $1 + 1 + 1 + 1 + 2 + 4 + 8 = 18$ and $1 + 8 = 9$ (check: $1{,}111{,}248 \div 9 = 123{,}472$).

4.   Which of the following numbers are divisible by 11? *Note:* Answers add up to 0 or a multiple of 11 for numbers divisible by 11:

   a.   **No,** because $+4 - 2 = 2$ (check: $42 \div 11 = 3$ r 9).

   b.   **Yes,** because $+1 - 8 + 7 = 0$ (check: $187 \div 11 = 17$).

   c.   **Yes,** because $+7 - 2 + 6 = 11$ (check: $726 \div 11 = 66$).

   d.   **Yes,** because $+1 - 9 + 6 - 9 = -11$ (check: $1{,}969 \div 11 = 179$).

Less is more: Checking divisibility by subtracting

The tests for divisibility by 4 and 8 are a little unusual, but with practice you may find them useful from time to time.

Divisible by 4

Remember: To find out whether a number is divisible by 4:

1.   **Drop all but the last two digits.**

2.   **Keep subtracting 20 until the result is 20 or less.**

If the result is divisible by 4 (that is, 4, 8, 12, 16, or 20), the number you started with is also divisible by 4.

Divisible by 8

Remember: To test a number for divisibility by 8:

1. **Drop all but the last three digits.**
2. **Keep subtracting 200 until the result is 200 or less.**
3. **Keep subtracting 40 until the result is 40 or less.**

If the result is divisible by 8 (that is, 8, 16, 24, 32, or 40), the original number is also divisible by 8.

Remember: Odd numbers are *never* divisible by 4 or 8.

Examples

Q.   Is 856 divisible by 4?

A.   **Yes.** Drop all but the last two digits. Then keep subtracting 20 until you know whether the number is divisible by 4: 856 → 56 → 36 → 16, and 16 is divisible by 4 (check: 856 ÷ 4 = 214).

Q.   Is 1,492 divisible by 8?

A.   **No.** Drop all but the last three digits. Then keep subtracting 200 and then 40 until you know whether the number is divisible by 8: 1,492 → 492 → 292 → 92 → 52 → 12, and 12 is not divisible by 8 (check: 1,492 ÷ 8 = 186 r 4).

Practice Questions

1.   Which of the following numbers are divisible by 4?

   a.   82
   b.   756
   c.   4,463
   d.   789,508

2.   Which of the following numbers are divisible by 8?

   a.   112
   b.   386
   c.   5,458
   d.   39,216

Practice Answers

1. Drop all but the last two digits. Then keep subtracting 20 until you know whether the number is divisible by 4. (***Remember:*** If the number is odd, it's *not* divisible by 4.)

    a. **No,** because $82 \rightarrow 62 \rightarrow 42 \rightarrow 22 \rightarrow 2$, and 2 is not divisible by 4 (check: $82 \div 4 = 20$ r 2).

    b. **Yes,** because $756 \rightarrow 56 \rightarrow 36 \rightarrow 16$, and 16 is divisible by 4 (check: $756 \div 4 = 189$).

    c. **No,** because 4,463 is an odd number (check: $4,463 \div 4 = 1,115$ r 3).

    d. **Yes,** because $789,508 \rightarrow 08$, and 8 is divisible by 4 (check: $789,508 \div 4 = 197,377$).

2. Drop all but the last three digits. Then keep subtracting 200, then 40, until you know whether the number is divisible by 8. (***Remember:*** If the number is odd, it's *not* divisible by 8.)

    a. **Yes,** because $112 \rightarrow 72 \rightarrow 32$, and 32 is divisible by 8 (check: $112 \div 8 = 14$).

    b. **No,** because $386 \rightarrow 186 \rightarrow 146 \rightarrow 106 \rightarrow 66 \rightarrow 26$, and 26 is not divisible by 8 (check: $386 \div 8 = 48$ r 2).

    c. **No,** because $5,458 \rightarrow 458 \rightarrow 258 \rightarrow 58 \rightarrow 18$, and 18 is not divisible by 8 (check: $5,458 \div 8 = 682$ r 2).

    d. **Yes,** because $39,216 \rightarrow 216 \rightarrow 16$, and 16 is divisible by 8 (check: $39,216 \div 8 = 4,902$).

Cross-checking: Using multiple tests

To test for divisibility by 6, 12, 15, and 18, use a pair of divisibility tests explained earlier in this chapter:

- **By 6:** Test for divisibility by 2 and 3: Any even number whose digital root is 3, 6, or 9 is divisible by 6; otherwise, the number isn't.

- **By 12:** Test for divisibility by 3 and 4: Any number whose digital root is 3, 6, or 9 *and* that can be decreased to a number that's divisible by 4 by repeatedly subtracting 20 is divisible by 12; all other numbers aren't.

- **By 15:** Test for divisibility by 3 and 5: Any number that ends in 5 or 0 and has a digital root of 3, 6, or 9 is divisible by 15; all other numbers aren't.

- **By 18:** Test for divisibility by 2 and 9: Any even number whose digital root is 9 is divisible by 18; other numbers aren't.

Examples

Q. Is 702 divisible by 6?

A. **Yes,** because 702 is even and $7 + 0 + 2 = 9$ (check: $702 \div 6 = 117$).

Q. Is 624 divisible by 12?

A. **Yes,** because 6 + 2 + 4 = 12 and 1 + 2 = 3, and 624 → 24 → 4, which is divisible by 4 (check: 624 ÷ 12 = 52).

Q. Is 2,160 divisible by 15?

A. **Yes,** because 2,160 ends in 0 and 2 + 1 + 6 + 0 = 9 (check: 2,160 ÷ 15 = 144).

Q. Is 8,142 divisible by 18?

A. **No,** because 8,142 is even, but 8 + 1 + 4 + 2 = 15 and 1 + 5 = 6 (check: 8,142 ÷ 18 = 452 r 6).

Practice Questions

1. Is 178 divisible by 6?

2. Is 338 divisible by 12?

3. Is 505 divisible by 15?

4. Is 1,656 divisible by 18?

Practice Answers

1. **No,** because 178 is even, but 1 + 7 + 8 = 16 and 1 + 6 = 7 (check: 178 ÷ 6 = 29 r 4).

2. **No,** because 3 + 3 + 8 = 14 and 1 + 4 = 5 (check: 178 ÷ 12 = 28 r 2).

3. **No,** because 505 ends in 5, but 5 + 0 + 5 = 10 and 1 + 0 = 1 (check: 505 ÷ 15 = 33 r 10).

4. **Yes,** because 1,656 is even, and 1 + 6 + 5 + 6 = 18 and 1 + 8 = 9 (check: 1,656 ÷ 18 = 92).

IDENTIFYING PRIME AND COMPOSITE NUMBERS

In the earlier section titled "Counting everyone in: Numbers you can divide everything by," I show you that every number (except 0 and 1) is divisible by at least two numbers: 1 and itself. In this section, I explore prime numbers and composite numbers (which I introduce you to in Chapter 1).

In Chapter 7, you need to know how to tell prime numbers from composite to break a number down into its prime factors. This tactic is important when you begin working with fractions.

Remember: A *prime number* is divisible by exactly two positive whole numbers: 1 and the number itself. A *composite number* is divisible by at least three numbers.

For example, 2 is a prime number because when you divide it by any number but 1 and 2, you get a remainder. So there's only one way to multiply two counting numbers and get 2 as a product:

$$1 \times 2 = 2$$

Similarly, 3 is prime because when you divide by any number but 1 or 3, you get a remainder. So the only way to multiply two numbers together and get 3 as a product is the following:

$$1 \times 3 = 3$$

On the other hand, 4 is a composite number because it's divisible by three numbers: 1, 2, and 4. In this case, you have two ways to multiply two counting numbers and get a product of 4:

$$1 \times 4 = 4$$
$$2 \times 2 = 4$$

But 5 is a prime number because it's divisible only by 1 and 5. Here's the only way to multiply two counting numbers and get 5 as a product:

$$1 \times 5 = 5$$

And 6 is a composite number because it's divisible by 1, 2, 3, and 6. Here are two ways to multiply two counting numbers and get a product of 6:

$$1 \times 6 = 6$$
$$2 \times 3 = 6$$

Remember: Every counting number except 1 is either prime or composite. The reason 1 is neither is that it's divisible by only *one* number, which is 1.

Here's a list of the prime numbers that are less than 30:

$$2, 3, 5, 7, 11, 13, 17, 19, 23, 29$$

Tip: When testing to see whether a number is prime or composite, perform divisibility tests in the following order (from easiest to hardest): 2, 5, 3, 11, 7, and 13. If you find that a number is divisible by one of these, you know that it's composite and you don't have to perform the remaining tests. Here's how you know which tests to perform:

- If a number less than 121 isn't divisible by 2, 3, 5, or 7, it's prime; otherwise, it's composite.

- If a number less than 289 isn't divisible by 2, 3, 5, 7, 11, or 13, it's prime; otherwise, it's composite.

Remember that 2 is the only prime number that's even. The next three odd numbers are prime — 3, 5, and 7. To keep the list going, think "lucky 7, lucky 11, unlucky 13" — they're all prime.

E

Example

Q. For each of the following numbers, tell which is prime and which is composite.

    a. 185

    b. 243

    c. 253

    d. 263

A. Check divisibility to identify prime and composite numbers:

    a. **185 is composite.** The number 185 ends in 5, so it's divisible by 5.

    b. **243 is composite.** The number 243 ends in an odd number, so it isn't divisible by 2. It doesn't end in 5 or 0, so it isn't divisible by 5. Its digital root is 9 (because 2 + 4 + 3 = 9), so it's divisible by 3. The math shows you that 243 ÷ 3 = 81.

    c. **253 is composite.** The number 253 ends in an odd number, so it isn't divisible by 2. It doesn't end in 5 or 0, so it isn't divisible by 5. Its digital root is 1 (because 2 + 5 + 3 = 10 and 1 + 0 = 1), so it isn't divisible by 3. But it is divisible by 11, because it passes the + and – test (+2 – 5 + 3 = 0). If you do the math, you find that 253 = 11 × 23.

    d. **263 is prime.** The number 263 ends in an odd number, so it isn't divisible by 2. It doesn't end in 5 or 0, so it isn't divisible by 5. Its digital root is 2 (because 2 + 6 + 3 = 11 and 1 + 1 = 2), so it isn't divisible by 3. It isn't divisible by 11, because it fails the + and – test (+2 – 6 + 3 = –1, which isn't 0 or divisible by 11). It isn't divisible by 7, because 263 ÷ 7 = 37 r 2. And it isn't divisible by 13, because 263 ÷ 13 = 20 r 3.

Q

Practice Questions

    1. Which of the following numbers are prime, and which are composite?

        a. 3

        b. 9

        c. 11

        d. 14

    2. Of the following numbers, tell which are prime and which are composite.

        a. 65

        b. 73

        c. 111

        d. 172

3. Find out whether each of these numbers is prime or composite.

    a. 23

    b. 51

    c. 91

    d. 113

4. Figure out which of the following are prime numbers and which are composite numbers.

    a. 143

    b. 169

    c. 187

    d. 283

Practice Answers

1. Which of the following numbers are prime, and which are composite?

    a. **3 is prime.** The only factors of 3 are 1 and 3.

    b. **9 is composite.** The factors of 9 are 1, 3, and 9.

    c. **11 is prime.** Eleven's only factors are 1 and 11.

    d. **14 is composite.** As an even number, 14 is also divisible by 2 and therefore can't be prime.

2. Of the following numbers, tell which are prime and which are composite.

    a. **65 is composite.** Because 65 ends in 5, it's divisible by 5.

    b. **73 is prime.** The number 73 isn't even, doesn't end in 5 or 0, and isn't a multiple of 7.

    c. **111 is composite.** The digital root of 111 is $1 + 1 + 1 = 3$, so it's divisible by 3 (check: $111 \div 3 = 37$).

    d. **172 is composite.** The number 172 is even, so it's divisible by 2.

3. Find out whether each of these numbers is prime or composite.

    a. **23 is prime.** The number 23 isn't even, doesn't end in 5 or 0, has a digital root of 5, and isn't a multiple of 7.

    b. **51 is composite.** The digital root of 51 is 6, so it's a multiple of 3 (check: $51 \div 3 = 17$).

    c. **91 is composite.** The number 91 is a multiple of 7: $7 \times 13 = 91$.

    d. **113 is prime.** The number 113 is odd, doesn't end in 5 or 0, and has a digital root of 5, so it's not divisible by 2, 5, or 3. It's also not a multiple of 7: $113 \div 7 = 16$ r 1.

4. Figure out which of the following are prime numbers and which are composite numbers:

    a. **143 is composite.** +1 – 4 + 3 = 0, so 143 is divisible by 11.

    b. **169 is composite.** You can evenly divide 13 into 169 to get 13.

    c. **187 is composite.** +1 – 8 + 7 = 0, so 187 is a multiple of 11.

    d. **283 is prime.** The number 283 is odd, doesn't end in 5 or 0, and has a digital root of 4; therefore, it's not divisible by 2, 5, or 3. It's not divisible by 11, because +2 – 8 + 3 = –3, which isn't a multiple of 11. It also isn't divisible by 7 (because 283 ÷ 7 = 40 r 3) or 13 (because 283 ÷ 13 = 21 r 10).

7

FABULOUS FACTORS AND MARVELOUS MULTIPLES

IN THIS CHAPTER

- Understanding how factors and multiples are related
- Listing all the factors of a number
- Breaking down a number into its prime factors
- Generating multiples of a number
- Finding the greatest common factor (GCF) and least common multiple (LCM)

In Chapter 1, I introduce you to sequences of numbers based on the multiplication table. In this chapter, I tell you about two important ways to think about these sequences: as *factors* and as *multiples*. Factors and multiples are really two sides of the same coin. Here I show you what you need to know about these two important concepts.

For starters, I discuss how factors and multiples are connected to multiplication and division. Then I show you how to find all the factor pairs of a number and how to decompose (split up) any number into its prime factors. To finish up on factors, I show you how to find the greatest common factor (GCF) of any set of numbers. After that, I tackle multiples, showing you how to generate the multiples of a number and then use this skill to find the least common multiple (LCM) of a set of numbers.

KNOWING SIX WAYS TO SAY THE SAME THING

In this section, I introduce you to factors and multiples, and I show you how these two important concepts are connected. As I discuss in Chapter 3, multiplication and division are inverse operations. For example, the following equation is true:

$$5 \times 4 = 20$$

So this equation using the inverse operation is also true:

$$20 \div 4 = 5$$

You may have noticed that, in math, you tend to run into the same ideas over and over again. For example, mathematicians have six different ways to talk about this relationship.

The following three statements all focus on the relationship between 5 and 20 from the perspective of multiplication:

- 5 *multiplied* by some number is 20.
- 5 is a *factor* of 20.
- 20 is a *multiple* of 5.

In two of the examples, you can see this relationship reflected in the words *multiplied* and *multiple*. For the remaining example, keep in mind that two factors are multiplied to equal a product.

Similarly, the following three statements all focus on the relationship between 5 and 20 from the perspective of division:

- 20 *divided* by some number is 5.
- 20 is *divisible by* 5.
- 5 is a *divisor* of 20.

Why do mathematicians need all these words for the same thing? Maybe for the same reason that Eskimos need a bunch of words for snow. In any case, in this chapter, I focus on the words *factor* and *multiple*. When you understand the concepts, which word you choose doesn't matter a whole lot.

UNDERSTANDING FACTORS AND MULTIPLES

In the preceding section, I connect the concept of divisibility (see Chapter 6) with factors and multiples.

For example, 12 is divisible by 3 because $12 \div 3 = 4$, with no remainder. You can also describe this relationship between 12 and 3 using the words *factor* and *multiple*. When you're working with positive numbers, the factor is always the *smaller number* and the multiple is the *bigger number*. For example, 12 is divisible by 3, so

- The number 3 is a *factor* of 12.
- The number 12 is a *multiple* of 3.

Examples

Q. The number 40 is divisible by 5, so which number is the factor and which is the multiple?

A. The number **5 is the factor and 40 is the multiple,** because 5 is smaller and 40 is larger.

Q. Which two of the following statements means the same thing as "18 is a multiple of 6"?

   a. 6 is a factor of 18.

   b. 18 is divisible by 6.

   c. 6 is divisible by 18.

   d. 18 is a factor of 6.

A. **Choices *a* and *b*.** The number 6 is the factor and 18 is the multiple, because 6 is smaller than 18, so *a* is correct. And $18 \div 6 = 3$, so 18 is divisible by 6; therefore, *b* is correct.

Practice Questions

1. Which of the following statements are true, and which are false?

   a. 5 is a factor of 15.

   b. 9 is a multiple of 3.

   c. 11 is a factor of 12.

   d. 7 is a multiple of 14.

2. Which two of these statements mean the same thing as "18 is divisible by 6"?

   a. 18 is a factor of 6.

   b. 18 is a multiple of 6.

   c. 6 is a factor of 18.

   d. 6 is a multiple of 18.

3. Which two of these statements mean the same thing as "10 is a factor of 50"?

   a. 10 is divisible by 50.

   b. 10 is a multiple of 50.

   c. 50 is divisible by 10.

   d. 50 is a multiple of 10.

4. Which of the following statements are true, and which are false?

   a. 3 is a factor of 42.

   b. 11 is a multiple of 121.

   c. 88 is a multiple of 9.

   d. 11 is a factor of 121.

Practice Answers

1. a. **True:** $5 \times 3 = 15$.

    b. **True:** $3 \times 3 = 9$.

    c. **False:** You can't multiply 11 by any whole number to get 12.

    d. **False:** The number 7 is a factor of 14.

2. **Choices *b* and *c*.** You're looking for something that says that the smaller number, 6, is a factor of the larger number (choice *c*) or one that says the larger number, 18, is a multiple of the smaller number (choice *b*).

3. **Choices *c* and *d*.** Factors are numbers you multiply to get larger ones, so you can say 50 is divisible by 10 (choice *c*). Multiples are the larger numbers, the products you get when you multiply two factors; you can say 50 is a multiple of 10 (choice *d*) because $10 \times 5 = 50$.

4. a. **True:** $3 \times 14 = 42$.

    b. **False:** The number 11 is a factor of 121.

    c. **False:** You can't multiply 9 by any whole numbers to get 88 because $88 \div 9 = 9 \text{ r } 7$.

    d. **True:** $11 \times 11 = 121$.

FINDING FABULOUS FACTORS

In this section, I introduce you to factors. First, I show you how to find out whether one number is a factor of another. Then I show you how to list all the factor pairs of a number. After that, I introduce the key idea of a number's prime factors. This information all leads up to an essential skill: finding the greatest common factor (GCF) of a set of numbers.

Deciding when one number is a factor of another

Remember: You can easily tell whether a number is a factor of a second number: Just divide the second number by the first. If it divides evenly (with no remainder), the number is a factor; otherwise, it's not a factor.

For example, suppose you want to know whether 7 is a factor of 56. Here's how you find out:

$$56 \div 7 = 8$$

Because 7 divides 56 without leaving a remainder, 7 is a factor of 56.

And here's how you find out whether 4 is a factor of 34:

$$34 \div 4 = 8 \text{ r } 2$$

Because 4 divides 34 with a remainder of 2, 4 isn't a factor of 34.

This method works no matter how large the numbers are.

Understanding factor pairs

Remember: A factor pair of a number is any pair of two numbers that, when multiplied together, equal that number. For example, 35 has two factor pairs — 1 × 35 and 5 × 7 because

$$1 \times 35 = 35$$
$$5 \times 7 = 35$$

Similarly, 24 has four factor pairs — $1 \times 24, 2 \times 12, 3 \times 8$, and 4×6 — because

$$1 \times 24 = 24$$
$$2 \times 12 = 24$$
$$3 \times 8 = 24$$
$$4 \times 6 = 24$$

Tip: Every positive integer has at least one factor pair: 1 times the number itself. For example:

$$1 \times 2 = 2 \quad 1 \times 11 = 11 \quad 1 \times 43 = 43$$

When a number greater than 1 has only one factor pair, it's a prime number (see Chapter 6 for more on prime numbers).

GENERATING A NUMBER'S FACTORS

When one number is divisible by a second number, that second number is a *factor* of the first. For example, 10 is divisible by 2, so 2 is a factor of 10.

Remember: A good way to find all the factors of a number is by finding all of that number's factor pairs. A *factor pair* of a number is any pair of two numbers that, when multiplied together, equal that number. For example, 30 has four factor pairs — 1 × 30, 2 × 15, 3 × 10, and 5 × 6 — because

$$1 \times 30 = 30$$
$$2 \times 15 = 30$$
$$3 \times 10 = 30$$
$$5 \times 6 = 30$$

Tip: Here's how to find *all the* factor pairs of a number:

1. **Begin the list with 1 times the number itself.**

2. **Try to find a factor pair that includes 2 — that is, see whether the number is divisible by 2 (for tricks on testing for divisibility, see Chapter 6).**

   If it is, list the factor pair that includes 2.

3. **Test the number 3 in the same way.**

4. **Continue testing numbers until you find no more factor pairs.**

Example

Q. Write down all the factor pairs of 18.

A. **1 × 18, 2 × 9, 3 × 6.** According to Step 1, begin with 1 × 18:

$$1 \times 18$$

The number 18 is even, so it's divisible by 2. And 18 ÷ 2 = 9, so the next factor pair is 2 × 9:

$$1 \times 18$$
$$2 \times 9$$

The digital root of 18 is 9 (because 1 + 8 = 9), so 18 is divisible by 3. And 18 ÷ 3 = 6, so the next factor pair is 3 × 6:

$$1 \times 18$$
$$2 \times 9$$
$$3 \times 6$$

The number 18 isn't divisible by 4, because 18 ÷ 4 = 4 r 2. And 18 isn't divisible by 5, because it doesn't end with 5 or 0. This list of factor pairs is complete because there are no more numbers between 3 and 6, the last factor pair on the list.

Practice Questions

1. Find all the factor pairs of 12.

2. Write down all the factor pairs of 28.

3. Figure out all the factor pairs of 40.

4. Find all the factor pairs of 66.

Ⓐ Practice Answers

1. **1 × 12, 2 × 6, and 3 × 4.** The first factor pair is 1 × 12. And 12 is divisible by 2 (12 ÷ 2 = 6), so the next factor pair is 2 × 6. Because 12 is divisible by 3 (12 ÷ 3 = 4), the next factor pair is 3 × 4.

2. **1 × 28, 2 × 14, and 4 × 7.** The first factor pair is 1 × 28. And 28 is divisible by 2 (28 ÷ 2 = 14), so the next factor pair is 2 × 14. Although 28 isn't divisible by 3, it's divisible by 4 (28 ÷ 4 = 7), so the next factor pair is 4 × 7. Finally, 28 isn't divisible by 5 or 6.

3. **1 × 40, 2 × 20, 4 × 10, and 5 × 8.** The first factor pair is 1 × 40. Because 40 is divisible by 2 (40 ÷ 2 = 20), the next factor pair is 2 × 20. Although 40 isn't divisible by 3, it's divisible by 4 (40 ÷ 4 = 10), so the next factor pair is 4 × 10. And 40 is divisible by 5 (40 ÷ 5 = 8), so the next factor pair is 5 × 8. Finally, 40 isn't divisible by 6 or 7.

4. **1 × 66, 2 × 33, 3 × 22, and 6 × 11.** The first factor pair is 1 × 66. The number 66 is divisible by 2 (66 ÷ 2 = 33), so the next factor pair is 2 × 33. It's also divisible by 3 (66 ÷ 3 = 22), so the next factor pair is 3 × 22. Although 66 isn't divisible by 4 or 5, it's divisible by 6 (66 ÷ 6 = 11), so the next factor pair is 6 × 11. Finally, 66 isn't divisible by 7, 8, 9, or 10.

DECOMPOSING A NUMBER INTO ITS PRIME FACTORS

Every number is the product of a unique set of *prime factors,* a group of prime numbers (including repeats) that, when multiplied together, equals that number. This section shows you how to find those prime factors for a given number, a process called *decomposition*.

Tip: An easy way to decompose a number is to make a factorization tree. Here's how:

1. **Find two numbers that multiply to equal the original number; write them as numbers that branch off the original one.**

   Knowing the multiplication table can often help you here.

2. **If either number is prime, circle it and end that branch.**

3. **Continue branching off non-prime numbers into two factors; whenever a branch reaches a prime number, circle it and close the branch.**

   When every branch ends in a circled number, you're finished — just gather up the circled numbers.

Example

Q. Decompose the number 48 into its prime factors.

A. **48 = 2 × 2 × 2 × 2 × 3.** Begin making a factorization tree by finding two numbers that multiply to equal 48:

Continue making branches of the tree by doing the same for 6 and 8:

Circle the prime numbers and close those branches. At this point, the only open branch is 4. Break it down into 2 and 2:

Every branch ends in a circled number, so you're finished. The prime factors are 2, 2, 2, 2, and 3.

Practice Questions

1. Decompose 18 into its prime factors.

2. Decompose 42 into its prime factors.

3. Decompose 81 into its prime factors.

4. Decompose 120 into its prime factors.

Practice Answers

1. **18 = 2 × 3 × 3.** Here's one possible factoring tree:

2. **42 = 2 × 3 × 7.** Here's one possible factoring tree:

3. **81 = 3 × 3 × 3 × 3.** Here's one possible factoring tree:

4. **120 = 2 × 2 × 2 × 3 × 5.** Here's one possible factoring tree:

FINDING THE GREATEST COMMON FACTOR

The *greatest common factor* (GCF) of a set of numbers is the largest number that's a factor of every number in that set. Finding the GCF is helpful when you want to reduce a fraction to its lowest terms (see Chapter 9).

You can find the GCF in two ways. The first option is to list all the factor pairs of the numbers and choose the largest factor that appears in both (or all) the lists. (For info on finding factor pairs, see the earlier section "Generating a Number's Factors.")

Tip: The other method uses prime factors, which I discuss in the preceding section. Here's how to find the GCF:

1. **Decompose the numbers into their prime factors.**

2. **Underline the factors that all the original numbers have in common.**

3. **Multiply the underlined numbers to get the GCF.**

Examples

Q.   Find the greatest common factor of 12 and 20.

A.   **4.** Write down all the factor pairs of 12 and 20:

Factor pairs of 12: $1 \times 12, 2 \times 6, 3 \times 4$

Factor pairs of 20: $1 \times 20, 2 \times 10, 4 \times 5$

The number 4 is the greatest number that appears in both lists of factor pairs, so it's the GCF.

Q.   Find the greatest common factor of 24, 36, and 42.

A.   **6.** Decompose all three numbers down to their prime factors:

$$24 = 2 \times 2 \times 2 \times 3$$
$$36 = 2 \times 2 \times 3 \times 3$$
$$42 = 2 \times 3 \times 7$$

Underline all factors that are common to all three numbers:

$$24 = \underline{2} \times 2 \times 2 \times \underline{3}$$
$$36 = \underline{2} \times 2 \times \underline{3} \times 3$$
$$42 = \underline{2} \times \underline{3} \times 7$$

Multiply those underlined numbers to get your answer:

$$2 \times 3 = 6$$

Practice Questions

1.   Find the greatest common factor of 10 and 22.

2.   What's the GCF of 8 and 32?

3.   Find the GCF of 30 and 45.

4.   Figure out the GCF of 27 and 72.

5.   Find the GCF of 15, 20, and 35.

6.   Figure out the GCF of 44, 56, and 72.

A Practice Answers

1. **2.** Write down all the factor pairs of 10 and 22:

    $$10: 1 \times 10, 2 \times 5$$
    $$22: 1 \times 22, 2 \times 11$$

    The number 2 is the greatest number that appears on both lists.

2. **8.** Write down all the factor pairs of 8 and 32:

    $$8: 1 \times 8, 2 \times 4$$
    $$32: 1 \times 32, 2 \times 16, 4 \times 8$$

    The greatest number that appears on both lists is 8.

3. **15.** Write down all the factor pairs of 30 and 45:

    $$30: 1 \times 30, 2 \times 15, 3 \times 10, 5 \times 6$$
    $$45: 1 \times 45, 3 \times 15, 5 \times 9$$

    The greatest number that appears on both lists is 15.

4. The GCF of 27 and 72 is **9.** Decompose 27 and 72 into their prime factors and underline every factor that's common to both:

    $$27 = \underline{3} \times \underline{3} \times 3$$
    $$72 = 2 \times 2 \times 2 \times \underline{3} \times \underline{3}$$

    Multiply those underlined numbers to get your answer: $3 \times 3 = 9$.

5. The GCF of 15, 20, and 35 is **5.** Decompose the three numbers into their prime factors and underline every factor that's common to all three:

    $$15 = 3 \times \underline{5}$$
    $$20 = 2 \times 2 \times \underline{5}$$
    $$35 = \underline{5} \times 7$$

    The only factor common to all three numbers is 5.

6. The GCF of 44, 56, and 72 is **4.** Decompose all three numbers to their prime factors and underline each factor that's common to all three:

    $$44 = \underline{2} \times \underline{2} \times 11$$
    $$56 = \underline{2} \times \underline{2} \times 2 \times 7$$
    $$72 = \underline{2} \times \underline{2} \times 2 \times 3 \times 3$$

    Multiply those underlined numbers to get your answer: $2 \times 2 = 4$.

GENERATING THE MULTIPLES OF A NUMBER

Generating the multiples of a number is easier than generating the factors: Just multiply the number by 1, 2, 3, and so forth. But unlike the factors of a number — which are always less than the number itself — in the positive numbers, the multiples of a number are greater than or equal to that number. Therefore, you can never write down all the multiples of any positive number.

Examples

Q. Find the first six (positive) multiples of 4.

A. **4, 8, 12, 16, 20, 24.** Write down the number 4 and keep adding 4 to it until you've written down six numbers.

Q. List the first six multiples of 12.

A. **12, 24, 36, 48, 60, 72.** Write down 12 and keep adding 12 to it until you've written down six numbers.

Practice Questions

1. Write the first six multiples of 5.

2. Generate the first six multiples of 7.

3. List the first ten multiples of 8.

4. Write the first six multiples of 15.

Practice Answers

1. **5, 10, 15, 20, 25, and 30.**

Write down the number 5 and keep adding 5 to it until you've written six numbers.

2. **7, 14, 21, 28, 35, and 42.**

3. **8, 16, 24, 32, 40, 48, 56, 64, 72, and 80.**

4. **15, 30, 45, 60, 75, and 90.**

FINDING THE LEAST COMMON MULTIPLE

The *least common multiple* (LCM) of a set of numbers is the smallest number that's a multiple of every number in that set. For small numbers, you can simply list the first several multiples of each number until you get a match.

When you're finding the LCM of two numbers, you may want to list the multiples of the larger number first, stopping when the number of multiples you've written down equals the smaller number. Then list the multiples of the smaller number and look for a match.

However, you may have to write down a lot of multiples with this method, and the disadvantage becomes even greater when you're trying to find the LCM of large numbers. I recommend a method that uses prime factors when you're facing big numbers or more than two numbers. Here's how:

1. **Write down the prime decompositions of all the numbers.**

    See the earlier "Decomposing a Number into Its Prime Factors" section for details.

2. **For each prime number you find, underline the *most repeated occurrences* of each.**

    In other words, compare the decompositions. If one breakdown contains two 2s and another contains three 2s, you'd underline the three 2s. If one decomposition contains one 7 and the rest don't have any, you'd underline the 7.

3. **Multiply the underlined numbers to get the LCM.**

Examples

Q. Find the LCM of 6 and 8.

A. **24.** Because 8 is the larger number, write down six multiples of 8:

Multiples of 8: 8, 16, 24, 32, 40, 48

Now, write down multiples of 6 until you find a matching number:

Multiples of 6: 6, 12, 18, 24

Q. Find the LCM of 12, 15, and 18.

A. **180.** Begin by writing the prime decompositions of all three numbers. Then, for each prime number you find, underline the most repeated occurrences of each:

$$12 = \underline{2} \times \underline{2} \times 3$$
$$15 = 3 \times \underline{5}$$
$$18 = 2 \times \underline{3} \times \underline{3}$$

Notice that 2 appears in the decomposition of 12 most often (twice), so I underline both of those 2s. Similarly, 3 appears in the decomposition of 18 most often (twice), and 5 appears in the decomposition of 15 most often (once). Now, multiply all the underlined numbers:

$$2 \times 2 \times 3 \times 3 \times 5 = 180$$

Practice Questions

1.   Find the LCM of 4 and 10.

2.   Find the LCM of 7 and 11.

3.   Find the LCM of 9 and 12.

4.   Find the LCM of 18 and 22.

Practice Answers

1.   The LCM of 4 and 10 is **20.** Write down four multiples of 10:

Multiples of 10: 10, 20, 30, 40

Next, generate multiples of 4 until you find a matching number:

Multiples of 4: 4, 8, 12, 16, 20

2.   The LCM of 7 and 11 is **77.** Write down seven multiples of 11:

Multiples of 11: 11, 22, 33, 44, 55, 66, 77

Next, generate multiples of 7 until you find a matching number:

Multiples of 7: 7, 14, 21, 28, 35, 42, 49, 56, 63, 70, 77

3.   The LCM of 9 and 12 is **36.** Write down nine multiples of 12:

Multiples of 12: 12, 24, 36, 48, 60, 72, 84, 96, 108

Next, generate multiples of 9 until you find a matching number:

Multiples of 9: 9, 18, 27, 36

4.   The LCM of 18 and 22 is **198.** First, decompose both numbers into their prime factors. Then underline the most frequent occurrences of each prime number:

$$18 = \underline{2} \times \underline{3} \times \underline{3}$$
$$22 = 2 \times \underline{11}$$

The factor 2 appears only once in any decomposition, so I underline a 2. The number 3 appears twice in the decomposition of 18, so I underline both of these. The number 11 appears only once, in the decomposition of 22, so I underline it. Now, multiply all the underlined numbers:

$$2 \times 3 \times 3 \times 11 = 198$$

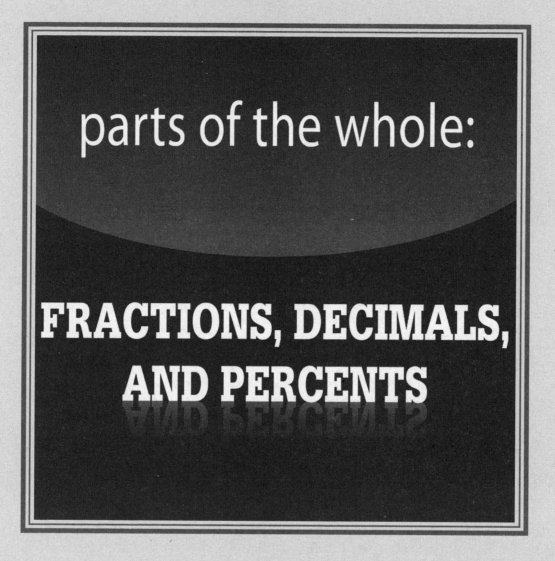

parts of the whole:

FRACTIONS, DECIMALS, AND PERCENTS

WEB EXTRA *To discover a great way to solve percent problems, go to www.dummies.com/extras/ ucanbasicmathprealgebra.*

IN THIS PART . . .

- Work with basic fractions, improper fractions, and mixed numbers.

- Add, subtract, multiply, and divide fractions, decimals, and percents.

- Convert the form of a rational number to a fraction, a decimal, or a percent.

- Use ratios and proportions.

- Solve word problems that involve fractions, decimals, percentages.

8

FOOLING WITH FRACTIONS

Suppose that today is your birthday and your friends are throwing you a surprise party. After opening all your presents, you finish blowing out the candles on your cake, but now you have a problem: Eight of you want some cake, but you have only *one cake.* Several solutions are proposed:

• You can all go into the kitchen and bake seven more cakes.

• Instead of eating cake, everyone can eat celery sticks.

• Because it's your birthday, you can eat the *whole* cake and everyone else can eat celery sticks. (That idea was yours.)

• You can cut the cake into eight equal slices so that everyone can enjoy it.

After careful consideration, you choose the last option. With that decision, you've opened the door to the exciting world of fractions. Fractions represent parts of a thing that can be cut into pieces. In this chapter, I give you some basic information about fractions that you need to know, including the three basic types of fractions: proper fractions, improper fractions, and mixed numbers.

I move on to increasing and reducing the terms of fractions, which you need when you begin applying the Big Four operations to fractions in Chapter 9. I also show you how to convert between improper fractions and mixed numbers. Finally, I show you how to compare fractions using cross-multiplication. By the time you're done with this chapter, you'll see how fractions really can be a piece of cake!

SLICING A CAKE INTO FRACTIONS

Here's a simple fact: When you cut a cake into two equal pieces, each piece is half of the cake. As a fraction, you write that as $\frac{1}{2}$. In Figure 8-1, the shaded piece is half of the cake.

Figure 8-1:
Two halves
of a cake.

Every fraction is made up of two numbers separated by a line, or a fraction bar. The line can be either diagonal or horizontal — so you can write this fraction in either of the following two ways:

$$\frac{1}{2} \qquad\qquad 1/2$$

The number above the line is called the numerator. The numerator tells you how many pieces you have. In this case, you have one dark-shaded piece of cake, so the numerator is 1.

The number below the line is called the denominator. The denominator tells you how many equal pieces the whole cake has been cut into. In this case, the denominator is 2.

Similarly, when you cut a cake into three equal slices, each piece is a third of the cake (see Figure 8-2).

Figure 8-2:
Cake cut
into thirds.

This time, the shaded piece is one-third — $\frac{1}{3}$ — of the cake. Again, the numerator tells you how many pieces you have, and the denominator tells you how many equal pieces the whole cake has been cut up into.

Figure 8-3 shows a few more examples of ways to represent parts of the whole with fractions.

Figure 8-3:
Cakes cut
and shaded
into
(A) $\frac{3}{4}$,
(B) $\frac{2}{5}$,
(C) $\frac{1}{6}$,
and (D) $\frac{7}{10}$.

In each case, the numerator tells you how many pieces are shaded, and the denominator tells how many pieces there are altogether.

The fraction bar can also mean a division sign. In other words, $\frac{3}{4}$ signifies 3 ÷ 4. If you take three cakes and divide them among four people, each person gets $\frac{3}{4}$ of a cake.

KNOWING THE FRACTION FACTS OF LIFE

Fractions have their own special vocabulary and a few important properties that are worth knowing right from the start. When you know them, you find working with fractions a lot easier.

Telling the numerator from the denominator

The top number in a fraction is called the *numerator,* and the bottom number is called the *denominator.* For example, look at the following fraction:

$$\frac{3}{4}$$

In this example, the number 3 is the numerator, and the number 4 is the denominator. Similarly, look at this fraction:

$$\frac{55}{89}$$

The number 55 is the numerator, and the number 89 is the denominator.

Flipping for reciprocals

When you flip over a fraction, you get its reciprocal. For example, the following numbers are reciprocals:

$$\frac{2}{3} \quad \text{and} \quad \frac{3}{2}$$

$$\frac{11}{14} \quad \text{and} \quad \frac{14}{11}$$

$$\frac{19}{19} \text{ is its own reciprocal}$$

Using ones and zeros

When the denominator (bottom number) of a fraction is 1, the fraction is equal to the numerator by itself. Conversely, you can turn any whole number into a fraction by drawing a line and placing the number 1 under it. For example,

$$\frac{2}{1} = 2 \qquad\qquad \frac{9}{1} = 9 \qquad\qquad \frac{157}{1} = 157$$

Remember: When the numerator and denominator match, the fraction equals 1. After all, if you cut a cake into eight pieces and you keep all eight of them, you have the entire cake. Here are some fractions that equal 1:

$$\frac{8}{8} = 1 \qquad\qquad \frac{11}{11} = 1 \qquad\qquad \frac{365}{365} = 1$$

When the numerator of a fraction is 0, the fraction is equal to 0. For example,

$$\frac{0}{1} = 0 \qquad\qquad \frac{0}{12} = 0 \qquad\qquad \frac{0}{113} = 0$$

Warning: The denominator of a fraction can never be 0. Fractions with 0 in the denominator are *undefined* — that is, they have no mathematical meaning.

Remember from earlier in this chapter that placing a number in the denominator is similar to cutting a cake into that number of pieces. You can cut a cake into two, or ten, or even a million pieces. You can even cut it into one piece (that is, don't cut it at all). But you can't cut a cake into zero pieces. For this reason, putting 0 in the denominator — much like lighting an entire book of matches on fire — is something you should never, never do.

Mixing things up

A mixed number is a combination of a whole number and a proper fraction added together. Here are some examples:

$$1\frac{1}{2} \qquad\qquad 5\frac{3}{4} \qquad\qquad 99\frac{44}{100}$$

A mixed number is always equal to the whole number plus the fraction attached to it. So $1\frac{1}{2}$ means $1 + \frac{1}{2}$, $5\frac{3}{4}$ means $5 + \frac{3}{4}$, and so on.

Knowing proper from improper

When the numerator (top number) is less than the denominator (bottom number), the fraction is less than 1:

$$\frac{1}{2} < 1 \qquad\qquad \frac{3}{5} < 1 \qquad\qquad \frac{63}{73} < 1$$

Fractions like these are called *proper fractions*. Positive proper fractions are always between 0 and 1. However, when the numerator is greater than the denominator, the fraction is greater than 1. Take a look:

$$\frac{3}{2} > 1 \qquad\qquad \frac{7}{4} > 1 \qquad\qquad \frac{98}{97} > 1$$

Any fraction that's greater than 1 is called an *improper fraction*. Converting an improper fraction to a mixed number is customary, especially when it's the final answer to a problem.

Remember: An improper fraction is always top heavy, as if it's unstable and wants to fall over. To stabilize it, convert it to a mixed number. Proper fractions are always stable.

Later in this chapter, I discuss improper fractions in more detail when I show you how to convert between improper fractions and mixed numbers.

Examples

Q. For each cake pictured below, identify the fraction of the cake that's shaded.

a.　　　　b.　　　　c.　　　　d.

A. Put the number of shaded slices over the number of total slices in each cake:

a. $\frac{2}{3}$

b. $\frac{1}{4}$

c. $\frac{5}{8}$

d. $\frac{7}{10}$

Q. What's the reciprocal of each of the following fractions?

a. $\frac{3}{4}$

b. $\frac{6}{11}$

c. $\frac{22}{7}$

d. $\frac{41}{48}$

A. To find the reciprocal, switch around the numerator and the denominator:

a. The reciprocal of $\frac{3}{4}$ is $\frac{4}{3}$.

b. The reciprocal of $\frac{6}{11}$ is $\frac{11}{6}$.

c. The reciprocal of $\frac{22}{7}$ is $\frac{7}{22}$.

d. The reciprocal of $\frac{41}{48}$ is $\frac{48}{41}$.

Practice Questions

1. For each cake pictured, identify the fraction of the cake that's shaded.

a.   b.   c.   d.

2. Which of the following fractions are proper? Which are improper?

a. $\frac{3}{2}$

b. $\frac{8}{9}$

c. $\frac{20}{23}$

d. $\frac{75}{51}$

3. Rewrite each of the following fractions as a whole number:

a. $\frac{3}{3}$

b. $\frac{10}{1}$

c. $\frac{10}{10}$

d. $\frac{81}{1}$

4. Find the reciprocal of the following fractions:

a. $\dfrac{5}{7}$

b. $\dfrac{10}{3}$

c. $\dfrac{12}{17}$

d. $\dfrac{80}{91}$

A Practice Answers

1. **a.** You have 1 shaded slice and 3 slices in total, so it's $\dfrac{1}{3}$.

b. You have 3 shaded slices and 4 slices in total, so it's $\dfrac{3}{4}$.

c. You have 5 shaded slices and 6 slices in total, so it's $\dfrac{5}{6}$.

d. You have 7 shaded slices and 12 slices in total, so it's $\dfrac{7}{12}$.

2. **a.** The numerator (3) is greater than the denominator (2), so $\dfrac{3}{2}$ is an **improper fraction.**

b. The numerator (8) is less than the denominator (9), so $\dfrac{8}{9}$ is a **proper fraction.**

b. The numerator (20) is less than the denominator (23), so $\dfrac{20}{23}$ is a **proper fraction.**

b. The numerator (75) is greater than the denominator (51), so $\dfrac{75}{51}$ is an **improper fraction.**

3. **a.** The numerator and denominator are the same, so $\dfrac{3}{3} = \mathbf{1.}$

b. The denominator is 1, so $\dfrac{10}{1} = \mathbf{10.}$

c. The numerator and denominator are the same, so $\dfrac{10}{10} = \mathbf{1.}$

d. The denominator is 1, so $\dfrac{81}{1} = \mathbf{81.}$

4. Find the reciprocal of the following fractions by switching the numerator and denominator.

a. The reciprocal of $\dfrac{5}{7}$ is $\mathbf{\dfrac{7}{5}}$.

b. The reciprocal of $\dfrac{10}{3}$ is $\mathbf{\dfrac{3}{10}}$.

c. The reciprocal of $\dfrac{12}{17}$ is $\mathbf{\dfrac{17}{12}}$.

d. The reciprocal of $\dfrac{80}{91}$ is $\mathbf{\dfrac{91}{80}}$.

INCREASING AND REDUCING TERMS OF FRACTIONS

Take a look at these three fractions:

$$\frac{1}{2} \qquad \frac{2}{4} \qquad \frac{3}{6}$$

If you cut three cakes (as I do earlier in this chapter) into these three fractions (see Figure 8-4), exactly half of the cake will be shaded, just like in Figure 8-1, no matter how you slice it. (Get it? No matter how you slice it? You may as well laugh at the bad jokes, too — they're free.) The important point here isn't the humor, or the lack of it, but the idea about fractions.

(A)

(B)

(C)

Figure 8-4:
Cakes
cut and
shaded into
(A) $\frac{1}{2}$,
(B) $\frac{2}{4}$,
and (C) $\frac{3}{6}$.

The fractions $\frac{1}{2}$, $\frac{2}{4}$, and $\frac{3}{6}$ are all equal in value. In fact, you can write a lot of fractions that are also equal to these. As long as the numerator is exactly half the denominator, the fractions are all equal to

$$\frac{1}{2}$$

— for example,

$$\frac{11}{22} \qquad\qquad \frac{100}{200} \qquad\qquad \frac{1,000,000}{2,000,000}$$

These fractions are equal to $\frac{1}{2}$, but their terms (the numerator and denominator) are different. In this section, I show you how to both increase and reduce the terms of a fraction without changing its value.

Increasing the terms of fractions

Remember: To increase the terms of a fraction by a certain number, multiply both the numerator and the denominator by that number.

For example, to increase the terms of the fraction $\frac{3}{4}$ by 2, multiply both the numerator and the denominator by 2:

$$\frac{3}{4} = \frac{3 \times 2}{4 \times 2} = \frac{6}{8}$$

Similarly, to increase the terms of the fraction $\frac{5}{11}$ by 7, multiply both the numerator and the denominator by 7:

$$\frac{5}{11} = \frac{5 \times 7}{11 \times 7} = \frac{35}{77}$$

Remember: Increasing the terms of a fraction doesn't change its value. Because you're multiplying the numerator and denominator by the same number, you're essentially multiplying the fraction by a fraction that equals 1.

One key point to know is how to increase the terms of a fraction so that the denominator becomes a preset number. Here's how you do it:

1. **Divide the new denominator by the old denominator.**

   To keep the fractions equal, you have to multiply the numerator and denominator of the old fraction by the same number. This first step tells you what the old denominator was multiplied by to get the new one.

   For example, suppose you want to raise the terms of the fraction $\frac{4}{7}$ so that the denominator is 35. You're trying to fill in the question mark here:

   $$\frac{4}{7} = \frac{?}{35}$$

   Divide 35 by 7, which tells you that the denominator was multiplied by 5.

2. **Multiply this result by the old numerator to get the new numerator.**

   You now know how the two denominators are related. The numerators need to have the same relationship, so multiply the old numerator by the number you found in Step 1.

   Multiply 5 by 4, which gives you 20. So here's the answer:

$$\frac{4}{7} = \frac{4 \times 5}{7 \times 5} = \frac{20}{35}$$

Reducing fractions to lowest terms

Reducing fractions is similar to increasing fractions, except that it involves division rather than multiplication. But because you can't always divide, reducing takes a bit more finesse.

In practice, reducing fractions is similar to factoring numbers. For this reason, if you're not up on factoring, you may want to review this topic in Chapter 7.

In this section, I show you the formal way to reduce fractions, which works in all cases. Then I show you a more informal way you can use when you're more comfortable.

Reducing fractions the formal way

Reducing fractions the formal way relies on understanding how to break down a number into its prime factors. I discuss this in detail in Chapter 7, so if you're shaky on this concept, you may want to review it first.

Here's how to reduce a fraction:

1. **Break down both the numerator (top number) and the denominator (bottom number) into their prime factors.**

   For example, suppose you want to reduce the fraction $\frac{12}{30}$. Break down both 12 and 30 into their prime factors:

$$\frac{12}{30} = \frac{2 \times 2 \times 3}{2 \times 3 \times 5}$$

2. **Cross out any common factors.**

   As you can see, I cross out a 2 and a 3 because they're common factors — that is, they appear in both the numerator and the denominator:

$$\frac{12}{30} = \frac{\cancel{2} \times 2 \times \cancel{3}}{\cancel{2} \times \cancel{3} \times 5}$$

3. **Multiply the remaining numbers to get the reduced numerator and denominator.**

   You can see now that the fraction $\frac{12}{30}$ reduces to $\frac{2}{5}$:

$$\frac{12}{30} = \frac{\cancel{2} \times 2 \times \cancel{3}}{\cancel{2} \times \cancel{3} \times 5} = \frac{2}{5}$$

As another example, here's how you reduce the fraction $\frac{32}{100}$:

$$\frac{32}{100} = \frac{\cancel{2} \times \cancel{2} \times 2 \times 2 \times 2}{\cancel{2} \times \cancel{2} \times 5 \times 5} = \frac{8}{25}$$

This time, cross out two 2s from both the top and the bottom as common factors. The remaining 2s on top and the 5s on the bottom aren't common factors. So the fraction $\frac{32}{100}$ reduces to $\frac{8}{25}$.

Reducing fractions the informal way

Here's an easier way to reduce fractions when you get comfortable with the concept:

1. **If the numerator (top number) and denominator (bottom number) are both divisible by 2 — that is, if they're both even — divide both by 2.**

   For example, suppose you want to reduce the fraction $\frac{36}{60}$. The numerator and the denominator are both even, so divide them both by 2:

   $$\frac{36}{60} = \frac{18}{30}$$

2. **Repeat Step 1 until the numerator or denominator (or both) is no longer divisible by 2.**

   In the resulting fraction, both numbers are still even, so repeat the first step again:

   $$\frac{18}{30} = \frac{9}{15}$$

3. **Repeat Step 1 using the number 3, and then 5, and then 7, continuing testing prime numbers until you're sure that the numerator and denominator have no common factors.**

   Now, the numerator and the denominator are both divisible by 3 (see Chapter 6 for easy ways to tell if one number is divisible by another), so divide both by 3:

   $$\frac{9}{15} = \frac{3}{5}$$

Neither the numerator nor the denominator is divisible by 3, so this step is complete. At this point, you can move on to test for divisibility by 5, 7, and so on, but you really don't need to. The numerator is 3, and it obviously isn't divisible by any larger number, so you know that the fraction $\frac{36}{60}$ reduces to $\frac{3}{5}$.

Examples

Q. Increase the terms of the fraction $\frac{4}{5}$ to a new fraction whose denominator is 15:

A. $\frac{12}{15}$. To start out, write the problem as follows:

$$\frac{4}{5} = \frac{?}{15}$$

The question mark stands for the numerator of the new fraction, which you want to fill in. Now divide the larger denominator (15) by the smaller denominator (5).

$$15 \div 5 = 3$$

Multiply this result by the numerator:

$$3 \times 4 = 12$$

Finally, take this number and use it to replace the question mark:

$$\frac{4}{5} = \frac{12}{15}$$

Examples

Q. Reduce the fraction $\frac{18}{42}$ to lowest terms.

A. $\frac{3}{7}$. The numerator and denominator aren't too large, so use the informal way: To start out, try to find a small number that the numerator and denominator are both divisible by. In this case, notice that the numerator and denominator are both divisible by 2, so divide both by 2:

$$\frac{18}{42} = \frac{(18 \div 2)}{(42 \div 2)} = \frac{9}{21}$$

Next, notice that the numerator and denominator are both divisible by 3 (see Chapter 6 for more on how to tell whether a number is divisible by 3), so divide both by 3:

$$\frac{9}{21} = \frac{(9 \div 3)}{(21 \div 3)} = \frac{3}{7}$$

At this point, there's no number (except for 1) that evenly divides both the numerator and denominator, so this is your answer.

Q. Reduce the fraction $\frac{135}{196}$ to lowest terms.

A. $\frac{135}{196}$. The numerator and denominator are both over 100, so use the formal way. First, decompose both the numerator and denominator down to their prime factors:

$$\frac{135}{196} = \frac{(3 \times 3 \times 3 \times 5)}{(2 \times 2 \times 7 \times 7)}$$

The numerator and denominator have no common factors, so the fraction is already in lowest terms.

Practice Questions

1. Increase the terms of the fraction $\frac{7}{10}$ to change the denominator is 120.

2. Increase the terms of $\frac{13}{15}$ so that the denominator is changed to 360.

3. Convert the mixed number $5\frac{1}{4}$ to an improper fraction.

4. Change $7\frac{2}{9}$ to an improper fraction.

5. Express the mixed number $10\frac{5}{12}$ as an improper fraction.

6. Convert the improper fraction $\frac{13}{4}$ to a mixed number.

7. Express the improper fraction $\frac{29}{10}$ as a mixed number.

8. Change $\frac{100}{7}$ to a mixed number.

Practice Answers

1. $\frac{7}{10} = \frac{84}{120}$. To start out, write the problem as follows:

$$\frac{7}{10} = \frac{?}{120}$$

Divide the larger denominator (120) by the smaller denominator (3) and then multiply this result by the numerator (7):

$$12 \times 7 = 84$$

Take this numerator and use it to replace the question mark; your answer is $\frac{84}{120}$

2. $\frac{13}{15} = \frac{312}{360}$. To start out, write the problem as follows:

$$\frac{13}{15} = \frac{?}{360}$$

Divide the larger denominator (360) by the smaller denominator (15) and then multiply this result by the numerator (7):

$$24 \times 13 = 312$$

Take this numerator and use it to replace the question mark; your answer is $\frac{312}{360}$

3. $\frac{2}{3} = \frac{12}{18}$. To start out, write the problem as follows:

$$\frac{2}{3} = \frac{?}{18}$$

Divide the larger denominator (18) by the smaller denominator (3) and then multiply this result by the numerator (2):

$$6 \times 2 = 12$$

Take this number and use it to replace the question mark; your answer is $\frac{12}{18}$.

4. $\frac{4}{9} = \frac{24}{54}$. Write the problem as follows:

$$\frac{4}{9} = \frac{?}{54}$$

Divide the larger denominator (54) by the smaller denominator (9) and then multiply this result by the numerator (4):

$$6 \times 4 = 24$$

Take this number and use it to replace the question mark; your answer is $\frac{24}{54}$.

5. $\frac{12}{60} = \frac{1}{5}$. The numerator (12) and the denominator (60) are both even, so divide both by 2:

$$\frac{12}{60} = \frac{6}{30}$$

They're still both even, so divide both by 2 again:

$$= \frac{3}{15}$$

Now the numerator and denominator are both divisible by 3, so divide both by 3:

$$= \frac{1}{5}$$

6. $\frac{45}{75} = \frac{3}{5}$. The numerator (45) and the denominator (75) are both divisible by 5, so divide both by 5:

$$\frac{45}{75} = \frac{9}{15}$$

Now the numerator and denominator are both divisible by 3, so divide both by 3:

$$= \frac{3}{5}$$

7. $\frac{135}{180} = \frac{3}{4}$. The numerator (135) and the denominator (180) are both divisible by 5, so divide both by 5:

$$\frac{135}{180} = \frac{27}{36}$$

Now the numerator and denominator are both divisible by 3, so divide both by 3:

$$= \frac{9}{12}$$

They're still both divisible by 3, so divide both by 3 again:

$$= \frac{3}{4}$$

8. $\frac{108}{217} = \mathbf{\frac{108}{217}}$. With a numerator and denominator this large, reduce using the formal way. First, decompose both the numerator and denominator down to their prime factors:

$$\frac{108}{217} = \frac{(2 \times 2 \times 3 \times 3 \times 3)}{(7 \times 31)}$$

The numerator and denominator have no common factors, so the fraction is already in lowest terms.

CONVERTING BETWEEN IMPROPER FRACTIONS AND MIXED NUMBERS

In "Knowing the Fraction Facts of Life," I tell you that any fraction whose numerator is greater than its denominator is an improper fraction. Improper fractions are useful and easy to work with, but for some reason, people just don't like them. (The word *improper* should've tipped you off.) Teachers especially don't like them, and they really don't like an improper fraction to appear as the answer to a problem. However, they love mixed numbers. One reason they love them is that estimating the approximate size of a mixed number is easy.

For example, if I tell you to put $\frac{31}{3}$ of a gallon of gasoline in my car, you probably find it hard to estimate roughly how much that is: 5 gallons, 10 gallons, 20 gallons?

But if I tell you to get $10\frac{1}{3}$ gallons of gasoline, you know immediately that this amount is a little more than 10 but less than 11 gallons. Although $10\frac{1}{3}$ is the same as $\frac{31}{3}$, knowing the mixed number is a lot more helpful in practice. For this reason, you often have to convert improper fractions to mixed numbers.

Knowing the parts of a mixed number

Every mixed number has both a whole number part and a fractional part. So the three numbers in a mixed number are

- The whole number
- The numerator
- The denominator

For example, in the mixed number $3\frac{1}{2}$, the whole number part is 3 and the fractional part is $\frac{1}{2}$. So this mixed number is made up of three numbers: the whole number (3), the numerator (1), and the denominator (2). Knowing these three parts of a mixed number is helpful for converting back and forth between mixed numbers and improper fractions.

Converting a mixed number to an improper fraction

To convert a mixed number to an improper fraction, follow these steps:

1. **Multiply the denominator of the fractional part by the whole number, and add the result to the numerator.**

   For example, suppose you want to convert the mixed number $5\frac{2}{3}$ to an improper fraction. First, multiply 3 by 5 and add 2:

   $$3 \times 5 + 2 = 17$$

2. **Use this result as your numerator, and place it over the denominator you already have.**

   Place this result over the denominator:

   $$\frac{17}{3}$$

   So the mixed number $5\frac{2}{3}$ equals the improper fraction $\frac{17}{3}$. This method works for all mixed numbers. Furthermore, if you start with the fractional part reduced, the answer is also reduced (see the earlier "Increasing and Reducing Terms of Fractions" section).

Converting an improper fraction to a mixed number

To convert an improper fraction to a mixed number, divide the numerator by the denominator. Then write the mixed number in this way:

- The quotient (answer) is the whole-number part.
- The remainder is the numerator.
- The denominator of the improper fraction is the denominator.

For example, suppose you want to write the improper fraction $\frac{19}{5}$ as a mixed number. First, divide 19 by 5:

$$19 \div 5 = 3 \text{ r } 4$$

Then write the mixed number as follows:

$$3\frac{4}{5}$$

This method works for all improper fractions. And as is true of conversions in the other direction, if you start with a reduced fraction, you don't have to reduce your answer (see "Increasing and Reducing Terms of Fractions").

Examples

Q. Convert the mixed number $2\frac{3}{4}$ to an improper fraction.

A. $\frac{11}{4}$. Multiply the whole number (2) by the denominator (4), and then add the numerator (3):

$$2 \times 4 + 3 = 11$$

Use this number as the numerator of your answer, keeping the same denominator:

$$\frac{11}{4}$$

Q. Convert the mixed number $3\frac{5}{7}$ to an improper fraction:

A. $\frac{26}{7}$. Multiply the whole number (3) by the denominator (7), and then add the numerator (5). This time, I do the whole process in one step:

$$3\frac{5}{7} = \frac{(3 \times 7 + 5)}{7} = \frac{26}{7}$$

Q. Convert the improper fraction $\frac{11}{2}$ to a mixed number:

A. $5\frac{1}{2}$. Divide the numerator (11) by the denominator (2):

$$\text{Quotient} \quad \rightarrow \quad \begin{array}{r} 5 \\ 2\overline{)\ 11} \\ -10 \\ \hline \end{array}$$
$$\text{Remainder} \quad \rightarrow \quad 1$$

Now build a mixed number using the quotient (5) as the whole number and the remainder (1) as the numerator, keeping the same denominator (2):

$$5\frac{1}{2}$$

Q. Convert the improper fraction $\frac{39}{5}$ to a mixed number:

A. $7\frac{4}{5}$. Divide the numerator (39) by the denominator (5):

$$\text{Quotient} \quad \rightarrow \quad \begin{array}{r} 7 \\ 5\overline{)\ 39} \\ -35 \\ \hline \end{array}$$
$$\text{Remainder} \quad \rightarrow \quad 4$$

Build your answer using the quotient (7) as the whole number and the remainder (4) as the numerator, keeping the same denominator (5):

$$7\frac{4}{5}$$

Practice Questions

1. Increase the terms of the fraction $\frac{2}{3}$ so that the denominator is 18.

2. Increase the terms of $\frac{4}{9}$, changing the denominator to 54.

3. Reduce the fraction $\frac{12}{60}$ to lowest terms.

4. Reduce $\frac{45}{75}$ to lowest terms.

5. Reduce the fraction $\frac{135}{180}$ to lowest terms.

6. Reduce $\frac{108}{217}$ to lowest terms.

Practice Answers

1. $5\frac{1}{4} = \frac{(5 \times 4 + 1)}{4} = \mathbf{\frac{21}{4}}$

2. $7\frac{2}{9} = \frac{(7 \times 9 + 2)}{9} = \mathbf{\frac{65}{9}}$

3. $10\frac{5}{12} = \frac{(10 \times 12 + 5)}{12} = \mathbf{\frac{125}{12}}$

4. $\frac{13}{4} = \mathbf{3\frac{1}{4}}$. Divide the numerator (13) by the denominator (4):

$$\text{Quotient} \rightarrow \quad \begin{array}{r} 3 \\ 4\overline{)\ 13} \\ -12 \\ \hline \end{array}$$
$$\text{Remainder} \rightarrow \quad 1$$

Build your answer using the quotient (3) as the whole number and the remainder (1) as the numerator, keeping the same denominator (4): $3\frac{1}{4}$.

5. $\frac{29}{10} = \mathbf{2\frac{9}{10}}$. Divide the numerator (29) by the denominator (10):

$$\text{Quotient} \rightarrow \quad \begin{array}{r} 2 \\ 10\overline{)\ 29} \\ -20 \\ \hline \end{array}$$
$$\text{Remainder} \rightarrow \quad 9$$

Build your answer using the quotient (2) as the whole number and the remainder (9) as the numerator, keeping the same denominator (10): $2\frac{9}{10}$.

6. $\frac{100}{7} = 14\frac{2}{7}$. Divide the numerator (100) by the denominator (7):

$$
\begin{array}{r}
\text{Quotient} \quad \rightarrow \quad 14 \\
7)\overline{100} \\
-7 \\
\hline
30 \\
-28 \\
\hline
\text{Remainder} \quad \rightarrow \quad 2
\end{array}
$$

Build your answer using the quotient (14) as the whole number and the remainder (2) as the numerator, keeping the same denominator (7): $14\frac{2}{7}$.

COMPARING FRACTIONS WITH CROSS-MULTIPLICATION

Cross-multiplication is a handy tool for getting a common denominator for two fractions, which is important for many operations involving fractions. In this section, I show you how to cross-multiply to compare a pair of fractions to find out which is greater or less. (In Chapter 9, I show you how to use cross-multiplication to add fractions, and in Chapter 19, I show you how to use it to help solve algebra equations.)

Here's how to cross-multiply two fractions:

1. **Multiply the numerator (top number) of the first fraction by the denominator (bottom number) of the second, writing the answer below the first fraction.**

2. **Multiply the numerator of the second fraction by the denominator of the first, writing the answer below the second fraction.**

The result is that each fraction now has a new number written underneath it. The larger number is below the larger fraction.

Tip: You can use cross-multiplication to rewrite a pair of fractions as two new fractions with a common denominator:

1. **Cross-multiply the two fractions to find the numerators of the new fractions.**

2. **Multiply the denominators of the two fractions to find the new denominators.**

When two fractions have the same denominator, the one with the greater numerator is the greater fraction.

Examples

Q. Which fraction is greater: $\frac{5}{8}$ or $\frac{6}{11}$?

A. $\frac{5}{8}$. Cross-multiply the two fractions:

$$\frac{5}{8} \qquad \frac{6}{11}$$
$$55 \qquad 48$$

$$\frac{5}{8} \diagdown \frac{6}{11}$$
$$55 \qquad 48$$

Because 55 is greater than 48, $\frac{5}{8}$ is greater than $\frac{6}{11}$.

Q. Which of these three fractions is the least: $\frac{3}{4}$, $\frac{7}{10}$, or $\frac{8}{11}$?

A. $\frac{7}{10}$. Cross-multiply the first two fractions:

$$\frac{3}{4} \diagdown \frac{7}{10}$$
$$30 \qquad 28$$

Because 28 is less than 30, $\frac{7}{10}$ is less than $\frac{3}{4}$, so you can rule out $\frac{3}{4}$. Now compare $\frac{7}{10}$ and $\frac{8}{11}$ similarly:

$$\frac{7}{10} \diagdown \frac{8}{11}$$
$$77 \qquad 80$$

Because 77 is less than 80, $\frac{7}{10}$ is less than $\frac{8}{11}$. Therefore, $\frac{7}{10}$ is the least of the three fractions.

Practice Questions

1. Which is the greater fraction: $\frac{1}{5}$ or $\frac{2}{9}$?

2. Find the lesser fraction: $\frac{3}{7}$ or $\frac{5}{12}$.

3. Among these three fractions, which is greatest: $\frac{1}{10}$, $\frac{2}{21}$, or $\frac{3}{29}$?

4. Figure out which of the following fractions is the least: $\frac{1}{3}$, $\frac{2}{7}$, $\frac{4}{13}$, or $\frac{8}{25}$.

Ⓐ Practice Answers

1. $\frac{2}{9}$ **is greater than** $\frac{1}{5}$. Cross-multiply to compare the two fractions:

$$\frac{2}{9} \times \frac{1}{5}$$
$$\quad 10 \qquad 9$$

Because 10 is greater than 9, $\frac{2}{9}$ is greater than $\frac{1}{5}$.

2. $\frac{5}{12}$ **is less than** $\frac{3}{7}$. Cross-multiply to compare the two fractions:

$$\frac{5}{12} \times \frac{3}{7}$$
$$\quad 35 \qquad 36$$

Because 35 is less than 36, $\frac{5}{12}$ is less than $\frac{3}{7}$.

3. $\frac{3}{29}$ **is greater than** $\frac{1}{10}$ **and** $\frac{2}{21}$. Use cross-multiplication to compare the first two fractions.

$$\frac{1}{10} \times \frac{2}{21}$$
$$\quad 21 \qquad 20$$

Because 21 is greater than 20, $\frac{1}{10}$ is greater than $\frac{2}{21}$, so you can rule out $\frac{2}{21}$. Next, compare $\frac{1}{10}$ and $\frac{3}{29}$ by cross-multiplying.

$$\frac{1}{10} \times \frac{3}{29}$$
$$\quad 29 \qquad 30$$

Because 30 is greater than 29, $\frac{3}{29}$ is greater than $\frac{1}{10}$. Therefore, $\frac{3}{29}$ is the greatest of the three fractions.

4. $\frac{2}{7}$ **is less than** $\frac{1}{3}$, $\frac{4}{13}$, **and** $\frac{8}{25}$. Cross-multiply to compare the first two fractions.

$$\frac{1}{3} \times \frac{2}{7}$$
$$\quad 7 \qquad 6$$

Because 6 is less than 7, $\frac{2}{7}$ is less than $\frac{1}{3}$, so you can rule out $\frac{1}{3}$. Next, compare $\frac{2}{7}$ and $\frac{4}{13}$:

$$\frac{2}{7} \times \frac{4}{13}$$
$$\quad 26 \qquad 28$$

Because 26 is less than 28, $\frac{2}{7}$ is less than $\frac{4}{13}$, so you can rule out $\frac{4}{13}$.

Finally, compare $\frac{2}{7}$ and $\frac{8}{25}$:

$$\frac{2}{7} \underset{50 \quad 56}{\times} \frac{8}{25}$$

Because 50 is less than 56, $\frac{2}{7}$ is less than $\frac{8}{25}$. Therefore, $\frac{2}{7}$ is the lowest of the four fractions.

WORKING WITH RATIOS AND PROPORTIONS

A *ratio* is a mathematical comparison of two numbers, based on division. For example, suppose you bring 3 shirts and 5 ties with you on a business trip. Here are a few ways to express the ratio of shirts to ties:

$$3:5 \qquad 3 \text{ to } 5 \qquad \frac{3}{5}$$

A good way to work with a ratio is to turn it into a fraction. Be sure to keep the order the same: The first number goes on top of the fraction, and the second number goes on the bottom.

You can use a ratio to solve problems by setting up a *proportion equation* — that is, an equation involving two ratios.

Typically, a proportion looks like a word equation, as follows:

$$\frac{\text{scarves}}{\text{caps}} = \frac{2}{3}$$

For example, suppose you know that both you and your friend Andrew brought the same proportion of scarves to caps. If you also know that Andrew brought 8 scarves, you can use this proportion to find out how many caps he brought. Just increase the terms of the fraction $\frac{2}{3}$ so that the numerator becomes 8. I do this in two steps:

$$\frac{\text{scarves}}{\text{caps}} = \frac{2 \times 4}{3 \times 4}$$

$$\frac{\text{scarves}}{\text{caps}} = \frac{8}{12}$$

As you can see, the ratio 8:12 is equivalent to the ratio 2:3 because the fractions $\frac{2}{3}$ and $\frac{8}{12}$ are equal. Therefore, Andrew brought 12 caps.

Examples

Q. Clarence has 1 daughter and 4 sons. Set up a proportion equation based on this ratio.

A. $\dfrac{\text{Daughters}}{\text{Sons}} = \dfrac{1}{4}$

Q. An English language school has 3:7 ratio of European to Asian students. If the school has 28 students from Asia, how many students in the school are from Europe?

A. **12 students.** To begin, set up a proportion based on the ratio of European to Asian students (be sure that the order of the two numbers is the same as the two attributes that they stand for — 3 Europeans and 7 Asians):

$$\frac{\text{Europe}}{\text{Asia}} = \frac{3}{7}$$

Now, increase the terms of the fraction $\dfrac{3}{7}$ so that the number representing the count of Asian students becomes 28:

$$\frac{\text{Europe}}{\text{Asia}} = \frac{3 \times 4}{7 \times 4} = \frac{12}{28}$$

Therefore, given that the school has 28 Asian students, it has 12 European students.

Practice Questions

1. A farmers market sells a 4 to 5 ratio of vegetables to fruit. If it sells 35 different types of fruits, how many different types of vegetables does it sell?

2. An art gallery is currently featuring an exhibition with a 2:7 ratio of sculpture to paintings. If the exhibition includes 18 sculptures, how many paintings does it include?

3. A summer camp has a 7 to 9 ratio of girls to boys. If it has 117 boys, what is the total number of children attending the summer camp?

4. The budget of a small town has a 3:8 ratio of state funding to municipal funding. If the town received $600,000 in municipal funding last year, what was its total budget from both state and municipal sources?

Ⓐ **Practice Answers**

1. **28.** To begin, set up a proportion based on the ratio of vegetables to fruit:

$$\frac{\text{Vegetables}}{\text{Fruit}} = \frac{4}{5}$$

Now, increase the terms of the fraction $\frac{4}{5}$ so that the number representing fruit becomes 35:

$$\frac{\text{Vegetables}}{\text{Fruit}} = \frac{4 \times 7}{5 \times 7} = \frac{28}{35}$$

Therefore, given that the farmers market has 35 varieties of fruit, it has 28 varieties of vegetables.

2. **63.** To begin, set up a proportion based on the ratio of sculptures to paintings:

$$\frac{\text{Sculptures}}{\text{Paintings}} = \frac{2}{7}$$

Now, increase the terms of the fraction $\frac{2}{7}$ so that the number representing sculptures becomes 18:

$$\frac{\text{Sculptures}}{\text{Paintings}} = \frac{2 \times 9}{7 \times 9} = \frac{18}{63}$$

Therefore, the gallery has 63 paintings.

3. **208.** To begin, set up a proportion based on the ratio of girls to boys:

$$\frac{\text{Girls}}{\text{Boys}} = \frac{7}{9}$$

Now, you want to increase the terms of the fraction $\frac{7}{9}$ so that the number representing boys becomes 117. (To do this, notice that $117 \div 9 = 13$, so $9 \times 13 = 117$):

$$\frac{\text{Girls}}{\text{Boys}} = \frac{7 \times 13}{9 \times 13} = \frac{91}{117}$$

Therefore, the camp has 91 girls and 117 boys, so the total number of children is 208.

4. **$825,000.** To begin, set up a proportion based on the ratio of state to municipal funding:

$$\frac{\text{State}}{\text{Municipal}} = \frac{3}{8}$$

Now, you want to increase the terms of the fraction $\frac{3}{8}$ so that the number representing municipal funding becomes 600,000. (To do this, notice that $600,000 \div 8 = 75,000$, so $8 \times 75,000 = 600,000$):

$$\frac{\text{State}}{\text{Municipal}} = \frac{3 \times 75,000}{8 \times 75,000} = \frac{225,000}{600,000}$$

Therefore, the town's budget includes $225,000 in state funding and $600,000 in municipal funding, so the total budget is $825,000.

9

PARTING WAYS: FRACTIONS AND THE BIG FOUR OPERATIONS

IN THIS CHAPTER

- Looking at multiplication and division of fractions
- Adding and subtracting fractions in a bunch of different ways
- Applying the four operations to mixed numbers

In this chapter, the focus is on applying the Big Four operations to fractions. I start by showing you how to multiply and divide fractions, which isn't much more difficult than multiplying whole numbers. Surprisingly, adding and subtracting fractions is a bit trickier. I show you a variety of methods, each with its own strengths and weaknesses, and I recommend how to choose which method will work best, depending on the problem you have to solve.

Later in the chapter, I move on to mixed numbers. Again, multiplication and division won't likely pose too much of a problem because the process in each case is almost the same as multiplying and dividing fractions. I save adding and subtracting mixed numbers for the very end. By then, you'll be much more comfortable with fractions and ready to tackle the challenge.

MULTIPLYING AND DIVIDING FRACTIONS

One of the odd little ironies of life is that multiplying and dividing fractions is easier than adding or subtracting them — just two easy steps and you're done! For this reason, I show you how to multiply and divide fractions before I show you how to add or subtract them. In fact, you may find multiplying fractions easier than multiplying whole numbers because the numbers you're working with are usually small. More good news is that dividing fractions is nearly as easy as multiplying them. So I'm not even wishing you good luck — you don't need it!

Multiplying numerators and denominators straight across

Everything in life should be as simple as multiplying fractions. All you need for multiplying fractions is a pen or pencil, something to write on (preferably not your hand), and a basic knowledge of the multiplication table. (See Chapter 3 for a multiplication refresher.)

Remember: Here's how to multiply two fractions:

1. **Multiply the numerators (the numbers on top) to get the numerator of the answer.**

2. **Multiply the denominators (the numbers on the bottom) to get the denominator of the answer.**

For example, here's how to multiply $\frac{2}{5} \times \frac{3}{7}$:

$$\frac{2}{5} \times \frac{3}{7} = \frac{2 \times 3}{5 \times 7} = \frac{6}{35}$$

Sometimes when you multiply fractions, you have an opportunity to reduce to lowest terms. (For more on when and how to reduce a fraction, see Chapter 8.) As a rule, math people are crazy about reduced fractions, and teachers sometimes take points off a right answer if you could've reduced it but didn't. Here's a multiplication problem that ends up with an answer that's not in its lowest terms:

$$\frac{4}{5} \times \frac{7}{8} = \frac{4 \times 7}{5 \times 8} = \frac{28}{40}$$

Because the numerator and the denominator are both even numbers, this fraction can be reduced. Start by dividing both numbers by 2:

$$\frac{28 \div 2}{40 \div 2} = \frac{14}{20}$$

Again, the numerator and the denominator are both even, so do it again:

$$\frac{14 \div 2}{20 \div 2} = \frac{7}{10}$$

This fraction is now fully reduced.

Tip: When multiplying fractions, you can often make your job easier by canceling out equal factors in the numerator and denominator. Canceling out equal factors makes the numbers that you're multiplying smaller and easier to work with, and it also saves you the trouble of reducing at the end. Here's how it works:

- When the numerator of one fraction and the denominator of the other are the same, change both of these numbers to 1s. (See the nearby sidebar for why this works.)

- When the numerator of one fraction and the denominator of the other are divisible by the same number, factor this number out of both. In other words, divide the numerator and denominator by that common factor. (For more on how to find factors, see Chapter 7.)

For example, suppose you want to multiply the following two numbers: $\frac{5}{13} \times \frac{13}{20}$

You can make this problem easier by canceling out the number 13, as follows:

$$\frac{5}{1\,\cancel{13}} \times \frac{\cancel{13}\,1}{20} = \frac{5 \times 1}{1 \times 20} = \frac{5}{20}$$

You can make it even easier by noticing that 20 and 5 are both divisible by 5, so you can also factor out the number 5 before multiplying:

$$\frac{1\,\cancel{5}}{1} \times \frac{1}{\cancel{20}\,4} = \frac{1}{1} \times \frac{1}{4} = \frac{1}{4}$$

One is the easiest number

With fractions, the relationship between the numbers, not the actual numbers themselves, is most important. Understanding how to multiply and divide fractions can give you a deeper understanding of why you can increase or decrease the numbers within a fraction without changing the value of the whole fraction.

When you multiply or divide any number by 1, the answer is the same number. This rule also goes for fractions, so

$$\frac{3}{8} \times 1 = \frac{3}{8} \text{ and } \frac{3}{8} \div 1 = \frac{3}{8}$$

$$\frac{5}{13} \times 1 = \frac{5}{13} \text{ and } \frac{5}{13} \div 1 = \frac{5}{13}$$

$$\frac{67}{70} \times 1 = \frac{67}{70} \text{ and } \frac{67}{70} \div 1 = \frac{67}{70}$$

And as I discuss in Chapter 8, when a fraction has the same number in both the numerator and the denominator, its value is 1. In other words, the fractions $\frac{2}{2}$, $\frac{3}{3}$, and $\frac{4}{4}$ are all equal to 1. Look what happens when you multiply the fraction $\frac{3}{4}$ by $\frac{2}{2}$:

$$\frac{3}{4} \times \frac{2}{2} = \frac{3 \times 2}{4 \times 2} = \frac{6}{8}$$

The net effect is that you've increased the terms of the original fraction by 2. But all you've done is multiply the fraction by 1, so the value of the fraction hasn't changed. The fraction $\frac{6}{8}$ is equal to $\frac{3}{4}$.

Similarly, reducing the fraction $\frac{6}{9}$ by a factor of 3 is the same as dividing that fraction by $\frac{3}{3}$ (which is equal to 1):

$$\frac{6}{9} \div \frac{3}{3} = \frac{6 \div 3}{9 \div 3} = \frac{2}{3}$$

So $\frac{6}{9}$ is equal to $\frac{2}{3}$.

Examples

Q. Multiply $\frac{2}{5}$ by $\frac{4}{9}$.

A. $\frac{8}{45}$. Multiply the two numerators (top numbers) to get the numerator of the answer. Then multiply the two denominators (bottom numbers) to get the denominator of the answer:

$$\frac{2}{5} \times \frac{4}{9} = \frac{(2 \times 4)}{(5 \times 9)} = \frac{8}{45}$$

In this case, you don't have to reduce the answer.

Q. Find $\frac{4}{7} \times \frac{5}{8}$.

A. $\frac{5}{14}$. Before you multiply, notice that the numerator 4 and the denominator 8 are both even. So, divide both of these numbers by 2 just as you would when reducing a fraction:

$$\frac{4}{7} \times \frac{5}{8} = \frac{2}{7} \times \frac{5}{4}$$

The numerator 2 and the denominator 4 are still even, so repeat the process:

$$= \frac{1}{7} \times \frac{5}{2}$$

At this point, neither numerator has a common factor with either denominator, so you're ready to multiply. Multiply the two numerators to get the numerator of the answer. Then multiply the two denominators to get the denominator of the answer:

$$\frac{(1 \times 5)}{(7 \times 2)} = \frac{5}{14}$$

Because you canceled all common factors before multiplying, this answer is in lowest terms.

Practice Questions

1. Multiply $\frac{2}{3}$ by $\frac{7}{9}$.

2. Find $\frac{3}{8} \times \frac{6}{11}$.

3. Multiply $\frac{2}{9}$ by $\frac{3}{10}$.

4. Figure out $\frac{9}{14} \times \frac{8}{15}$.

Practice Answers

1. $\frac{2}{3} \times \frac{7}{9} = \frac{(2 \times 7)}{(3 \times 9)} = \frac{14}{27}$.

2. $\frac{3}{8} \times \frac{6}{11} = \frac{9}{44}$. Multiply the numerator by the numerator and the denominator by the denominator:

    $$\frac{3}{8} \times \frac{6}{11} = \frac{(3 \times 6)}{(8 \times 11)} = \frac{18}{88}$$

    The numerator and denominator are both even, so both can be reduced by a factor of 2:

    $$= \frac{9}{44}$$

3. $\frac{2}{9} \times \frac{3}{10} = \frac{1}{15}$. Begin by canceling out common factors. The numerator 2 and the denominator 10 are both even, so divide both by 2:

    $$\frac{2}{9} \times \frac{3}{10} = \frac{1}{9} \times \frac{3}{5}$$

    Next, the numerator 3 and the denominator 9 are both divisible by 3, so divide both by 3:

    $$= \frac{1}{3} \times \frac{1}{5}$$

    Now multiply straight across:

    $$= \frac{(1 \times 1)}{(3 \times 5)} = \frac{1}{15}$$

    Because you canceled out all common factors before multiplying, this answer is already reduced.

4. $\frac{9}{14} \times \frac{8}{15} = \frac{12}{35}$. Start by canceling out common factors. The numbers 14 and 8 are both divisible by 2, and 9 and 15 are divisible by 3:

    $$\frac{9}{14} \times \frac{8}{15}$$
    $$= \frac{9}{7} \times \frac{4}{15}$$
    $$= \frac{3}{7} \times \frac{4}{5}$$

    Now multiply:

    $$= \frac{(3 \times 4)}{(7 \times 5)} = \frac{12}{35}$$

Doing a flip to divide fractions

Dividing fractions is just as easy as multiplying them. In fact, when you divide fractions, you really turn the problem into multiplication.

Remember: To divide one fraction by another, multiply the first fraction by the reciprocal of the second. (As I discuss in Chapter 8, the *reciprocal* of a fraction is simply that fraction turned upside down.)

For example, here's how you turn fraction division into multiplication:

$$\frac{1}{3} \div \frac{4}{5} = \frac{1}{3} \times \frac{5}{4}$$

As you can see, I turn $\frac{4}{5}$ into its reciprocal — $\frac{5}{4}$ — and change the division sign to a multiplication sign. After that, just multiply the fractions as I describe in "Multiplying numerators and denominators straight across":

$$\frac{1}{3} \times \frac{5}{4} = \frac{1 \times 5}{3 \times 4} = \frac{5}{12}$$

Remember: As with multiplication, in some cases, you may have to reduce your result at the end. But you can also make your job easier by canceling out equal factors. (See the preceding section.)

Examples

Q. Divide $\frac{5}{8}$ by $\frac{3}{7}$.

A. $1\frac{11}{24}$. Change the division to multiplication:

$$\frac{5}{8} \div \frac{3}{7} = \frac{5}{8} \times \frac{7}{3}$$

Solve the problem using fraction multiplication:

$$= \frac{(5 \times 7)}{(8 \times 3)} = \frac{35}{24}$$

The answer is an improper fraction (because the numerator is greater than the denominator), so change it to a mixed number. Divide the numerator by the denominator and put the remainder over the denominator:

$$= 1\frac{11}{24}$$

Q. Calculate $\frac{7}{10} \div \frac{2}{5}$.

A. $1\frac{3}{4}$. Change the division to multiplication:

$$\frac{7}{10} \div \frac{2}{5} = \frac{7}{10} \times \frac{5}{2}$$

Notice that you have a 5 in one of the numerators and a 10 in the other fraction's denominator, so you can cancel out the common factor, which is 5; that would change your problem to $\frac{7}{2} \times \frac{1}{2}$. Or you can simply do your calculations and reduce the fraction later, as I do here. Solve by multiplying these two fractions:

$$= \frac{(7 \times 5)}{(10 \times 2)} = \frac{35}{20}$$

This time, the numerator and denominator are both divisible by 5, so you can reduce them:

$$= \frac{7}{4}$$

Because the numerator is greater than the denominator, the fraction is improper, so change it to a mixed number:

$$1\frac{3}{4}$$

Practice Questions

1. Divide $\frac{1}{4}$ by $\frac{6}{7}$.

2. Find $\frac{3}{5} \div \frac{9}{10}$.

3. Divide $\frac{8}{9}$ by $\frac{3}{10}$.

4. Solve $\frac{14}{15} \div \frac{7}{12}$.

Practice Answers

1. $\frac{1}{4} \div \frac{6}{7} = \frac{7}{24}$. First, change the problem to multiplication, multiplying by the reciprocal of the second fraction:

$$\frac{1}{4} \div \frac{6}{7} = \frac{1}{4} \times \frac{7}{6}$$

Now complete the problem using fraction multiplication:

$$= \frac{(1 \times 7)}{(4 \times 6)} = \frac{7}{24}$$

2. $\frac{3}{5} \div \frac{9}{10} = \frac{2}{3}$. Change the problem to multiplication, using the reciprocal of the second fraction:

$$\frac{3}{5} \div \frac{9}{10} = \frac{3}{5} \times \frac{10}{9}$$

Complete the problem using fraction multiplication:

$$= \frac{(3 \times 10)}{(5 \times 9)} = \frac{30}{45}$$

Both the numerator and denominator are divisible by 5, so reduce the fraction by this factor:

$$= \frac{6}{9}$$

They're still both divisible by 3, so reduce the fraction by this factor:

$$= \frac{2}{3}$$

3. $\frac{8}{9} \div \frac{3}{10} = 2\frac{26}{27}$. Change the problem to multiplication, using the reciprocal of the second fraction:

$$\frac{8}{9} \div \frac{3}{10} = \frac{8}{9} \times \frac{10}{3}$$

Complete the problem using fraction multiplication:

$$= \frac{(8 \times 10)}{(9 \times 3)} = \frac{80}{27}$$

The numerator is greater than the denominator, so change this improper fraction to a mixed number:

$$= 2\frac{26}{27}$$

4. $\frac{14}{15} \div \frac{7}{12} = 1\frac{3}{5}$. Change the problem to multiplication, using the reciprocal of the second fraction:

$$\frac{14}{15} \div \frac{7}{12} = \frac{14}{15} \times \frac{12}{7}$$

Complete the problem using fraction multiplication:

$$= \frac{(14 \times 12)}{(15 \times 7)} = \frac{168}{105}$$

The numerator is greater than the denominator, so change this improper fraction to a mixed number:

$$= 1\frac{63}{105}$$

Now the numerator and the denominator are both divisible by 3, so reduce the fractional part of this mixed number by a factor of 3:

$$= 1\frac{21}{35}$$

The numerator and denominator are now both divisible by 7, so reduce the fractional part by this factor:

$$= 1\frac{3}{5}$$

ADDING AND SUBTRACTING FRACTIONS WITH THE SAME DENOMINATOR

When you add and subtract fractions, one important item to notice is whether their denominators (the numbers on the bottom) are the same. If they're the same — woo-hoo! Adding and subtracting fractions that have the same denominator is a walk in the park.

Remember: To add two fractions that have the same denominator (bottom number), add the numerators (top numbers) and leave the denominator unchanged.

For example, consider the following problem:

$$\frac{1}{5} + \frac{2}{5} = \frac{1+2}{5} = \frac{3}{5}$$

As you can see, to add these two fractions, you add the numerators (1 + 2) and keep the denominator (5).

Why does this work? Chapter 8 tells you that you can think about fractions as pieces of cake. The denominator in this case tells you that the entire cake has been cut into five pieces. So when you add $\frac{1}{5} + \frac{2}{5}$, you're really adding one piece plus two pieces. The answer, of course, is three pieces — that is, $\frac{3}{5}$.

Even if you have to add more than two fractions, as long as the denominators are all the same, you just add the numerators and leave the denominator unchanged:

$$\frac{1}{17} + \frac{3}{17} + \frac{4}{17} + \frac{6}{17} = \frac{1+3+4+6}{17} = \frac{14}{17}$$

Sometimes when you add fractions with the same denominator, you have to reduce the result to lowest terms (to find out more about reducing, flip to Chapter 8). Take this problem, for example:

$$\frac{1}{4} + \frac{1}{4} = \frac{1+1}{4} = \frac{2}{4}$$

The numerator and the denominator are both even, so you know they can be reduced:

$$\frac{2}{4} = \frac{1}{2}$$

In other cases, the sum of two proper fractions is an improper fraction. You get a numerator that's larger than the denominator when the two fractions add up to more than 1, as in this case:

$$\frac{3}{7} + \frac{5}{7} = \frac{8}{7}$$

If you have more work to do with this fraction, leave it as an improper fraction so that it's easier to work with. But if this is your final answer, you may need to turn it into a mixed number (I cover mixed numbers in Chapter 8):

$$\frac{8}{7} = 8 \div 7 = 1r1 = 1\frac{1}{7}$$

Warning: When two fractions have the same numerator, don't add them by adding the denominators and leaving the numerator unchanged.

As with addition, subtracting fractions with the same denominator is always easy. When the denominators are the same, you can just think of the fractions as pieces of cake.

To subtract one fraction from another when they both have the same denominator (bottom number), subtract the numerator (top number) of the second from the numerator of the first and keep the denominator the same. For example:

$$\frac{3}{5} - \frac{2}{5} = \frac{3-2}{5} = \frac{1}{5}$$

Sometimes, as when you add fractions, you have to reduce:

$$\frac{3}{10} - \frac{1}{10} = \frac{3-1}{10} = \frac{2}{10}$$

Because the numerator and denominator are both even, you can reduce this fraction by a factor of 2:

$$\frac{2}{10} = \frac{2 \div 2}{10 \div 2} = \frac{1}{5}$$

Unlike with addition, when you subtract one proper fraction from another, you never get an improper fraction.

Examples

Q. Find $\frac{5}{8} + \frac{7}{8}$.

A. $1\frac{1}{2}$. The denominators are both 8, so add the numerators (5 and 7) to get the new numerator and keep the denominator the same:

$$\frac{5}{8} + \frac{7}{8} = \frac{(5+7)}{8} = \frac{12}{8}$$

The numerator is greater than the denominator, so the answer is an improper fraction. Change it to a mixed number and then reduce (as I show you in Chapter 8):

$$= 1\frac{2}{4} = 1\frac{1}{2}$$

Q. What is $\frac{11}{12} - \frac{5}{12}$.

A. The denominators are both 12, so subtract the numerators (11 and 5) to get the new numerator and keep the denominator the same:

$$\frac{11}{12} - \frac{5}{12} = \frac{(11-5)}{12} = \frac{6}{12}$$

The numerator is a factor of the denominator, so reduce it (as I show you in Chapter 8):

$$= \frac{1}{2}$$

Practice Questions

1. What is $\frac{3}{11} + \frac{7}{11}$?

2. Find $\frac{11}{15} - \frac{4}{15}$.

3. What is the sum of $\frac{5}{9}$ and $\frac{8}{9}$?

4. Subtract $\frac{13}{20}$ minus $\frac{9}{20}$

5. Add $\frac{7}{9}$ and $\frac{8}{9}$.

6. What is $\frac{19}{24} - \frac{11}{24}$?

A **Practice Answers**

1. $\frac{10}{11}$. The denominators are the same, so add numerators and keep the denominator the same:

$$\frac{3}{11}+\frac{7}{11}=\frac{10}{11}$$

2. $\frac{7}{15}$. The denominators are the same, so subtract the first numerator minus the second and keep the denominator the same:

$$\frac{11}{15}-\frac{4}{15}=\frac{7}{15}$$

3. $1\frac{4}{9}$. The denominators are the same, so add numerators and keep the denominator the same:

$$\frac{5}{9}+\frac{8}{9}=\frac{13}{9}$$

The result is an improper fraction, so convert it to a mixed number:

$$=1\frac{4}{9}$$

4. $\frac{1}{5}$. The denominators are the same, so subtract the first numerator minus the second and keep the denominator the same:

$$\frac{13}{20}-\frac{9}{20}=\frac{4}{20}$$

Both the numerator and denominator are divisible by 4, so reduce the fraction by 4:

$$=\frac{1}{5}$$

5. $\frac{7}{9}+\frac{8}{9}=1\frac{2}{3}$. The denominators are the same, so add the numerators:

$$\frac{7}{9}+\frac{8}{9}=\frac{15}{9}$$

Both the numerator and denominator are divisible by 3, so reduce the fraction by 3:

$$=\frac{5}{3}$$

The result is an improper fraction, so convert it to a mixed number:

$$=1\frac{2}{3}$$

6. $\frac{1}{3}$. The denominators are the same, so subtract the first numerator minus the second:

$$\frac{19}{24} - \frac{11}{24} = \frac{8}{24}$$

Both the numerator and denominator are divisible by 8, so reduce the fraction by 8:

$$= \frac{1}{3}$$

Adding fractions with different denominators

When the fractions that you want to add have different denominators, adding them isn't quite as easy. At the same time, it doesn't have to be as hard as most teachers make it.

Now, I'm shimmying out onto a brittle limb here, but this needs to be said: Fractions can be added in a very simple way. It always works. It makes adding fractions only a little more difficult than multiplying them. And as you move up the math food chain into algebra, it becomes the most useful method.

So why doesn't anybody talk about it? I think it's a clear case of tradition being stronger than common sense. The traditional way to add fractions is more difficult, more time-consuming, and more likely to cause an error. But generation after generation has been taught that it's the right way to add fractions. It's a vicious cycle.

But in this book, I'm breaking with tradition. I first show you the easy way to add fractions. Then I show you a quick trick that works in a few special cases. Finally, I show you the traditional way to add fractions.

Tip: At some point in your life, I bet some teacher somewhere told you these golden words of wisdom: "You can't add two fractions with different denominators." Your teacher was wrong! Here's the way to do it:

1. **Cross-multiply the two fractions and add the results together to get the numerator of the answer.**

   Suppose you want to add the fractions $\frac{1}{3}$ and $\frac{2}{5}$. To get the numerator of the answer, cross-multiply. In other words, multiply the numerator of each fraction by the denominator of the other:

   $$\frac{1}{3} + \frac{2}{5}$$
   $$1 \times 5 = 5$$
   $$2 \times 3 = 6$$

   Add the results to get the numerator of the answer:

   $$5 + 6 = 11$$

2. **Multiply the two denominators to get the denominator of the answer.**

   To get the denominator, just multiply the denominators of the two fractions:

   $$3 \times 5 = 15$$

   The denominator of the answer is 15.

3. **Write your answer as a fraction.**

   $$\frac{1}{3} + \frac{2}{5} = \frac{11}{15}$$

As you discover in the earlier section "Finding the sum of fractions with the same denominator," when you add fractions, you sometimes need to reduce the answer you get. Here's an example:

$$\frac{5}{8} + \frac{3}{10} = \frac{(5 \times 10) + (3 \times 8)}{8 \times 10} = \frac{50 + 24}{80} = \frac{74}{80}$$

Because the numerator and the denominator are both even numbers, you know that the fraction can be reduced. So try dividing both numbers by 2:

$$\frac{74 \div 2}{80 \div 2} = \frac{37}{40}$$

This fraction can't be reduced further, so $\frac{37}{40}$ is the final answer.

As you also discover in "Finding the sum of fractions with the same denominator," sometimes when you add two proper fractions, your answer is an improper fraction:

$$\frac{4}{5} + \frac{3}{7} = \frac{(4 \times 7) + (3 \times 5)}{5 \times 7} = \frac{28 + 15}{35} = \frac{43}{35}$$

If you have more work to do with this fraction, leave it as an improper fraction so that it's easier to work with. But if this is your final answer, you may need to turn it into a mixed number (see Chapter 8 for details).

$$\frac{43}{35} = 43 \div 35 = 1 \text{ r } 8 = 1\frac{8}{35}$$

Tip: In some cases, you have to add more than one fraction. The method is similar, with one small tweak. For example, suppose you want to add $\frac{1}{2} + \frac{3}{5} + \frac{4}{7}$:

1. **Start by multiplying the *numerator* of the first fraction by the *denominators* of all the other fractions.**

   $$\frac{1}{2} + \frac{3}{5} + \frac{4}{7}$$
   $$(1 \times 5 \times 7) = 35$$

2. **Do the same with the second fraction, and add this value to the first.**

   $$\frac{1}{2} + \frac{3}{5} + \frac{4}{7}$$
   $$35 + (3 \times 2 \times 7) = 35 + 42$$

3. **Do the same with the remaining fraction(s).**

$$\frac{1}{2}+\frac{3}{5}+\frac{4}{7}$$

$$35+42+(4\times2\times5)=35+42+40=117$$

When you're done, you have the numerator of the answer.

4. **To get the denominator, just multiply all the denominators together:**

$$\frac{1}{2}+\frac{3}{5}+\frac{4}{7}$$

$$=\frac{35+42+40}{2\times5\times7}=\frac{117}{70}$$

As usual, you may need to reduce or change an improper fraction to a mixed number. In this example, you just need to change to a mixed number (see Chapter 8 for details):

$$\frac{117}{70}=117\div70=1\ r\ 47=1\frac{47}{70}$$

Examples

Q. $\frac{4}{5}+\frac{1}{3}=?$

A. $1\frac{2}{15}$ Cross-multiply and add the results to find the numerator of the answer, and multiply the denominators to find the denominator:

$$\frac{4}{5}+\frac{1}{3}=\frac{12+5}{15}=\frac{17}{15}$$

The result is an improper fraction, so change it to a mixed number:

$$=1\frac{2}{15}$$

Practice Questions

1. What is $\frac{2}{9}$ plus $\frac{3}{4}$?

2. Add $\frac{3}{8}+\frac{6}{7}$

3. $\frac{5}{12}+\frac{1}{10}=?$

4. What is the sum of $\frac{5}{6}$ and $\frac{7}{9}$?

Practice Answers

1. $\frac{35}{36}$. Cross-multiply and add the results to find the numerator of the answer, and multiply the denominators to find the denominator:

$$\frac{2}{9}+\frac{3}{4}=\frac{8+27}{36}=\frac{35}{36}$$

2. $1\frac{13}{56}$. Cross-multiply and add the results to find the numerator of the answer, and multiply the denominators to find the denominator:

$$\frac{3}{8}+\frac{6}{7}=\frac{21+48}{56}=\frac{69}{56}$$

The result is an improper fraction, so change it to a mixed number:

$$=1\frac{13}{56}$$

3. $\frac{31}{60}$. Cross-multiply and add the results to find the numerator of the answer, and multiply the denominators to find the denominator:

$$\frac{5}{12}+\frac{1}{10}=\frac{12+50}{120}=\frac{62}{120}$$

This fraction can be reduced by a factor of 2:

$$=\frac{31}{60}$$

4. $1\frac{11}{18}$. Cross-multiply and add the results to find the numerator of the answer, and multiply the denominators to find the denominator:

$$\frac{5}{6}+\frac{7}{9}=\frac{45+42}{54}=\frac{87}{54}$$

The result is an improper fraction, so change it to a mixed number:

$$=1\frac{33}{54}$$

This fraction can be reduced by a factor of 3:

$$=1\frac{11}{18}$$

Trying a quick trick

I show you a way to add fractions with different denominators in the preceding section. It always works, and it's easy. So why do I want to show you another way? It feels like déjà vu.

In some cases, though, you can save yourself a lot of effort with a little bit of smart thinking. You can't always use this method, but you can use it when one denominator is a multiple of the other. (For more on multiples, see Chapter 7.) Look at the following problem:

$$\frac{11}{12} + \frac{19}{24}$$

First, I solve it the way I show you in the preceding section:

$$\frac{11}{12} + \frac{19}{24} = \frac{(11 \times 24) + (19 \times 12)}{12 \times 24} = \frac{264 + 228}{288} = \frac{492}{288}$$

Those numbers are pretty big, and I'm still not done because the numerator is larger than the denominator. The answer is an improper fraction. Worse yet, the numerator and denominator are both even numbers, so the answer still needs to be reduced.

With certain fraction addition problems, I can give you a smarter way to work. The trick is to turn a problem with different denominators into a much easier problem with the same denominator.

Remember: Before you add two fractions with different denominators, check the denominators to see whether one is a multiple of the other (for more on multiples, flip to Chapter 7). If it is, you can use the quick trick:

1. **Increase the terms of the fraction with the smaller denominator so that it has the larger denominator.**

    Look at the earlier problem in this new way:

    $$\frac{11}{12} + \frac{19}{24}$$

    As you can see, 12 divides into 24 without a remainder. In this case, you want to raise the terms of $\frac{11}{12}$ so that the denominator is 24:

    $$\frac{11}{12} = \frac{?}{24}$$

    I show you how to do this kind of problem in Chapter 8. To fill in the question mark, the trick is to divide 24 by 12 to find out how the denominators are related; then multiply the result by 11:

    $$? = (24 \div 12) \times 11 = 22$$
    $$\text{So } \frac{11}{12} = \frac{22}{24}$$

2. **Rewrite the problem, substituting this increased version of the fraction, and add as I show you earlier in "Finding the sum of fractions with the same denominator."**

    Now you can rewrite the problem this way:

    $$\frac{22}{24} + \frac{19}{24} = \frac{41}{24}$$

As you can see, the numbers in this case are much smaller and easier to work with. The answer here is an improper fraction; changing it to a mixed number is easy:

$$\frac{41}{24} = 41 \div 24 = 1 \text{ r } 17 = 1\frac{17}{24}$$

Example

Q. $\frac{7}{8} + \frac{13}{16} = ?$

A. $1\frac{11}{16}$ The number 8 is a factor of 16, so increase the fraction $\frac{7}{8}$ to a fraction with 16 in the denominator, then add the fractions:

$$\frac{7}{8} + \frac{13}{16} = \frac{14}{16} + \frac{13}{16} = \frac{27}{16}$$

The result is an improper fraction, so change it to a mixed number:

$$= 1\frac{11}{16}$$

Practice Questions

1. What is the sum of the fractions $\frac{13}{20}$ and $\frac{3}{10}$?

2. $\frac{5}{18} + \frac{8}{9} = ?$

3. What is $\frac{11}{24}$ plus $\frac{7}{8}$?

4. Add $\frac{47}{100} + \frac{19}{25}$.

Practice Answers

1. $\frac{19}{20}$. The number 10 is a factor of 20, so increase the fraction $\frac{3}{10}$ to a fraction with 20 in the denominator, then add the fractions:

$$\frac{13}{20} + \frac{3}{10} = \frac{13}{20} + \frac{6}{20} = \frac{19}{20}$$

2. $1\frac{1}{6}$. The number 9 is a factor of 18, so increase the fraction $\frac{8}{9}$ to a fraction with 18 in the denominator, then add the fractions:

$$\frac{5}{18} + \frac{8}{9} = \frac{5}{18} + \frac{16}{18} = \frac{21}{18}$$

The result is an improper fraction, so change it to a mixed number:

$$= 1\frac{3}{18}$$

This fraction can be reduced by a factor of 3:

$$= 1\frac{1}{6}$$

3. $1\frac{1}{3}$. The number 8 is a factor of 24, so increase the fraction $\frac{7}{8}$ to a fraction with 24 in the denominator, then add the fractions:

$$\frac{11}{24} + \frac{7}{8} = \frac{11}{24} + \frac{21}{24} = \frac{32}{24}$$

The result is an improper fraction, so change it to a mixed number:

$$= 1\frac{8}{24}$$

This fraction can be reduced by a factor of 8:

$$= 1\frac{1}{3}$$

4. $1\frac{23}{100}$. The number 25 is a factor of 100, so increase the fraction $\frac{19}{25}$ to a fraction with 100 in the denominator, then add the fractions:

$$\frac{47}{100} + \frac{19}{25} = \frac{47}{100} + \frac{76}{100} = \frac{123}{100}$$

The result is an improper fraction, so change it to a mixed number:

$$= 1\frac{23}{100}$$

Relying on the traditional way

In the two preceding sections, I show you two ways to add fractions with different denominators. They both work great, depending on the circumstances. So why do I want to show you yet a third way? It feels like déjà vu all over again.

The truth is that I don't want to show you this way. But they're *forcing* me to. And you know who *they* are, don't you? The man — the system — the powers that be. The ones who want to keep you down in the mud, groveling at their feet. Okay, so I'm exaggerating a little. But let me impress on you that you don't have to add fractions this way unless you really want to (or unless your teacher insists on it).

Remember: Here's the traditional way to add fractions with two different denominators:

1. **Find the least common multiple (LCM) of the two denominators (for more on finding the LCM of two numbers, see Chapter 7).**

   Suppose you want to add the fractions $\frac{3}{4} + \frac{7}{10}$. First find the LCM of the two denominators, 4 and 10. Here's how to find the LCM using the multiplication table method:

   • **Multiples of 10:** 10, 20, 30, 40

   • **Multiples of 4:** 4, 8, 12, 16, 20

   So the LCM of 4 and 10 is 20.

2. **Increase the terms of each fraction so that the denominator of each equals the LCM (for more on how to do this, see Chapter 8).**

   Increase each fraction to higher terms so that the denominator of each is 20.

   $$\frac{3}{4} = \frac{3 \times 5}{4 \times 5} = \frac{15}{20} \text{ and } \frac{7}{10} = \frac{7 \times 2}{10 \times 2} = \frac{14}{20}$$

3. **Substitute these two new fractions for the original ones and add as I show you earlier in "Finding the sum of fractions with the same denominator."**

   At this point, you have two fractions that have the same denominator:

   $$\frac{15}{20} + \frac{14}{20} = \frac{29}{20}$$

   When the answer is an improper fraction, you still need to change it to a mixed number:

   $$\frac{29}{20} = 29 \div 20 = 1 \text{ r } 9 = 1\frac{9}{20}$$

As another example, suppose you want to add the numbers $\frac{5}{6} + \frac{3}{10} + \frac{2}{15}$.

1. **Find the LCM of 6, 10, and 15.**

   This time, I use the prime factorization method (see Chapter 7 for details on how to do this). Start by decomposing the three denominators to their prime factors:

   $$6 = 2 \times 3$$
   $$10 = 2 \times 5$$
   $$15 = 3 \times 5$$

These denominators have a total of three different prime factors — 2, 3, and 5. Each prime factor appears only once in any decomposition, so the LCM of 6, 10, and 15 is

$$2 \times 3 \times 5 = 30$$

2. **You need to increase the terms of all three fractions so that their denominators are 30:**

$$\frac{5}{6} = \frac{5 \times 5}{6 \times 5} = \frac{25}{30}$$

$$\frac{3}{10} = \frac{3 \times 3}{10 \times 3} = \frac{9}{30}$$

$$\frac{2}{15} = \frac{2 \times 2}{15 \times 2} = \frac{4}{30}$$

3. **Simply add the three new fractions:**

$$\frac{25}{30} + \frac{9}{30} + \frac{4}{30} = \frac{38}{30}$$

Again, you need to change this improper fraction to a mixed number:

$$\frac{38}{30} = 38 \div 30 = 1 \text{ r } 8 = 1\frac{8}{30}$$

Because both the numerator and the denominator are divisible by 2, you can reduce the fraction:

$$1\frac{8}{30} = 1\frac{4}{15}$$

E

Example

Q. Add $\frac{7}{8} + \frac{5}{6}$.

A. $1\frac{17}{24}$. To find the LCM of 8 and 6, generate the multiples of these two numbers:

Multiples of 8: 8, 16, **24**, 32, 40, 48

Multiples of 6: 6, 12, 18, **24**

Increase the terms of both fractions so that each has a denominator of 24:

$$\frac{7}{8} + \frac{5}{6} = \frac{21}{24} + \frac{20}{24} = \frac{41}{24}$$

Convert this improper fraction to a mixed number:

$$= 1\frac{17}{24}$$

Practice Questions

1. What is $\frac{3}{4}$ plus $\frac{1}{6}$?

2. Add the fractions $\frac{5}{6}$ and $\frac{7}{9}$.

3. $\frac{3}{10} + \frac{11}{15} = ?$

4. What is the sum of $\frac{9}{16}$ and $\frac{5}{12}$?

Practice Answers

1. $\frac{11}{12}$. To find the LCM of 4 and 6, generate the multiples of these two numbers:

   Multiples of 6: 6, 12, 18, 24

   Multiples of 4: 4, 8, 12

   Increase the terms of both fractions so that each has a denominator of 12:

   $$\frac{3}{4} + \frac{1}{6} = \frac{9}{12} + \frac{2}{12} = \frac{11}{12}$$

2. $1\frac{11}{18}$. To find the LCM of 6 and 9, generate the multiples of these two numbers:

   Multiples of 9: 9, 18, 27, 36, 45, 54

   Multiples of 6: 6, 12, 18

   Increase the terms of both fractions so that each has a denominator of 18:

   $$\frac{5}{6} + \frac{7}{9} = \frac{15}{18} + \frac{14}{18} = \frac{29}{18}$$

   Change this improper fraction to a mixed number:

   $$= 1\frac{11}{18}$$

3. $1\frac{1}{30}$. To find the LCM of 10 and 15, generate the multiples of these two numbers:

   Multiples of 15: 15, 30

   Multiples of 10: 10, 20, 30

   Increase the terms of both fractions so that each has a denominator of 30:

   $$\frac{3}{10} + \frac{11}{15} = \frac{9}{30} + \frac{22}{30} = \frac{31}{30}$$

   Change this improper fraction to a mixed number:

   $$= 1\frac{1}{30}$$

4. $\dfrac{47}{48}$. To find the LCM of 12 and 16, find the prime factorizations of these two numbers:

$$12 = 2 \times 2 \times 3$$
$$16 = 2 \times 2 \times 2 \times 2$$

The number 2 appears four times in the prime factorization of 16, and 3 appears once in the prime factorization of 12. Therefore, the least common multiple of 12 and 16 is:

$$2 \times 2 \times 2 \times 2 \times 3 = 48$$

Increase the terms of the two fractions to fractions with denominators of 48:

$$\frac{9}{16} + \frac{5}{12} = \frac{27}{48} + \frac{20}{48} = \frac{47}{48}$$

TAKING IT AWAY: SUBTRACTING FRACTIONS

Subtracting fractions isn't really much different than adding them. As with addition, when the denominators are the same, subtraction is easy. And when the denominators are different, the methods I show you for adding fractions can be tweaked for subtracting them.

So to figure out how to subtract fractions, you can read the section "All Together Now: Adding Fractions" and substitute a minus sign (–) for every plus sign (+). But it'd be just a little chintzy if I expected you to do that. So in this section, I show you four ways to subtract fractions that mirror what I discuss earlier in this chapter about adding them.

Subtracting fractions with different denominators

Just as with addition, you have a choice of methods when subtracting fractions. These three methods are similar to the methods I show you for adding fractions: the easy way, the quick trick, and the traditional way.

The easy way always works, and I recommend this method for most of your fraction subtracting needs. The quick trick is a great timesaver, so use it when you can. And as for the traditional way — well, even if I don't like it, your teacher and other math purists probably do.

Knowing the easy way

This way of subtracting fractions works in all cases, and it's easy. (In the next section, I show you a quick way to subtract fractions when one denominator is a multiple of the other.) Here's the easy way to subtract fractions that have different denominators:

1. **Cross-multiply the two fractions and subtract the second number from the first to get the numerator of the answer.**

   For example, suppose you want to subtract $\dfrac{6}{7} - \dfrac{2}{5}$. To get the numerator, cross-multiply the two fractions and then subtract the second number from the first number (see Chapter 8 for info on cross-multiplication):

$$\frac{6}{7} - \frac{2}{5}$$
$$(6 \times 5) - (2 \times 7) = 30 - 14 = 16$$

Remember: After you cross-multiply, be sure to subtract in the correct order. (The first number is the numerator of the first fraction times the denominator of the second.)

2. **Multiply the two denominators to get the denominator of the answer.**

$$7 \times 5 = 35$$

$$\frac{16}{35}$$

Here's another example to work with:

$$\frac{9}{10} - \frac{5}{6}$$

This time, I put all the steps together:

$$\frac{9}{10} - \frac{5}{6} = \frac{(9 \times 6) - (5 \times 10)}{10 \times 6}$$

With the problem set up like this, you just have to simplify the result:

$$= \frac{54 - 50}{60} = \frac{4}{60}$$

In this case, you can reduce the fraction:

$$\frac{4}{60} = \frac{1}{15}$$

Example

Q. $\frac{3}{5} - \frac{1}{4} = ?$

A. $\frac{7}{20}$. Cross-multiply and subtract to find the numerator of the answer, and multiply the denominators to find the denominator:

$$\frac{3}{5} - \frac{1}{4} = \frac{12 - 5}{20} = \frac{7}{20}$$

Practice Questions

1. Subtract $\frac{1}{2} - \frac{1}{10}$

2. What is $\frac{5}{7}$ minus $\frac{3}{10}$?

3. What is the result when you subtract $\frac{6}{11}$ minus $\frac{3}{8}$?

4. $\frac{7}{10} - \frac{7}{12} = ?$

Practice Answers

1. $\frac{2}{5}$. Cross-multiply and subtract to find the numerator of the answer, and multiply the denominators to find the denominator:

$$\frac{1}{2} - \frac{1}{10} = \frac{10-2}{20} = \frac{8}{20}$$

This fraction can be reduced by a factor of 4:

$$= \frac{2}{5}$$

2. $\frac{29}{70}$. Cross-multiply and subtract to find the numerator of the answer, and multiply the denominators to find the denominator:

$$\frac{5}{7} - \frac{3}{10} = \frac{50-21}{70} = \frac{29}{70}$$

3. $\frac{15}{88}$. Cross-multiply and subtract to find the numerator of the answer, and multiply the denominators to find the denominator:

$$\frac{6}{11} - \frac{3}{8} = \frac{48-33}{88} = \frac{15}{88}$$

4. $\frac{7}{60}$. Cross-multiply and subtract to find the numerator of the answer, and multiply the denominators to find the denominator:

$$\frac{7}{10} - \frac{7}{12} = \frac{84-70}{120} = \frac{14}{120}$$

This fraction can be reduced by a factor of 2:

$$= \frac{7}{60}$$

Cutting it short with a quick trick

The easy way I show you in the preceding section works best when the numerators and denominators are small. When they're larger, you may be able to take a shortcut.

Before you subtract fractions with different denominators, check the denominators to see whether one is a multiple of the other (for more on multiples, see Chapter 7). If it is, you can use the quick trick:

1. **Increase the terms of the fraction with the smaller denominator so that it has the larger denominator.**

   For example, suppose you want to find $\frac{17}{20} - \frac{31}{80}$. If you cross-multiply these fractions, your results are going to be much bigger than you want to work with. But fortunately, 80 is a multiple of 20, so you can use the quick way.

First, increase the terms of $\frac{17}{20}$ so that the denominator is 80 (for more on increasing the terms of fractions, see Chapter 8):

$$\frac{17}{20} = \frac{?}{80}$$
$$? = 80 \div 20 \times 17 = 68$$
$$\text{So } \frac{17}{20} = \frac{68}{80}$$

2. **Rewrite the problem, substituting this increased version of the fraction, and subtract as I show you earlier in "Subtracting fractions with the same denominator."**

Here's the problem as a subtraction of fractions with the same denominator, which is much easier to solve:

$$\frac{68}{80} - \frac{31}{80} = \frac{37}{80}$$

In this case, you don't have to reduce to lowest terms, although you may have to in other problems. (See Chapter 8 for more on reducing fractions.)

Example

Q. $\frac{13}{24} - \frac{1}{8} = ?$

A. $\frac{5}{12}$. The number 8 is a factor of 24, so increase the fraction $\frac{1}{8}$ to a fraction with 24 in the denominator, then subtract the fractions:

$$\frac{13}{24} - \frac{1}{8} = \frac{13}{24} - \frac{3}{24} = \frac{10}{24}$$

Reduce this fraction by a factor of 2:

$$= \frac{5}{12}$$

Practice Questions

1. $\frac{9}{10} - \frac{1}{2} = ?$

2. What is the difference between $\frac{8}{9}$ and $\frac{11}{18}$?

3. What is $\frac{13}{21}$ minus $\frac{1}{3}$?

4. Subtract $\frac{29}{30} - \frac{1}{6}$.

A Practice Answers

1. $\frac{2}{5}$. The number 2 is a factor of 10, so increase the fraction $\frac{1}{2}$ to a fraction with 10 in the denominator, then subtract:

$$\frac{9}{10} - \frac{1}{2} = \frac{9}{10} - \frac{5}{10} = \frac{4}{10}$$

Reduce this fraction by a factor of 2:

$$= \frac{2}{5}$$

2. $\frac{5}{18}$. The number 9 is a factor of 18, so increase the fraction $\frac{8}{9}$ to a fraction with 18 in the denominator, then subtract:

$$\frac{8}{9} - \frac{11}{18} = \frac{16}{18} - \frac{11}{18} = \frac{5}{18}$$

3. $\frac{2}{7}$. The number 3 is a factor of 21, so increase the fraction $\frac{1}{3}$ to a fraction with 21 in the denominator, then subtract:

$$\frac{13}{21} - \frac{1}{3} = \frac{13}{21} - \frac{7}{21} = \frac{6}{21}$$

This fraction can be reduced by a factor of 3:

$$= \frac{2}{7}$$

4. $\frac{4}{5}$. The number 6 is a factor of 30, so increase the fraction $\frac{1}{6}$ to a fraction with 30 in the denominator, then subtract:

$$\frac{29}{30} - \frac{1}{6} = \frac{29}{30} - \frac{5}{30} = \frac{24}{30}$$

This fraction can be reduced by a factor of 6:

$$= \frac{4}{5}$$

Keeping your teacher happy with the traditional way

As I describe earlier in this chapter in "All Together Now: Adding Fractions," you want to use the traditional way only as a last resort. I recommend that you use it only when the numerator and denominator are too large to use the easy way and when you can't use the quick trick.

To use the traditional way to subtract fractions with two different denominators, follow these steps:

1. **Find the least common multiple (LCM) of the two denominators (for more on finding the LCM of two numbers, see Chapter 7).**

   For example, suppose you want to subtract $\frac{7}{8} - \frac{11}{14}$. Here's how to find the LCM of 8 and 14:

   Multiples of 8 : 8, 16, 24, 32, 40, 48, **56**

   Multiples of 14 : 14, 28, 42, **56**

   So the LCM of 8 and 14 is 56.

2. **Increase each fraction to higher terms so that the denominator of each equals the LCM (for more on how to do this, see Chapter 7).**

   The denominators of both now are 56:

   $$\frac{7}{8} = \frac{7 \times 7}{8 \times 7} = \frac{49}{56}$$
   $$\frac{11}{14} = \frac{11 \times 4}{14 \times 4} = \frac{44}{56}$$

3. **Substitute these two new fractions for the original ones and subtract as I show you earlier in "Subtracting fractions with the same denominator."**

   $$\frac{49}{56} - \frac{44}{56} = \frac{5}{56}$$

This time, you don't need to reduce because 5 is a prime number and 56 isn't divisible by 5. In some cases, however, you have to reduce the answer to lowest terms.

E

Example

Q. Subtract $\frac{11}{12} - \frac{7}{8}$.

A. $\frac{1}{24}$. To find the LCM of 12 and 8 generate the multiples of these two numbers:

Multiples of 12: 12, **24**, 36, 48, 60, 72

Multiples of 8: 8, 16, **24**

Increase the terms of both fractions so that each has a denominator of 24:

$$\frac{11}{12} - \frac{7}{8} = \frac{22}{24} - \frac{21}{24} = \frac{1}{24}$$

Practice Questions

1.   Subtract $\frac{5}{6}$ minus $\frac{3}{10}$.

2.   What is $\frac{7}{9}$ minus $\frac{1}{12}$?

3. $\frac{14}{15} - \frac{17}{20} = ?$

4.   What is the difference between $\frac{13}{20}$ and $\frac{7}{24}$?

Practice Answers

1. $\frac{8}{15}$. To find the LCM of 6 and 10, generate the multiples of these two numbers:

Multiples of 6: 6, 12, 18, 24, **30**

Multiples of 10: 10, 20, **30**

Increase the terms of both fractions so that each has a denominator of 30:

$$\frac{5}{6} - \frac{3}{10} = \frac{25}{30} - \frac{9}{30} = \frac{16}{30}$$

Reduce this fraction by a factor of 2:

$$= \frac{8}{15}$$

2. $\frac{25}{36}$. To find the LCM of 9 and 12, generate the multiples of these two numbers:

Multiples of 12: 12, 24, **36,** 48, 60, 72

Multiples of 9: 9, 18, 27, **36**

Increase the terms of both fractions so that each has a denominator of 36:

$$\frac{7}{9} - \frac{1}{12} = \frac{28}{36} - \frac{3}{36} = \frac{25}{36}$$

3. $\frac{1}{12}$. To find the LCM of 20 and 15, generate the multiples of these two numbers:

Multiples of 15: 15, 30, 45, **60**

Multiples of 10: 10, 20, 30, 40, 50, **60**

Increase the terms of both fractions so that each has a denominator of 60:

$$\frac{14}{15} - \frac{17}{20} = \frac{56}{60} - \frac{51}{60} = \frac{5}{60}$$

Now, reduce this result to lowest terms:

$$= \frac{1}{12}$$

4. $\frac{43}{120}$. To find the LCM of 20 and 24, find the prime factorizations of these two numbers:

$$20 = 2 \times 2 \times 5$$
$$24 = 2 \times 2 \times 2 \times 3$$

The number 2 appears three times in the prime factorization of 24, 3 appears once in the prime factorization of 24, and the number 5 appears once in the prime factorization of 20. Therefore, the least common multiple of 20 and 24 is:

$$2 \times 2 \times 2 \times 3 \times 5 = 120$$

Increase the terms of the two fractions to fractions with denominators of 120:

$$\frac{13}{20} - \frac{7}{24} = \frac{78}{120} - \frac{35}{120} = \frac{43}{120}$$

WORKING PROPERLY WITH MIXED NUMBERS

All the methods I describe earlier in this chapter work for both proper and improper fractions. Unfortunately, mixed numbers are ornery little critters, and you need to figure out how to deal with them on their own terms. (For more on mixed numbers, flip to Chapter 8.)

Multiplying and dividing mixed numbers

I can't give you a direct method for multiplying and dividing mixed numbers. The only way is to convert the mixed numbers to improper fractions and multiply or divide as usual. Here's how to multiply or divide mixed numbers:

1. **Convert all mixed numbers to improper fractions (see Chapter 8 for details).**

   For example, suppose you want to multiply $1\frac{3}{5} \times 2\frac{1}{3}$. First convert $1\frac{3}{5}$ and $2\frac{1}{3}$ to improper fractions:

   $$1\frac{3}{5} = \frac{5 \times 1 + 3}{5} = \frac{8}{5}$$
   $$2\frac{1}{3} = \frac{3 \times 2 + 1}{3} = \frac{7}{3}$$

2. **Multiply these improper fractions (as I show you earlier in this chapter, in "Multiplying and Dividing Fractions").**

   $$\frac{8}{5} \times \frac{7}{3} = \frac{8 \times 7}{5 \times 3} = \frac{56}{15}$$

3. **If the answer is an improper fraction, convert it back to a mixed number (see Chapter 8).**

   $$\frac{56}{15} = 56 \div 15 = 3r11 = 3\frac{11}{15}$$

   In this case, the answer is already in lowest terms, so you don't have to reduce it.

As a second example, suppose you want to divide $3\frac{2}{3}$ by $1\frac{4}{7}$.

1. Convert $3\frac{2}{3}$ and $1\frac{4}{7}$ to improper fractions:

$$3\frac{2}{3} = \frac{3 \times 3 + 2}{3} = \frac{11}{3}$$
$$1\frac{4}{7} = \frac{7 \times 1 + 4}{7} = \frac{11}{7}$$

2. **Divide these improper fractions.**

   Divide fractions by multiplying the first fraction by the reciprocal of the second (see the earlier "Multiplying and Dividing Fractions" section):

$$\frac{11}{3} \div \frac{11}{7} = \frac{11}{3} \times \frac{7}{11}$$

   In this case, before you multiply, you can cancel out factors of 11 in the numerator and denominator:

$$\frac{1\cancel{1}}{3} \times \frac{7}{\cancel{1}1} = \frac{1 \times 7}{3 \times 1} = \frac{7}{3}$$

3. **Convert the answer to a mixed number.**

$$\frac{7}{3} = 7 \div 3 = 2\text{r}1 = 2\frac{1}{3}$$

Examples

Q. What is $2\frac{1}{5} \times 3\frac{1}{4}$?

A. $7\frac{3}{20}$. First, convert both mixed numbers to improper fractions. Multiply the whole number by the denominator and add the numerator; then place your answer over the original denominator:

$$2\frac{1}{5} = \frac{(2 \times 5 + 1)}{5} = \frac{11}{5}$$
$$3\frac{1}{4} = \frac{(3 \times 4 + 1)}{4} = \frac{13}{4}$$

Now multiply the two fractions:

$$\frac{11}{5} \times \frac{13}{4} = \frac{(11 \times 13)}{5 \times 4} = \frac{143}{20}$$

Because the answer is an improper fraction, change it back to a mixed number:

$$\begin{array}{r} 7 \\ 20\overline{)143} \\ -140 \\ \hline 3 \end{array}$$

The final answer is $7\frac{3}{20}$.

Q.   What is $3\frac{1}{2} \div 1\frac{1}{7}$?

A. $3\frac{1}{16}$. First, convert both mixed numbers to improper fractions:

$$3\frac{1}{2} = \frac{(3 \times 2 + 1)}{2} = \frac{7}{2}$$
$$1\frac{1}{7} = \frac{(1 \times 7 + 1)}{7} = \frac{8}{7}$$

Now divide the two fractions:

$$\frac{7}{2} \div \frac{8}{7} = \frac{7}{2} \times \frac{7}{8} = \frac{49}{16}$$

Because the answer is an improper fraction, convert it to a mixed number:

$$\begin{array}{r} 3 \\ 16\overline{)49} \\ -48 \\ \hline 1 \end{array}$$

The final answer is $3\frac{1}{16}$.

Practice Questions

1. Multiply $2\frac{1}{3}$ by $1\frac{3}{7}$.

2. Find $2\frac{2}{5} \times 1\frac{5}{6}$.

3. Multiply $4\frac{4}{5}$ by $3\frac{1}{8}$.

4. Calculate $4\frac{1}{2} \div 1\frac{5}{8}$.

5. Divide $2\frac{1}{10}$ by $2\frac{1}{4}$.

6. Find $1\frac{2}{7} \div 6\frac{3}{10}$.

A

Practice Answers

1. $2\frac{1}{3} \times 1\frac{3}{7} = 3\frac{1}{3}$. Change both mixed numbers to improper fractions:

$$2\frac{1}{3} = \frac{(2 \times 3 + 1)}{3} = \frac{7}{3}$$
$$1\frac{3}{7} = \frac{(1 \times 7 + 3)}{7} = \frac{10}{7}$$

Set up the multiplication:

$$\frac{7}{3} \times \frac{10}{7}$$

Before you multiply, you can cancel out 7s in the numerator and denominator:

$$= \frac{1}{3} \times \frac{10}{1} = \frac{10}{3}$$

Because the answer is an improper fraction, change it to a mixed number:

$$\begin{array}{r} 3 \\ 3{\overline{\smash{)}10}} \\ \underline{-9} \\ 1 \end{array}$$

So, the final is $3\frac{1}{3}$.

2. $2\frac{2}{5} \times 1\frac{5}{6} = 4\frac{2}{5}$. Change both mixed numbers to improper fractions:

$$2\frac{2}{5} = \frac{(2 \times 5 + 2)}{5} = \frac{12}{5}$$
$$1\frac{5}{6} = \frac{(1 \times 6 + 5)}{6} = \frac{11}{6}$$

Set up the multiplication:

$$\frac{12}{5} \times \frac{11}{6}$$

Before you multiply, you can cancel out 6s in the numerator and denominator:

$$= \frac{2}{5} \times \frac{11}{1} = \frac{22}{5}$$

Because the answer is an improper fraction, change it to a mixed number:

$$\begin{array}{r} 4 \\ 5{\overline{\smash{)}22}} \\ \underline{-20} \\ 2 \end{array}$$

The final answer is $4\frac{2}{5}$.

3. $4\frac{4}{5} \times 3\frac{1}{8} = 15$. Change both mixed numbers to improper fractions:

$$4\frac{4}{5} = \frac{(4 \times 5 + 4)}{5} = \frac{24}{5}$$

$$3\frac{1}{8} = \frac{(3 \times 8 + 1)}{8} = \frac{25}{8}$$

Set up the multiplication:

$$\frac{24}{5} \times \frac{25}{8}$$

Before you multiply, cancel out both 5s and 8s in the numerator and denominator:

$$= \frac{24}{1} \times \frac{5}{8} = \frac{3}{1} \times \frac{5}{1} = 15$$

4. $4\frac{1}{2} \div 1\frac{5}{8} = 2\frac{10}{13}$. Change both mixed numbers to improper fractions:

$$4\frac{1}{2} = \frac{(4 \times 2 \times 1)}{2} = \frac{9}{2}$$

$$1\frac{5}{8} = \frac{(1 \times 8 + 5)}{8} = \frac{13}{8}$$

Set up the division:

$$\frac{9}{2} \div \frac{13}{8}$$

Change the problem to multiplication using the reciprocal of the second fraction:

$$= \frac{9}{2} \times \frac{8}{13}$$

Cancel out a factor of 2 and multiply:

$$= \frac{9}{1} \times \frac{4}{13} = \frac{36}{13}$$

Because the answer is an improper fraction, change it to a mixed number:

$$= 2\frac{10}{13}$$

5. $2\frac{1}{10} \div 2\frac{1}{4} = \frac{14}{15}$. Change both mixed numbers to improper fractions:

$$2\frac{1}{10} = \frac{(2 \times 10 + 1)}{10} = \frac{21}{10}$$

$$2\frac{1}{4} = \frac{(2 \times 4 + 1)}{4} = \frac{9}{4}$$

Set up the division:

$$\frac{21}{10} \div \frac{9}{4}$$

Change the problem to multiplication using the reciprocal of the second fraction:

$$= \frac{21}{10} \times \frac{4}{9}$$

Before you multiply, cancel 2s and 3s from the numerator and denominator:

$$= \frac{21}{5} \times \frac{2}{9} = \frac{7}{5} \times \frac{2}{3} = \frac{14}{15}$$

6. $1\frac{2}{7} \div 6\frac{3}{10} = \frac{10}{49}$. Change both mixed numbers to improper fractions:

$$1\frac{2}{7} = \frac{(1 \times 7 + 2)}{7} = \frac{9}{7}$$
$$6\frac{3}{10} = \frac{(6 \times 10 + 3)}{10} = \frac{63}{10}$$

Set up the division:

$$\frac{9}{7} \div \frac{63}{10}$$

Change the problem to multiplication using the reciprocal of the second fraction:

$$= \frac{9}{7} \times \frac{10}{63}$$

Before you multiply, cancel 9s from the numerator and denominator:

$$= \frac{1}{7} \times \frac{10}{7} = \frac{10}{49}$$

Adding mixed numbers

One way to add and subtract mixed numbers is to convert them to improper fractions, much as I describe earlier in this chapter in "Multiplying and dividing mixed numbers," and then to add or subtract them using a method from the "All Together Now: Adding Fractions" or "Take It Away: Subtracting Fractions" sections. Doing so is a perfectly valid way of getting the right answer without learning a new method.

Unfortunately, teachers just love to make people add and subtract mixed numbers in their own special way. The good news is that a lot of folks find this way easier than all the converting stuff.

Working in pairs: Adding two mixed numbers

Adding mixed numbers looks a lot like adding whole numbers: You stack them one on top of the other, draw a line, and add. For this reason, some students feel more comfortable adding mixed numbers than adding fractions. Here's how to add two mixed numbers:

1. **Add the fractional parts using any method you like; if necessary, change this sum to a mixed number and reduce it.**

2. **If the answer you found in Step 1 is an improper fraction, change it to a mixed number, write down the fractional part, and carry the whole number part to the whole number column.**

3. **Add the whole number parts (including any number carried).**

You may also need to reduce your answer to lowest terms (see Chapter 8). In the examples that follow, I show you everything you need to know.

Summing up mixed numbers when the denominators are the same

As with any problem involving fractions, adding is always easier when the denominators are the same. For example, suppose you want to add $3\frac{1}{3} + 5\frac{1}{3}$. Doing mixed number problems is often easier if you place one number above the other:

$$3\frac{1}{3}$$
$$+5\frac{1}{3}$$

As you can see, this arrangement is similar to how you add whole numbers, but it includes an extra column for fractions. Here's how you add these two mixed numbers step by step:

1. **Add the fractions.**

$$\frac{1}{3} + \frac{1}{3} = \frac{2}{3}$$

2. **Switch improper fractions to mixed numbers; write down your answer.**

   Because $\frac{2}{3}$ is a proper fraction, you don't have to change it.

3. **Add the whole number parts.**

$$3 + 5 = 8$$

Here's how your problem looks in column form:

$$3\frac{1}{3}$$
$$+5\frac{1}{3}$$
$$\overline{8\frac{2}{3}}$$

This problem is about as simple as they get. In this case, all three steps are pretty easy. But sometimes, Step 2 requires more attention. For example, suppose you want to add $8\frac{3}{5} + 6\frac{4}{5}$. Here's how you do it:

1. Add the fractions.

$$\frac{3}{5} + \frac{4}{5} = \frac{7}{5}$$

2. Switch improper fractions to mixed numbers, write down the fractional part, and carry over the whole number.

Because the sum is an improper fraction, convert it to the mixed number $1\frac{2}{5}$ (flip to Chapter 8 for more on converting improper fractions to mixed numbers). Write down $\frac{2}{5}$ and carry the 1 over to the whole number column.

3. Add the whole number parts, including any whole numbers you carried over when you switched to a mixed number.

$$1 + 8 + 6 = 15$$

Here's how the solved problem looks in column form. (Be sure to line up the whole numbers in one column and the fractions in another.)

$$
\begin{array}{r}
1 \phantom{\frac{3}{5}} \\
8\frac{3}{5} \\
+6\frac{4}{5} \\
\hline
15\frac{2}{5}
\end{array}
$$

As with any other problems involving fractions, sometimes you need to reduce at the end of Step 1.

The same basic idea works no matter how many mixed numbers you want to add. For example, suppose you want to add $5\frac{4}{9} + 11\frac{7}{9} + 3\frac{8}{9} + 1\frac{5}{9}$:

1. Add the fractions.

$$\frac{4}{9} + \frac{7}{9} + \frac{8}{9} + \frac{5}{9} = \frac{24}{9}$$

2. Switch improper fractions to mixed numbers, write down the fractional part, and carry over the whole number.

Because the result is an improper fraction, convert it to the mixed number $2\frac{6}{9}$ and then reduce it to $2\frac{2}{3}$ (for more on converting and reducing fractions, see Chapter 8). I recommend doing these calculations on a piece of scrap paper.

Write down $\frac{2}{3}$ and carry the 2 to the whole number column.

3. Add the whole numbers.

$$2 + 5 + 11 + 3 + 1 = 22$$

Here's how the problem looks after you solve it:

$$2$$

$$5\frac{4}{9}$$

$$11\frac{7}{9}$$

$$3\frac{8}{9}$$

$$+1\frac{5}{9}$$

$$\overline{22\frac{2}{3}}$$

Summing up mixed numbers when the denominators are different

The most difficult type of mixed number addition is when the denominators of the fractions are different. This difference doesn't change Steps 2 or 3, but it does make Step 1 tougher.

For example, suppose you want to add $16\frac{3}{5}$ and $7\frac{7}{9}$.

1. Add the fractions.

Add $\frac{3}{5}$ and $\frac{7}{9}$. You can use any method from earlier in this chapter. Here, I use the easy way:

$$\frac{3}{5} + \frac{7}{9} = \frac{(3 \times 9) + (7 \times 5)}{5 \times 9} = \frac{27 + 35}{45} = \frac{62}{45}$$

2. Switch improper fractions to mixed numbers, write down the fractional part, and carry over the whole number.

This fraction is improper, so change it to the mixed number $1\frac{17}{45}$. Fortunately, the fractional part of this mixed number isn't reducible.

Write down the $\frac{17}{45}$ and carry over the 1 to the whole number column.

3. Add the whole numbers.

$$1 + 16 + 7 = 24$$

Here's how the completed problem looks:

$$1$$

$$16\frac{3}{5}$$

$$+7\frac{7}{9}$$

$$\overline{24\frac{17}{45}}$$

Examples

Q. Add $4\frac{1}{8} + 2\frac{3}{8}$.

A. $6\frac{1}{2}$. To start out, set up the problem in column form:

$$4\frac{1}{8}$$
$$+2\frac{3}{8}$$

Add the fractions and reduce the result:

$$\frac{1}{8} + \frac{3}{8} = \frac{4}{8} = \frac{1}{2}$$

Because this result is a proper fraction, you don't have to worry about carrying. Next, add the whole-number parts:

$$4 + 2 = 6$$

Here's how the problem looks in column form:

$$4\frac{1}{8}$$
$$+2\frac{3}{8}$$
$$\overline{6\frac{1}{2}}$$

Q. Add $5\frac{3}{4} + 4\frac{7}{9}$.

A. $10\frac{19}{36}$. To start out, set up the problem in column form:

$$5\frac{3}{4}$$
$$+4\frac{7}{9}$$

To add the fractional parts, change the two denominators to a common denominator using cross-multiplication. The new numerators are 3 × 9 = 27 and 7 × 4 = 28, and the new denominators are 4 × 9 = 36:

$$\frac{3}{4} \qquad\qquad\qquad \frac{7}{9}$$
$$\downarrow \qquad\qquad\qquad \downarrow$$
$$\frac{27}{36} \qquad\qquad\qquad \frac{28}{36}$$

Now you can add:

$$\frac{27}{36} + \frac{28}{36} = \frac{55}{36}$$

Because this result is an improper fraction, change it to a mixed number.

$$= 1\frac{19}{36}$$

Carry the 1 from this mixed number into the whole-number column and add:

$$1 + 5 + 4 = 10$$

Here's how the problem looks in column form:

$$\begin{array}{r} 1 \\ 5\frac{27}{36} \\ +4\frac{28}{36} \\ \hline 10\frac{19}{36} \end{array}$$

Practice Questions

1. Add $3\frac{1}{5}$ and $4\frac{2}{5}$.

2. Find $7\frac{1}{3} + 1\frac{1}{6}$.

3. Add $12\frac{4}{9}$ and $7\frac{8}{9}$.

4. Find the sum of $5\frac{2}{3}$ and $9\frac{3}{5}$.

5. Add $13\frac{6}{7} + 2\frac{5}{14}$.

6. Find $21\frac{9}{10} + 38\frac{3}{4}$.

Practice Answers

1. $3\frac{1}{5} + 4\frac{2}{5} = 7\frac{3}{5}$. Set up the problem in column form:

$$\begin{array}{r} 3\frac{1}{5} \\ +4\frac{2}{5} \\ \hline \end{array}$$

Add the fractional parts:

$$\frac{1}{5} + \frac{2}{5} = \frac{3}{5}$$

Because this result is a proper fraction, you don't have to worry about carrying. Next, add the whole-number parts:

$$3 + 4 = 7$$

Here's how the completed problem looks:

$$3\frac{1}{5}$$
$$+4\frac{2}{5}$$
$$\overline{7\frac{3}{5}}$$

2. $7\frac{1}{3} + 1\frac{1}{6} = 8\frac{1}{2}$. To start out, set up the problem in column form:

$$7\frac{1}{3}$$
$$+1\frac{1}{6}$$
$$\overline{\phantom{+1\frac{1}{6}}}$$

Next, add the fractional parts. The denominators are different, but 3 is a factor of 6, so you can use the quick trick. Increase the terms of $\frac{1}{3}$ so that the denominator is 6 by multiplying the numerator and denominator by 2:

$$\frac{1}{3} = \frac{2}{6}$$

Now you can add and reduce the result:

$$\frac{2}{6} + \frac{1}{6} = \frac{3}{6} = \frac{1}{2}$$

Because this result is a proper fraction, you don't have to worry about carrying. Next, add the whole-number parts:

$$7 + 1 = 8$$

Here's how the completed problem looks:

$$7\frac{2}{6}$$
$$+1\frac{1}{6}$$
$$\overline{8\frac{1}{2}}$$

3. $12\frac{4}{9} + 7\frac{8}{9} = 20\frac{1}{3}$. Set up the problem in column form:

$$12\frac{4}{9}$$
$$+7\frac{8}{9}$$
$$\overline{\phantom{+7\frac{8}{9}}}$$

Add the fractional parts and reduce the result:

$$\frac{4}{9} + \frac{8}{9} = \frac{12}{9} = \frac{4}{3}$$

Because this result is an improper fraction, convert it to a mixed number:

$$= 1\frac{1}{3}$$

Carry the 1 from this mixed number into the whole-number column and add:

$$1 + 12 + 7 = 20$$

Here's how the completed problem looks:

$$12\overset{1}{}\frac{4}{9}$$

$$+7\frac{8}{9}$$

$$20\frac{1}{3}$$

4. $5\frac{2}{3} + 9\frac{3}{5} = 15\frac{4}{15}$. Set up the problem in column form:

$$5\frac{2}{3}$$

$$+9\frac{3}{5}$$

Start by adding the fractional parts. Because the denominators are different, change them to a common denominator by cross-multiplying. The new numerators are $2 \times 5 = 10$ and $3 \times 3 = 9$, and the new denominators are $3 \times 5 = 15$:

$$\frac{2}{3} \qquad\qquad\qquad \frac{3}{5}$$

$$\downarrow \qquad\qquad\qquad \downarrow$$

$$\frac{10}{15} \qquad\qquad\qquad \frac{9}{15}$$

Now you can add:

$$\frac{10}{15} + \frac{9}{15} = \frac{19}{15}$$

Because this result is an improper fraction, convert it to a mixed number:

$$= 1\frac{4}{15}$$

Carry the 1 from this mixed number into the whole-number column and add:

$$1 + 5 + 9 = 15$$

Here's how the completed problem looks:

$$5\overset{1}{}\frac{10}{15}$$

$$+9\frac{9}{15}$$

$$15\frac{4}{15}$$

5. $13\frac{6}{7} + 2\frac{5}{14} = 16\frac{3}{14}$. Set up the problem in column form:

$$13\frac{6}{7}$$

$$+2\frac{5}{14}$$

Begin by adding the fractional parts. Because the denominator 7 is a factor of the denominator 14, you can use the quick trick. Increase the terms of $\frac{6}{7}$ so that the denominator is 14 by multiplying the numerator and denominator by 2:

$$\frac{6}{7} = \frac{12}{14}$$

Now you can add:

$$\frac{12}{14} + \frac{5}{14} = \frac{17}{14}$$

Because this result is an improper fraction, convert it to a mixed number:

$$= 1\frac{3}{14}$$

Carry the 1 from this mixed number into the whole-number column and add:

$$1 + 13 + 2 = 16$$

Here's how the completed problem looks:

$$13\overset{1}{}\frac{12}{14}$$

$$+2\frac{5}{14}$$

$$16\frac{3}{14}$$

6. $21\frac{9}{10} + 38\frac{3}{4} = 60\frac{13}{20}$. Set up the problem in column form:

$$21\frac{9}{10}$$

$$+38\frac{3}{4}$$

To add the fractional parts, change the denominators to a common denominator by using cross-multiplication. The new numerators are 9 × 4 = 36 and 3 × 10 = 30, and the new denominators are 10 × 4 = 40:

$$\frac{9}{10}$$
$$\downarrow$$
$$\frac{36}{40}$$

$$\frac{3}{4}$$
$$\downarrow$$
$$\frac{30}{40}$$

Now you can add:

$$\frac{36}{40} + \frac{30}{40} = \frac{66}{40}$$

The numerator and denominator are both even, so reduce this fraction by a factor of 2:

$$= \frac{33}{20}$$

Because this result is an improper fraction, convert it to a mixed number:

$$= 1\frac{13}{20}$$

Carry the 1 from this mixed number into the whole-number column and add:

$$1 + 21 + 38 = 60$$

Here's how the completed problem looks:

$$2\overset{1}{1}\frac{36}{40}$$
$$+38\frac{30}{40}$$
$$\overline{60\frac{13}{20}}$$

Subtracting mixed numbers

The basic way to subtract mixed numbers is close to the way you add them. Again, the subtraction looks more like what you're used to with whole numbers. Here's how to subtract two mixed numbers:

1. **Find the difference of the fractional parts using any method you like.**

2. **Find the difference of the two whole number parts.**

Along the way, though, you may encounter a couple more twists and turns. I keep you on track so that, by the end of this section, you can do any mixed-number subtraction problem.

Taking away mixed numbers when the denominators are the same

As with addition, subtraction is much easier when the denominators are the same. For example, suppose you want to subtract $7\frac{3}{5} - 3\frac{1}{5}$. Here's what the problem looks like in column form:

$$7\frac{3}{5}$$
$$-3\frac{1}{5}$$
$$\overline{4\frac{2}{5}}$$

In this problem, I subtract $\frac{3}{5} - \frac{1}{5} = \frac{2}{5}$. Then I subtract $7 - 3 = 4$. Not too terrible, agreed?

One complication arises when you try to subtract a larger fractional part from a smaller one. Suppose you want to find $11\frac{1}{6} - 2\frac{5}{6}$. This time, if you try to subtract the fractions, you get

$$\frac{1}{6} - \frac{5}{6} = -\frac{4}{6}$$

Obviously, you don't want to end up with a negative number in your answer. You can handle this problem by borrowing from the column to the left. This idea is similar to the borrowing that you use in regular subtraction, with one key difference.

When borrowing in mixed-number subtraction,

1. **Borrow 1 from the whole-number portion and add it to the fractional portion, turning the fraction into a mixed number.**

   To find $11\frac{1}{6} - 2\frac{5}{6}$, borrow 1 from the 11 and add it to $\frac{1}{6}$, making it the mixed number $1\frac{1}{6}$:

   $$11\frac{1}{6} = 10 + 1\frac{1}{6}$$

2. **Change this new mixed number into an improper fraction.**

   Here's what you get when you change $1\frac{1}{6}$ into an improper fraction: $10 + 1\frac{1}{6} = 10\frac{7}{6}$ The result is $10\frac{7}{6}$. This answer is a weird cross between a mixed number and an improper fraction, but it's what you need to handle the job.

3. **Use the result in your subtraction.**

   $$10\frac{7}{6}$$
   $$-2\frac{5}{6}$$
   $$\overline{8\frac{2}{6}}$$

In this case, you have to reduce the fractional part of the answer:

$$8\frac{2}{6} = 8\frac{1}{3}$$

Subtracting mixed numbers when the denominators are different

Subtracting mixed numbers when the denominators are different is just about the hairiest thing you're ever going to have to do in pre-algebra. Fortunately, though, if you work through this chapter, you acquire all the skills you need.

Suppose you want to subtract $15\frac{4}{11} - 12\frac{3}{7}$. Because the denominators are different, subtracting the fractions becomes more difficult. But you have another question to think about: In this problem, do you need to borrow? If $\frac{4}{11}$ is greater than $\frac{3}{7}$, you don't have to borrow. But if $\frac{4}{11}$ is less than $\frac{3}{7}$, you do. (For more on borrowing in mixed-number subtraction, see the preceding section.)

In Chapter 8, I show you how to test two fractions to see which is greater by cross-multiplying:

$$4 \times 7 = 28$$
$$3 \times 11 = 33$$

Because 28 is less than 33, $\frac{4}{11}$ is less than $\frac{3}{7}$, so you do have to borrow. I get the borrowing out of the way first:

$$15\frac{4}{11} = 14 + 1\frac{4}{11} = 14\frac{15}{11}$$

Now the problem looks like this:

$$14\frac{15}{11} - 12\frac{3}{7}$$

The first step, subtracting the fractions, is the most time-consuming, so as I show you earlier in "Subtracting fractions with different denominators," you can take care of that on the side:

$$\frac{15}{11} - \frac{3}{7} = \frac{(15 \times 7) - (3 \times 11)}{11 \times 7} = \frac{105 - 33}{77} = \frac{72}{77}$$

The good news is that this fraction can't be reduced (72 and 77 have no common factors: $72 = 2 \times 2 \times 2 \times 3 \times 3$ and $77 = 7 \times 11$). So the hard part of the problem is done, and the rest follows easily:

$$14\frac{15}{11}$$
$$-12\frac{3}{7}$$
$$\overline{2\frac{72}{77}}$$

This problem is about as difficult as a mixed-number subtraction problem gets. Take a look at it step by step. Better yet, copy the problem and then close the book and try to work through the steps on your own. If you get stuck, that's okay: Better now than on an exam!

Examples

Q.   Subtract $8\frac{4}{5} - 6\frac{3}{5}$.

A. $2\frac{1}{5}$. To start out, set up the problem in column form:

$$8\frac{4}{5}$$
$$-6\frac{3}{5}$$

The fractional parts already have a common denominator, so subtract:

$$\frac{4}{5} - \frac{3}{5} = \frac{1}{5}$$

Next, subtract the whole-number parts:

$$8 - 6 = 2$$

Here's how the completed problem looks:

$$8\frac{4}{5}$$
$$-6\frac{3}{5}$$
$$\overline{2\frac{1}{5}}$$

Q.   Subtract $9\frac{1}{6} - 3\frac{5}{6}$.

A. $5\frac{1}{3}$. To start out, set up the problem in column form:

$$9\frac{1}{6}$$
$$-3\frac{5}{6}$$

The fractional parts already have a common denominator, so subtract. Notice that $\frac{1}{6}$ is less than $\frac{5}{6}$, so you need to borrow 1 from 9:

$$8\cancel{9}1\frac{1}{6}$$
$$-3\frac{5}{6}$$

Now convert the mixed number $1\frac{1}{6}$ into an improper fraction:

$$8\frac{7}{6}$$

$$-3\frac{5}{6}$$

At this point, you can subtract the fractional parts and reduce:

$$\frac{7}{6}-\frac{5}{6}=\frac{2}{6}=\frac{1}{3}$$

Next, subtract the whole-number parts:

$$8-3=5$$

Here's how the completed problem looks:

$$8\frac{7}{6}$$

$$-3\frac{5}{6}$$

$$5\frac{1}{3}$$

Q. Subtract $19\frac{4}{11}-6\frac{3}{8}$.

A. $12\frac{87}{88}$. To start out, set up the problem in column form:

$$19\frac{4}{11}$$

$$-6\frac{3}{8}$$

The fractional parts have different denominators, so change them to a common denominator using cross-multiplication. The new numerators are $4 \times 8 = 32$ and $3 \times 11 = 33$, and the new denominators are $11 \times 8 = 88$:

$$\frac{4}{11} \qquad\qquad \frac{3}{8}$$

$$\downarrow \qquad\qquad \downarrow$$

$$\frac{32}{88} \qquad\qquad \frac{33}{88}$$

Here's how the problem looks now:

$$19\frac{32}{88}$$

$$-6\frac{33}{88}$$

Because $\frac{32}{88}$ is less than $\frac{33}{88}$, you need to borrow before you can subtract:

$$18\cancel{19}1\frac{32}{88}$$
$$-6\frac{33}{88}$$

Now turn the mixed number $1\frac{32}{88}$ into an improper fraction:

$$18\frac{120}{88}$$
$$-6\frac{33}{88}$$

At this point, you can subtract the fractional parts and the whole-number parts:

$$18\frac{120}{88}$$
$$-6\frac{33}{88}$$
$$\overline{12\frac{87}{88}}$$

Practice Questions

1. Subtract $5\frac{7}{9}-2\frac{4}{9}$.

2. Find $9\frac{1}{8}-7\frac{5}{8}$.

3. Subtract $11\frac{3}{4}-4\frac{2}{3}$.

4. Figure out $16\frac{2}{5}-8\frac{4}{9}$.

Practice Answers

1. $5\frac{7}{9}-2\frac{4}{9}=3\frac{1}{3}$. Set up the problem in column form:

$$5\frac{7}{9}$$
$$-2\frac{4}{9}$$

Subtract the fractional parts and reduce:

$$\frac{7}{9}-\frac{4}{9}=\frac{3}{9}=\frac{1}{3}$$

Subtract the whole-number parts:

$$5 - 2 = 3$$

Here's how the completed problem looks:

$$
\begin{array}{r}
5\frac{7}{9} \\
-2\frac{4}{9} \\
\hline
3\frac{1}{3}
\end{array}
$$

2. $9\frac{1}{8} - 7\frac{5}{8} = 1\frac{1}{2}$. Set up the problem in column form:

$$
\begin{array}{r}
9\frac{1}{8} \\
-7\frac{5}{8} \\
\hline
\end{array}
$$

The first fraction $\left(\frac{1}{8}\right)$ is less than the second fraction $\left(\frac{5}{8}\right)$, so you need to borrow 1 from 9 before you can subtract:

$$
\begin{array}{r}
-7\frac{5}{8} \\
\hline
\end{array}
$$

Change the mixed number $1\frac{1}{8}$ to an improper fraction:

$$
\begin{array}{r}
8\frac{9}{8} \\
-7\frac{5}{8} \\
\hline
\end{array}
$$

Now you can subtract the fractional parts and reduce:

$$\frac{9}{8} - \frac{5}{8} = \frac{4}{8} = \frac{1}{2}$$

Subtract the whole-number parts:

$$8 - 7 = 1$$

Here's how the completed problem looks:

$$
\begin{array}{r}
8\frac{9}{8} \\
-7\frac{5}{8} \\
\hline
1\frac{1}{2}
\end{array}
$$

3. $11\frac{3}{4} - 4\frac{2}{3} = 7\frac{1}{12}$. Set up the problem in column form:

$$11\frac{3}{4}$$
$$-4\frac{2}{3}$$

The denominators are different, so get a common denominator using cross-multiplication. The new numerators are $3 \times 3 = 9$ and $2 \times 4 = 8$, and the new denominators are $4 \times 3 = 12$:

$$\frac{3}{4} \qquad\qquad\qquad \frac{2}{3}$$
$$\downarrow \qquad\qquad\qquad \downarrow$$
$$\frac{9}{12} \qquad\qquad\qquad \frac{8}{12}$$

Because $\frac{9}{12}$ is greater than $\frac{8}{12}$, you don't need to borrow before you can subtract fractions:

$$11\frac{9}{12}$$
$$-4\frac{8}{12}$$
$$7\frac{1}{12}$$

4. $16\frac{2}{5} - 8\frac{4}{9} = 7\frac{43}{45}$. To start out, set up the problem in column form:

$$16\frac{2}{5}$$
$$-8\frac{4}{9}$$

The denominators are different, so find a common denominator using cross-multiplication. The new numerators are $2 \times 9 = 18$ and $4 \times 5 = 20$, and the new denominators are $5 \times 9 = 45$:

$$\frac{2}{5} \qquad\qquad\qquad \frac{4}{9}$$
$$\downarrow \qquad\qquad\qquad \downarrow$$
$$\frac{18}{45} \qquad\qquad\qquad \frac{20}{45}$$

Because $\frac{18}{45}$ is less than $\frac{20}{45}$, you need to borrow 1 from 16 before you can subtract fractions:

$$15\,\cancel{16}1\frac{18}{45}$$
$$-8\frac{20}{45}$$

Change the mixed number $1\frac{18}{45}$ to an improper fraction:

$$15\frac{63}{45}$$
$$-8\frac{20}{45}$$

Now you can subtract:

$$15\frac{63}{45}$$
$$-8\frac{20}{45}$$
$$7\frac{43}{45}$$

10

DALLYING WITH DECIMALS

IN THIS CHAPTER

- Understanding the decimal basics
- Applying decimals to the Big Four operations
- Looking at decimal and fraction conversions
- Making sense of repeating decimals

Because early humans used their fingers for counting, the number system is based on the number 10. So numbers come in ones, tens, hundreds, thousands, and so on. A *decimal* — with its handy decimal point — allows people to work with numbers smaller than 1: tenths, hundredths, thousandths, and the like.

Here's some lovely news: Decimals are much easier to work with than fractions (which I discuss in Chapters 8 and 9). Decimals look and feel more like whole numbers than fractions do, so when you're working with decimals, you don't have to worry about reducing and increasing terms, improper fractions, mixed numbers, and a lot of other stuff.

Performing the Big Four operations — addition, subtraction, multiplication, and division — on decimals is very close to performing them on whole numbers (which I cover in Part II of the book). The numerals 0 through 9 work just like they usually do. As long as you get the decimal point in the right place, you're home free.

In this chapter, I show you all about working with decimals. I also show you how to convert fractions to decimals and decimals to fractions. Finally, I give you a peek into the strange world of repeating decimals.

UNDERSTANDING BASIC DECIMAL STUFF

The good news about decimals is that they look a lot more like whole numbers than fractions do. So a lot of what you find out about whole numbers in Chapter 2 applies to decimals as well. In this section, I introduce you to decimals, starting with place value.

When you understand place value of decimals, a lot falls into place. Then I discuss trailing zeros and what happens when you move the decimal point either to the left or to the right.

Counting dollars and decimals

You use decimals all the time when you count money. And a great way to begin thinking about decimals is as dollars and cents. For example, you know that $0.50 is half of a dollar (see Figure 10-1), so this information tells you:

$$0.5 = \frac{1}{2}$$

Notice that, in the decimal 0.5, I drop the zero at the end. This practice is common with decimals.

You also know that $0.25 is a quarter — that is, one-fourth of a dollar (see Figure 10-2) — so:

$$0.25 = \frac{1}{4}$$

Similarly, you know that $0.75 is three quarters, or three-fourths, of a dollar (see Figure 10-3), so:

$$0.75 = \frac{3}{4}$$

Taking this idea even further, you can use the remaining denominations of coins — dimes, nickels, and pennies — to make further connections between decimals and fractions.

$$\text{A dime} = \$0.10 = \frac{1}{10} \text{ of a dollar, so } \frac{1}{10} = 0.1$$

$$\text{A nickel} = \$0.05 = \frac{1}{20} \text{ of a dollar, so } \frac{1}{20} = 0.05$$

$$\text{A penny} = \$0.01 = \frac{1}{100} \text{ of a dollar, so } \frac{1}{100} = 0.01$$

Figure 10-1: One-half (0.5) of a dollar bill.

F 9 7 8 1 1 1 9 0 6 B

1

Figure 10-2:
One-fourth
(0.25) of a
dollar bill.

© John Wiley & Sons, Inc.

Figure 10-3:
Three-
fourths
(0.75) of a
dollar bill.

© John Wiley & Sons, Inc.

Remember: Notice that I again drop the final zero in the decimal 0.1, but I keep the zeros in the decimals 0.05 and 0.01. You can drop zeros from the right end of a decimal, but you can't drop zeros that fall between the decimal point and another digit.

Decimals are just as good for cutting up cake as for cutting up money. Figure 10-4 gives you a look at the four cut-up cakes that I show you in Chapter 8. This time, I give you the decimals that tell you how much cake you have. Fractions and decimals accomplish the same task: allowing you to cut a whole object into pieces and talk about how much you have.

Figure 10-4:
Cakes cut
and shaded
into (A) 0.75,
(B) 0.4,
(C) 0.1, and
(D) 0.7.

© John Wiley & Sons, Inc.

Identifying the place value of decimals

In Chapter 2, you find out about the place value of whole numbers. Table 10-1 shows how the whole number 4,672 breaks down in terms of place value.

Table 10-1   Breaking Down 4,672 in Terms of Place Value

Thousands	Hundreds	Tens	Ones
4	6	7	2

This number means 4,000 + 600 + 70 + 2.

With decimals, this idea is extended. First, a decimal point is placed to the right of the ones place in a whole number. Then more numbers are appended to the right of the decimal point.

Table 10-2   Breaking Down the Decimal 4,672.389

Thousands	Hundreds	Tens	Ones	Decimal Point	Tenths	Hundredths	Thousandths
4	6	7	2	.	3	8	9

For example, the decimal 4,672.389 breaks down as shown in Table 10-2.

This decimal means $4,000 + 600 + 70 + 2 + \frac{3}{10} + \frac{8}{100} + \frac{9}{1000}$.

The connection between fractions and decimals becomes obvious when you look at place value. Decimals really are a shorthand notation for fractions. You can represent any fraction as a decimal.

Knowing the decimal facts of life

When you understand how place value works in decimals (as I explain in the preceding section), a whole lot of facts about decimals begin to make sense. Two key ideas are trailing zeros and what happens when you move a decimal point left or right.

Understanding trailing zeros

You probably know that you can attach zeros to the beginning of a whole number without changing its value. For example, these three numbers are all equal in value:

27   027   0,000,027

The reason becomes clear when you know about place value of whole numbers. See Table 10-3.

Table 10-3   Example of Attaching Leading Zeros

Millions	Hundred Thousands	Ten Thousands	Thousands	Hundreds	Tens	Ones
0	0	0	0	0	2	7

As you can see, 0,000,027 simply means 0 + 0 + 0 + 0 + 0 + 20 + 7. No matter how many zeros you add to the beginning of a number, the number 27 doesn't change.

Zeros attached to the beginning of a number in this way are called *leading zeros*.

In decimals, this idea of zeros that don't add value to a number can be extended to trailing zeros.

Remember: A *trailing zero* is any zero that appears to the right of both the decimal point and every digit other than zero.

For example:

34.8   34.80   34.8000

All three of these numbers are the same. The reason becomes clear when you understand how place value works in decimals. See Table 10-4.

In this example, 34.8000 means $30 + 4 + \frac{8}{10} + \frac{0}{100} + \frac{0}{1000} + \frac{0}{10000}$.

Table 10-4   Example of Attaching Trailing Zeros

Tens	Ones	Decimal Point	Tenths	Hundredths	Thousandths	Ten Thousandths
3	4	.	8	0	0	0

Remember: You can attach or remove as many trailing zeros as you want without changing the value of a number.

When you understand trailing zeros, you can see that every whole number can easily be changed to a decimal. Just attach a decimal point and a 0 to the end of it. For example:

$$4 = 4.0$$
$$20 = 20.0$$
$$971 = 971.0$$

Warning: Make sure that you don't attach or remove any non-leading or nontrailing zeros — it changes the value of the decimal.

For example, look at this number:

$$0450.0070$$

In this number, you can remove the leading and trailing zeros without changing the value, as follows:

$$450.007$$

The remaining zeros, however, need to stay where they are as *placeholders* between the decimal point and digits other than zero. See Table 10-5.

Table 10-5   Example of Zeros as Placeholders

Thousands	Hundreds	Tens	Ones	Decimal Point	Tenths	Hundredths	Thousandths	Ten Thousandths
0	4	5	0	.	0	0	7	0

I continue to discuss zeros as placeholders in the next section.

Moving the decimal point

When you're working with whole numbers, you can multiply any number by 10 just by adding a zero to the end of it. For example:

$$45,971 \times 10 = 459,710$$

To see why this answer is so, again think about the place value of digits and look at Table 10-6.

Table 10-6 Example, Decimal Points and Place Value of Digits

Millions	Hundred Thousands	Ten Thousands	Thousands	Hundreds	Tens	Ones
		4	5	9	7	1
	4	5	9	7	1	0

Here's what these two numbers really mean:

$$45,971 = 40,000 + 5,000 + 900 + 70 + 1$$

$$459,710 = 400,000 + 50,000 + 9,000 + 700 + 10 + 0$$

As you can see, that little zero makes a big difference: It causes the rest of the numbers to shift one place.

This concept makes even more sense when you think about the decimal point. See Table 10-7.

Table 10-7 Example, Numbers Shifting One Place

Hundred Thousands	Ten Thousands	Thousands	Hundreds	Tens	Ones	Decimal Point	Tenths	Hundredths
	4	5	9	7	1	.	0	0
4	5	9	7	1	0	.	0	0

In effect, adding a 0 to the end of a whole number moves the decimal point one place to the right. So for any decimal, when you move the decimal point one place to the right, you multiply that number by 10. This fact becomes clear when you start with a simple number like 7:

7.0
70.0
700.0
7,000.0

In this case, the net effect is that you moved the decimal point three places to the right, which is the same as multiplying 7 by 1,000.

Similarly, to divide any number by 10, move the decimal point one place to the left. For example:

7.0

0.7

0.07

0.007

This time, the net effect is that you moved the decimal point three places to the left, which is the same as dividing 7 by 1,000.

Tip: You can express any whole number as a decimal simply by attaching a decimal point and a trailing zero to the end of it. For example,

$$7 = 7.0 \quad 12 = 12.0 \quad 1,568 = 1,568.0$$

To sum up, moving the decimal point to the *right* is the same as multiplying that decimal by a power of 10. For example,

- Moving the decimal point one place to the right is the same as multiplying by 10.

- Moving the decimal point two places to the right is the same as multiplying by 100.

- Moving the decimal point three places to the right is the same as multiplying by 1,000.

Similarly, moving the decimal point to the *left* is the same as dividing that decimal by a power of 10. For example,

- Moving the decimal point one place to the left is the same as dividing by 10.

- Moving the decimal point two places to the left is the same as dividing by 100.

- Moving the decimal point three places to the left is the same as dividing by 1,000.

Tip: To multiply a decimal by any power of 10, count the number of zeros and move the decimal point that many places to the right. To divide a decimal by any power of 10, count the number of zeros and move the decimal point that many places to the left.

Rounding decimals is similar to rounding whole numbers (if you need a refresher, see Chapter 2). Generally speaking, to round a number to a given decimal place, focus on that decimal place and the place to its immediate right; then round as you would with whole numbers:

- **Rounding down:** If the digit on the right is 0, 1, 2, 3, or 4, drop this digit and every digit to its right.

- **Rounding up:** If the digit on the right is 5, 6, 7, 8, or 9, add 1 to the digit you're rounding to and then drop every digit to its right.

Remember: When rounding, people often refer to the first three decimal places in two different ways — by the number of the decimal place and by the name:

- Rounding to *one decimal place* is the same as rounding *to the nearest tenth*.

- Rounding to *two decimal places* is the same as rounding *to the nearest hundredth*.

- Rounding to *three decimal places* is the same as rounding *to the nearest thousandth*.

When rounding to four or more decimal places, the names get longer, so they're usually not used.

Examples

Q. Expand the decimal 7,358.293.

A. $7,358.293 = 7,000 + 300 + 50 + 8 + \dfrac{2}{10} + \dfrac{9}{100} + \dfrac{3}{1,000}$

Q. Simplify the decimal 0400.0600 by removing all leading and trailing zeros, without removing placeholding zeros.

A. **400.06.** The first zero is a leading zero because it appears to the left of all nonzero digits. The last two zeros are trailing zeros because they appear to the right of all nonzero digits. The remaining three zeros are placeholding zeros.

Q. Multiply 3.458 × 100.

A. **345.8.** The number 100 has two zeros, so to multiply by 100, move the decimal point two places to the right.

Q. Divide 29.81 ÷ 10,000.

A. **0.002981.** The number 10,000 has four zeros, so to divide by 10,000, move the decimal point four places to the left.

Practice Questions

1. Expand the following decimals:

    a. 2.7

    b. 31.4

    c. 86.52

    d. 103.759

    e. 1,040.0005

    f. 16,821.1384

2. Simplify each of the following decimals by removing all leading and trailing zeros whenever possible, without removing placeholding zeros:

   a. 5.80

   b. 7.030

   c. 90.0400

   d. 9,000.005

   e. 0108.0060

   f. 00100.0102000

3. Do the following decimal multiplication and division problems by moving the decimal point the correct number of places:

   a. 7.32×10

   b. 9.04×100

   c. $51.6 \times 100,000$

   d. $2.786 \div 1,000$

   e. $943.812 \div 1,000,000$

4. Round each of the following decimals to the number of places indicated:

   a. Round 4.777 to one decimal place.

   b. Round 52.305 to the nearest tenth.

   c. Round 191.2839 to two decimal places.

   d. Round 99.995 to the nearest hundredth.

   e. Round 0.00791 to three decimal places.

   f. Round 909.9996 to the nearest thousandth.

A Practice Answers

1. a. $2.7 = 2 + \dfrac{7}{10}$

   b. $31.4 = 30 + 1 + \dfrac{4}{10}$

   c. $86.52 = 80 + 6 + \dfrac{5}{10} + \dfrac{2}{100}$

   d. $103.759 = 100 + 3 + \dfrac{7}{10} + \dfrac{5}{100} + \dfrac{9}{1,000}$

   e. $1,040.0005 = 1,000 + 40 + \dfrac{5}{10,000}$

   f. $16,821.1384 = 10,000 + 6,000 + 800 + 20 + 1 + \dfrac{1}{10} + \dfrac{3}{100} + \dfrac{8}{1,000} + \dfrac{4}{10,000}$

2.   **a.**   **5.8**

       b.   **7.03**

       c.   **90.04**

       d.   **9,000.005**

       e.   **108.006**

       f.   **100.0102**

3.   **a.**   **73.2**

       b.   **904**

       c.   **5,160,000**

       d.   **0.002786**

       e.   **0.000943812**

4.   **a.**   **4.8**

       b.   **52.3**

       c.   **191.28**

       d.   99.99$\underline{95}$ → **100.00**

       e.   0.00$\underline{791}$ → **0.008**

       f.   909.99$\underline{96}$ → **910.000**

PERFORMING THE BIG FOUR WITH DECIMALS

Everything you already know about adding, subtracting, multiplying, and dividing whole numbers (see Chapter 3) carries over when you work with decimals. In fact, in each case, there's really only one key difference: how to handle that pesky little decimal point. In this section, I show you how to perform the Big Four math operations with decimals.

The most common use of adding and subtracting decimals is working with money — for example, balancing your checkbook. Later in this book, you find that multiplying and dividing by decimals is useful for calculating percentages (see Chapter 11), using scientific notation (see Chapter 13), and measuring with the metric system (see Chapter 14).

Adding decimals

Adding decimals is almost as easy as adding whole numbers. As long as you set up the problem correctly, you're in good shape. To add decimals, follow these steps:

1. **Arrange the numbers in a column and line up the decimal points vertically.**

2. **Add as usual, column by column, from right to left.**

   3. **Place the decimal point in the answer in line with the other decimal points in the problem.**

For example, suppose you want to add the numbers 14.5 and 1.89. Line up the decimal points neatly, as follows:

$$14.5$$
$$+\,1.89$$

Begin adding from the right-most column. Treat the blank space after 14.5 as a 0 — you can write this in as a trailing 0 (see earlier in this chapter to see why adding zeros to the end of a decimal doesn't change its value). Adding this column gives you $0 + 9 = 9$:

$$14.50$$
$$+\,1.89$$
$$\overline{9}$$

Continuing to the left, $5 + 8 = 13$, so put down the 3 and carry the 1:

$$\overset{1}{14.50}$$
$$+\,1.89$$
$$\overline{39}$$

Complete the problem column by column, and at the end, put the decimal point directly below the others in the problem:

$$14.50$$
$$+\,1.89$$
$$\overline{16.39}$$

When adding more than one decimal, the same rules apply. For example, suppose you want to add $15.1 + 0.005 + 800 + 1.2345$. The most important idea is lining up the decimal points correctly:

$$15.1$$
$$0.005$$
$$800.0$$
$$+\,1.2345$$

Tip: To avoid mistakes, be especially neat when adding a lot of decimals.

Because the number 800 isn't a decimal, I place a decimal point and a 0 at the end of it, to be clear about how to line it up. If you like, you can make sure all numbers have the same number of decimal places (in this case, four) by adding trailing zeros. When you properly set up the problem, the addition is no more difficult than in any other addition problem:

$$
\begin{array}{r}
15.1000 \\
0.0050 \\
800.0000 \\
+1.2345 \\
\hline
816.3395
\end{array}
$$

Subtracting decimals

Subtracting decimals uses the same trick as adding them (which I talk about in the preceding section). Here's how you subtract decimals:

1. **Arrange the numbers in a column and line up the decimal points.**

2. **Subtract as usual, column by column from right to left.**

3. **When you're done, place the decimal point in the answer in line with the other decimal points in the problem.**

For example, suppose you want to figure out 144.87 – 0.321. First, line up the decimal points:

$$
\begin{array}{r}
144.870 \\
-0.321 \\
\hline
\end{array}
$$

In this case, I add a zero at the end of the first decimal. This placeholder reminds you that, in the right-most column, you need to borrow to get the answer to 0 – 1:

$$
\begin{array}{r}
144.8\,\overset{6}{\cancel{7}}\,10 \\
-\ 0.32\ \ \ 1 \\
\hline
4\ \ 9
\end{array}
$$

The rest of the problem is straightforward. Just finish the subtraction and drop the decimal point straight down:

$$
\begin{array}{r}
144.8\,\overset{6}{\cancel{7}}\,10 \\
-\ 0.32\ \ \ 1 \\
\hline
144.54\ \ 9
\end{array}
$$

As with addition, the decimal point in the answer goes directly below where it appears in the problem.

Examples

Q. Add the following decimals: 321.81 + 24.5 + 0.006 = ?

A. **346.316.** Place the decimal numbers in a column (as you would for column addition) with the decimal points lined up:

$$
\begin{array}{r}
321.810 \\
24.500 \\
+0.006 \\
\end{array}
$$

Notice that the decimal point in the answer lines up with the others. As you can see, I've also filled out the columns with trailing zeros. This is optional, but do it if it helps you to see how the columns line up.

Now add as you would when adding whole numbers, carrying when necessary (see Chapter 3 for more on carrying in addition):

$$
\begin{array}{r}
3\overset{1}{2}1.810 \\
24.500 \\
+0.006 \\
\hline
346.316 \\
\end{array}
$$

Q. Subtract the following decimals: 978.245 - 29.03 = ?

A. **949.215.** Place the decimals one on top of the other with the decimal points lined up, dropping the decimal point straight down in the answer:

$$
\begin{array}{r}
978.245 \\
-29.030 \\
\end{array}
$$

Now subtract as you would when subtracting whole numbers, borrowing when necessary (see Chapter 3 for more on borrowing in subtraction):

$$
\begin{array}{r}
9\overset{6}{7}\overset{1}{8}.245 \\
-29.030 \\
\hline
949.215 \\
\end{array}
$$

Practice Questions

1. Add these decimals: 17.4 + 2.18 = ?

2. Compute the following decimal addition: 0.0098 + 10.101 + 0.07 + 33 =

3. Add the following decimals: 1,000.001 + 75 + 0.03 + 800.2 = ?

4. Subtract these decimals: 0.748 - 0.23 = ?

5. Compute the following: 674.9 - 5.0001.

6. Find the solution to this decimal subtraction problem: 100.009 - 0.68 = ?

Practice Answers

1. 17.4 + 2.18 = **19.58**. Place the numbers in a column as you would for addition with whole numbers, but with the decimal points lined up. I've filled out the columns with trailing zeros to help show how the columns line up:

$$
\begin{array}{r}
17.40 \\
+2.18 \\
\hline
19.58
\end{array}
$$

   Notice that the decimal point in the answer lines up with the others.

2. 0.0098 + 10.101 + 0.07 + 33 = **43.1808**. Line up the decimal points and do column addition:

$$
\begin{array}{r}
.0\overset{1}{0}98 \\
10.1010 \\
.0700 \\
+33.0000 \\
\hline
43.1808
\end{array}
$$

3. 1,000.001 + 75 + 0.03 + 800.2 = **1,875.231**. Place the decimal numbers in a column, lining up the decimal points:

$$
\begin{array}{r}
1000.001 \\
75.000 \\
0.030 \\
+800.200 \\
\hline
1875.231
\end{array}
$$

4. 0.748 - 0.23 = **0.518**. Place the first number on top of the second number, with the decimal points lined up. I've also added a trailing 0 to the second number to fill out the right-hand column and emphasize how the columns line up:

$$
\begin{array}{r}
0.748 \\
-0.230 \\
\hline
0.518
\end{array}
$$

5. 674.9 - 5.0001 = **669.8999**. Place the first number on top of the second number, with the decimal points lined up. I've filled out the right-hand column with trailing zeros so I can complete the math:

$$
\begin{array}{r}
6\,7\,4.\,9\,0\,0\,0 \\
-5.\,0\,0\,0\,1 \\
\hline
6\,6\,9.\,8\,9\,9\,9
\end{array}
$$

6. 100.009 - 0.68 = **99.329**. Place the first number on top of the second number, with the decimal points lined up:

$$
\begin{array}{r}
1\,0\,0.\,0\,0\,9 \\
-0.\,6\,8\,0 \\
\hline
9\,9.\,3\,2\,9
\end{array}
$$

Multiplying decimals

Multiplying decimals is different from adding and subtracting them, in that you don't have to worry about lining up the decimal points (see the preceding sections). In fact, the only difference between multiplying whole numbers and decimals comes at the end.

Here's how to multiply decimals:

1. **Perform the multiplication as you do for whole numbers.**

2. **When you're done, count the number of digits to the right of the decimal point in each factor, and add the result.**

3. **Place the decimal point in your answer so that your answer has the same number of digits after the decimal point.**

This process sounds tricky, but multiplying decimals can actually be simpler than adding or subtracting them. Suppose, for instance, that you want to multiply 23.5 by 0.16. The first step is to pretend that you're multiplying numbers without decimal points:

$$
\begin{array}{r}
23.5 \\
\times 0.16 \\
\hline
1410 \\
2350 \\
\hline
3760
\end{array}
$$

This answer isn't complete, though, because you still need to find out where the decimal point goes in the answer. To do this, notice that 23.5 has one digit after the decimal point and that 0.16 has two digits after the decimal point. Because 1 + 2 = 3, place the decimal point in the answer so that it has three digits after the decimal point. (You can put your pencil at the 0 at the end of 3760 and move the decimal point three places to the left.)

$$
\begin{array}{rl}
23.5 & \text{1 digit after the decimal point} \\
\times 0.16 & \text{2 digits after the decimal point} \\
\hline
1410 & \\
2350 & \\
\hline
3760 & 1+2=3 \text{ digits after the decimal point}
\end{array}
$$

Warning: Even though the last digit in the answer is a 0, you still need to count this as a digit when placing the decimal point. When the decimal point is in place, you can drop trailing zeros (flip to "Understanding Basic Decimal Stuff," earlier in this chapter, to see why the zeros at the end of a decimal don't change the value of the number).

So the answer is 3.760, which is equal to 3.76.

Example

Q. Multiply the following decimals: 74.2 × 0.35 = ?

A. **25.97.** Ignoring the decimal points, perform the multiplication just as you would for whole numbers:

$$
\begin{array}{r}
74.2 \\
\times 0.35 \\
\hline
3710 \\
+22260 \\
\hline
25970
\end{array}
$$

At this point, you're ready to find out where the decimal point goes in the answer. Count the number of decimal places in the two factors (74.2 and 0.35), add these two numbers together (1 + 2 = 3), and place the decimal point in the answer so that it has three digits to the right of the decimal point:

$$
\begin{array}{rl}
74.2 & \leftarrow \text{1 digit after decimal} \\
\times 0.35 & \leftarrow \text{2 digits after decimal} \\
\hline
3710 & \\
+22260 & \\
\hline
25.970 & \leftarrow 1+2=3 \text{ digits after decimal}
\end{array}
$$

Practice Questions

1. Multiply these decimals: $0.635 \times 0.42 = ?$

2. Perform the following decimal multiplication: $0.675 \times 34.8 = ?$

3. Solve the following multiplication problem: $943 \times 0.0012 = ?$

4. Find the solution to this decimal multiplication: $1.006 \times 0.0807 = ?$

Practice Answers

1. **0.2667**. Place the first number on top of the second number, ignoring the decimal points. Complete the multiplication as you would for whole numbers:

$$0.635 \leftarrow 3 \text{ digits after decimal}$$
$$\underline{\times 0.42} \leftarrow 2 \text{ digits after decimal}$$
$$1270$$
$$\underline{+25400}$$
$$0.26670 \leftarrow 3 + 2 = 5 \text{ digits after decimal}$$

At this point, you're ready to find out where the decimal point goes in the answer. Count the number of decimal places in the two factors, add these two numbers together ($3 + 2 = 5$), and place the decimal point in the answer so that it has five digits after the decimal point. After you place the decimal point (but not before!), you can drop the trailing zero.

2. **23.49**. Ignore the decimal points and simply place the first number on top of the second. Complete the multiplication as you would for whole numbers:

$$0.675$$
$$\underline{\times 34.8}$$
$$5400$$
$$27000$$
$$\underline{+202500}$$
$$23.4900$$

Count the number of decimal places in the two factors, add these two numbers together ($3 + 1 = 4$), and place the decimal point in the answer so that it has four digits after the decimal point. Last, you can drop the trailing zeros.

3. **1.1316**. Complete the multiplication as you would for whole numbers:

$$943 \leftarrow 0 \text{ digit after decimal}$$
$$\underline{\times 0.0012} \leftarrow 4 \text{ digit after decimal}$$
$$1886$$
$$\underline{+9430}$$
$$1.1316 \leftarrow 0 + 4 = 4 \text{ digit after decimal}$$

Zero digits come after the decimal point in the first factor, and you have four after-decimal digits in the second factor, for a total of 4 (0 + 4 = 4); place the decimal point in the answer so that it has four digits after the decimal point.

4. **0.0811842**. Complete the multiplication as you would for whole numbers:

$$
\begin{array}{r}
1.006 \leftarrow 3 \text{ digits after decimal} \\
\times 0.0807 \leftarrow 4 \text{ digits after decimal} \\
\hline
7042 \\
+804800 \\
\hline
0.0811842 \leftarrow 3+4 = 7 \text{ digits after decimal}
\end{array}
$$

You have a total of seven digits after the decimal points in the two factors — three in the first factor and four in the second (3 + 4 = 7) — so place the decimal point in the answer so that it has seven digits after the decimal point. Notice that I need to create an extra decimal place in this case by attaching an additional non-trailing 0.

Dividing decimals

Long division has never been a crowd pleaser. Dividing decimals is almost the same as dividing whole numbers, which is why a lot of people don't particularly like dividing decimals, either.

But at least you can take comfort in the fact that, when you know how to do long division, figuring out how to divide decimals is easy. The main difference comes at the beginning, before you start dividing.

Here's how to divide decimals:

1. **Turn the *divisor* (the number you're dividing by) into a whole number by moving the decimal point all the way to the right; at the same time, move the decimal point in the *dividend* (the number you're dividing) the same number of places to the right.**

   For example, suppose you want to divide 10.274 by 0.11. Write the problem as usual:

   $$0.11\overline{)10.274}$$

   Turn 0.11 into a whole number by moving the decimal point in 0.11 two places to the right, giving you 11. At the same time, move the decimal point in 10.274 two places to the right, giving you 1,027.4:

   $$11.\overline{)1027.4}$$

2. **Place a decimal point in the *quotient* (the answer) directly above where the decimal point now appears in the dividend.**

   Here's what this step looks like:

   $$11.\overline{)1027.4}$$

3. **Divide as usual, being careful to line up the quotient properly so that the decimal point falls into place.**

To start out, notice that 11 is too large to go into either 1 or 10. However, 11 does go into 102 (nine times). So write the first digit of the quotient just above the 2 and continue:

$$
\begin{array}{r}
9 . \\
11.\overline{)1027.4} \\
\underline{99} \\
37
\end{array}
$$

I paused after bringing down the next number, 7. This time, 11 goes into 37 three times. The important point is to place the next digit in the answer just above the 7:

$$
\begin{array}{r}
93. \\
11.\overline{)1027.4} \\
\underline{99} \\
37 \\
\underline{33} \\
44
\end{array}
$$

I paused after bringing down the next number, 4. Now, 11 goes into 44 four times. Again, be careful to place the next digit in the quotient just above the 4, and complete the division:

$$
\begin{array}{r}
93.4 \\
11.\overline{)1027.4} \\
\underline{99} \\
37 \\
\underline{33} \\
44 \\
\underline{44} \\
0
\end{array}
$$

So the answer is 93.4. As you can see, as long as you're careful when placing the decimal point and the digits, the correct answer appears with the decimal point in the right position.

Dealing with more zeros in the dividend

Sometimes you have to add one or more trailing zeros to the dividend. As I discuss earlier in this chapter, you can add as many trailing zeros as you like to a decimal without changing its value. For example, suppose you want to divide 67.8 by 0.333:

$$
0.333\overline{)67.8}
$$

Follow these steps:

1. **Change 0.333 into a whole number by moving the decimal point three places to the right; at the same time, move the decimal point in 67.8 three places to the right:**

$$333.\overline{)67800.}$$

In this case, when you move the decimal point in 67.8, you run out of room, so you have to add a couple zeros to the dividend. This step is perfectly valid, and you need to do this whenever the divisor has more decimal places than the dividend.

2. **Place the decimal point in the quotient directly above where it appears in the dividend:**

$$333.\overline{)67800.}$$

3. **Divide as usual, being careful to correctly line up the numbers in the quotient. This time, 333 doesn't go into 6 or 67, but it does go into 678 (two times). So place the first digit of the quotient directly above the 8:**

$$
\begin{array}{r}
2 \\
333.\overline{)67800.} \\
\underline{666} \\
120
\end{array}
$$

I've jumped forward in the division to the place where I bring down the first 0. At this point, 333 doesn't go into 120, so you need to put a 0 above the first 0 in 67,800 and bring down the second 0. Now, 333 does go into 1,200, so place the next digit in the answer (3) over the second 0:

$$
\begin{array}{r}
203. \\
333.\overline{)67800.} \\
\underline{666} \\
1200 \\
\underline{999} \\
201
\end{array}
$$

This time, the division doesn't work out evenly. If this were a problem with whole numbers, you'd finish by writing down a remainder of 201. (For more on remainders in division, see Chapter 3.) But decimals are a different story. The next section explains why, with decimals, the show must go on.

Completing decimal division

When you're dividing whole numbers, you can complete the problem simply by writing down the remainder. But remainders are *never* allowed in decimal division.

A common way to complete a problem in decimal division is to round off the answer. In most cases, you're instructed to round your answer to the nearest whole number or to one or two decimal places (see earlier in this chapter to find out how to round off decimals).

To complete a decimal division problem by rounding it off, you need to add at least one trailing zero to the dividend:

- To round a decimal to a whole number, add one trailing zero.
- To round a decimal to one decimal place, add two trailing zeros.
- To round a decimal to two decimal places, add three trailing zeros.

Here's what the problem looks like with a trailing zero attached:

$$
\begin{array}{r}
203. \\
333.\overline{)67800.0} \\
\underline{666} \\
1200 \\
\underline{999} \\
2010
\end{array}
$$

Attaching a trailing zero doesn't change a decimal, but it does allow you to bring down one more number, changing 201 into 2,010. Now you can divide 333 into 2,010:

$$
\begin{array}{r}
203.6 \\
333.\overline{)67800.0} \\
\underline{666} \\
1200 \\
\underline{999} \\
2010 \\
\underline{1998} \\
12
\end{array}
$$

At this point, you can round the answer to the nearest whole number, 204. I give you more practice dividing decimals later in this chapter.

Examples

Q. Divide the following: $9.152 \div 0.8 = ?$

A. **11.44.** To start out, write the problem as usual:

$$0.8\overline{)9.152}$$

Turn 0.8 into the whole number 8 by moving the decimal point one place to the right. At the same time, move the decimal point in 9.152 one place to the right. Put your decimal point in the quotient directly above where it falls in 91.52:

$$8.\overline{)9\overset{\textbf{.}}{1.52}}$$

Now you're ready to divide. Just be careful to line up the quotient properly so that the decimal point falls into place.

$$
\begin{array}{r}
11.44 \\
8.\overline{)91.52} \\
-8 \\
\hline
11 \\
-8 \\
\hline
35 \\
-32 \\
\hline
32 \\
-32 \\
\hline
0
\end{array}
$$

Q. Q. Divide the following: 21.9 ÷ 0.015 = ?

A. **1,460.** Set up the problem as usual:

$$0.015\overline{)21.900}$$

Notice that I attach two trailing zeros to the dividend. I do this because I need to move the decimal points in each number three places to the right. Again, place the decimal point in the quotient directly above where it now appears in the dividend, 21900:

$$15.\overline{)21900.}$$

Now you're ready to divide. Line up the quotient carefully so the decimal point falls into place:

$$
\begin{array}{r}
1460. \\
15.\overline{)21900.} \\
-15 \\
\hline
69 \\
-60 \\
\hline
90 \\
-90 \\
\hline
0
\end{array}
$$

Even though the division comes out even after you write the digit 6 in the quotient, you still need to add a placeholding zero so that the decimal point appears in the correct place.

Practice Questions

1. Divide these two decimals: $9.345 \div 0.05 = ?$

2. Solve the following division: $3.15 \div .021 = ?$

3. Perform the following decimal division, rounding to one decimal place: $6.7 \div 10.1$.

4. Find the solution, rounding to the nearest hundredth: $9.13 \div 4.25$.

Practice Answers

1. **186.9.** To start out, write the problem as usual:

$$0.05 \overline{)9.345}$$

Turn the divisor (0.05) into a whole number by moving the decimal point two places to the right. At the same time, move the decimal point in the dividend (9.345) two places to the right. Place the decimal point in the quotient directly above where it now appears in the dividend:

$$5.\overline{)934.5}$$

Now you're ready to divide. Be careful to line up the quotient properly so that the decimal point falls into place.

$$
\begin{array}{r}
186.9 \\
5.\overline{)934.5} \\
\underline{-5} \\
43 \\
\underline{-40} \\
34 \\
-30 \\
45 \\
\underline{-45} \\
0
\end{array}
$$

2. **150**. Write the problem as usual:

$$0.021 \overline{)3.15}$$

You need to move the decimal point in the divisor (0.021) three places to the right, so attach an additional trailing zero to the dividend (3.15) to extend it to three decimal places:

$$0.021 \overline{)3.150}$$

Now you can move both decimal points three places to the right. Place the decimal point in the quotient above the decimal point in the dividend:

$$21.\overline{)3150.}$$

Divide, being careful to line up the quotient properly:

$$
\begin{array}{r}
150. \\
21\overline{)3150.} \\
-21 \\
\hline
105 \\
-105 \\
\hline
0
\end{array}
$$

Remember to insert a placeholding zero in the quotient so that the decimal point ends up in the correct place.

3. **0.7.** To start out, write the problem as usual:

$$10.1\overline{)6.7}$$

Turn the divisor (10.1) into a whole number by moving the decimal point one place to the right. At the same time, move the decimal point in the dividend (6.7) one place to the right:

$$101.\overline{)67.}$$

The problem asks you to round the quotient to one decimal place, so fill out the dividend with trailing zeros to two decimal places:

$$101.\overline{)67.00}$$

Now you're ready to divide:

$$
\begin{array}{r}
0.66 \\
101.\overline{)67.00} \\
-606 \\
\hline
640 \\
-606 \\
\hline
34
\end{array}
$$

Round the quotient to one decimal place:

$$0.6\underline{6} \to 0.7$$

4. **2.15.** First, write the problem as usual:

$$4.25\overline{)9.13}$$

Turn the divisor (4.25) into a whole number by moving the decimal point two places to the right. At the same time, move the decimal point in the dividend (9.13) two places to the right:

$$425\overline{)913.}$$

The problem asks you to round the quotient to the nearest hundredth, so fill out the dividend with trailing zeros to three decimal places:

$$425.\overline{)913.000}$$

Now divide, carefully lining up the quotient:

$$
\begin{array}{r}
2.148 \\
425.\overline{)913.000} \\
-850 \\
\hline
630 \\
-425 \\
\hline
2050 \\
-1700 \\
\hline
3500 \\
-3400 \\
\hline
100
\end{array}
$$

Round the quotient to the nearest hundredth:

$$2.1\underline{48} \rightarrow 2.15$$

CONVERTING BETWEEN DECIMALS AND FRACTIONS

Fractions (see Chapters 8 and 9) and decimals are similar, in that they both allow you to represent parts of the whole — that is, these numbers fall on the number line *between* whole numbers.

In practice, though, sometimes one of these options is more desirable than the other. For example, calculators love decimals but aren't so crazy about fractions. To use your calculator, you may have to change fractions into decimals.

As another example, some units of measurement (such as inches) use fractions, whereas others (such as meters) use decimals. To change units, you may need to convert between fractions and decimals.

In this section, I show you how to convert back and forth between fractions and decimals. (If you need a refresher on fractions, review Chapters 8 and 9 before proceeding.)

SIMPLE DECIMAL-FRACTION CONVERSIONS

Some decimals are so common that you can memorize how to represent them as fractions. Here's how to convert all the one-place decimals to fractions:

.1	.2	.3	.4	.5	.6	.7	.8	.9
$\frac{1}{10}$	$\frac{1}{5}$	$\frac{3}{10}$	$\frac{2}{5}$	$\frac{1}{2}$	$\frac{3}{5}$	$\frac{7}{10}$	$\frac{4}{5}$	$\frac{9}{10}$

And, here are few more common decimals that translate easily to fractions:

.125	.25	.375	.625	.75	.875
$\frac{1}{8}$	$\frac{1}{4}$	$\frac{3}{8}$	$\frac{5}{8}$	$\frac{3}{4}$	$\frac{7}{8}$

Examples

Q. Convert 13.7 to a mixed number.

A. $13\frac{7}{10}$. The whole-number part of the decimal (13) becomes the whole-number part of the mixed number. Use the conversion chart to change the rest of the decimal (0.7) to a fraction.

Q. Change $9\frac{4}{5}$ to a decimal.

A. **9.8.** The whole-number part of the mixed number (9) becomes the whole-number part of the decimal. Use the conversion chart to change the fractional part of the mixed number $\left(\frac{4}{5}\right)$ to a decimal.

Practice Questions

1. Convert the following decimals into fractions:

   a. 0.7

   b. 0.4

   c. 0.25

   d. 0.125

   e. 0.1

   f. 0.75

2. Change these fractions to decimals:

   a. $\frac{9}{10}$

   b. $\frac{2}{5}$

   c. $\frac{3}{4}$

   d. $\frac{3}{8}$

   e. $\frac{7}{8}$

   f. $\frac{1}{2}$

3. Change these decimals to mixed numbers:

   a. 1.6

   b. 3.3

   c. 14.5

   d. 20.75

   e. 100.625

   f. 375.375

4. Change these mixed numbers to decimals:

   a. $1\frac{1}{5}$

   b. $2\frac{1}{10}$

   c. $3\frac{1}{2}$

   d. $5\frac{1}{4}$

   e. $7\frac{1}{8}$

   f. $12\frac{5}{8}$

Practice Answers

1. **a.** $0.7 = \frac{7}{10}$

   b. $0.4 = \frac{2}{5}$

   c. $0.25 = \frac{1}{4}$

   d. $0.125 = \frac{1}{8}$

   e. $0.1 = \frac{1}{10}$

   f. $0.75 = \frac{3}{4}$

2. **a.** $\frac{9}{10} = 0.9$

   b. $\frac{2}{5} = 0.4$

   c. $\frac{3}{4} = 0.75$

   d. $\frac{3}{8} = 0.375$

   e. $\frac{7}{8} = 0.875$

   f. $\frac{1}{2} = 0.5$

3. **a.** $1.6 = 1\frac{3}{5}$

   b. $3.3 = 3\frac{3}{10}$

   c. $14.5 = 14\frac{1}{2}$

   d. $20.75 = 20\frac{3}{4}$

   e. $100.625 = 100\frac{5}{8}$

   f. $375.375 = 375\frac{3}{8}$

4. **a.** $1\dfrac{1}{5} = 1.2$

     b. $2\dfrac{1}{10} = 2.1$

     c. $3\dfrac{1}{2} = 3.5$

     d. $5\dfrac{1}{4} = 5.25$

     e. $7\dfrac{1}{8} = 7.125$

     f. $12\dfrac{5}{8} = 12.625$

Changing decimals to fractions

Converting a decimal to a fraction is pretty simple. The only tricky part comes in when you have to reduce the fraction or change it to a mixed number.

In this section, I first show you the easy case, when no further work is necessary. Then I show you the harder case, when you need to tweak the fraction. I also show you a great time-saving trick.

Doing a basic decimal-to-fraction conversion

Here's how to convert a decimal to a fraction:

1. **Draw a line (fraction bar) under the decimal and place a 1 underneath it.**

   Suppose you want to turn the decimal 0.3763 into a fraction. Draw a line under 0.3763 and place a 1 underneath it:

   $$\frac{0.3763}{1}$$

   This number looks like a fraction, but technically it isn't one because the top number (the numerator) is a decimal.

2. **Move the decimal point one place to the right and add a 0 after the 1.**

   $$= \frac{3.763}{10}$$

3. **Repeat Step 2 until the decimal point moves all the way to the right so you can drop the decimal point entirely.**

   In this case, this is a three-step process:

   $$\frac{37.63}{100} = \frac{376.3}{1000} = \frac{3763}{10000}$$

As you can see on the last step, the decimal point in the numerator moves all the way to the end of the number, so dropping the decimal point is okay.

Note: Moving a decimal point one place to the right is the same thing as multiplying a number by 10. When you move the decimal point four places in this problem, you're essentially multiplying the 0.3763 and the 1 by 10,000. Notice that the number of digits after the decimal point in the original decimal is equal to the number of 0s that end up following the 1.

Tip: A quick way to make a fraction out of a decimal is to use the name of the smallest decimal place in that decimal. For example,

- In the decimal 0.3, the smallest decimal place is the tenths place, so the equivalent fraction is $\frac{3}{10}$.

- In the decimal 0.29, the smallest decimal place is the hundredths place, so the equivalent fraction is $\frac{29}{100}$.

- In the decimal 0.817, the smallest decimal place is the thousandths place, so the equivalent fraction is $\frac{817}{1,000}$.

In the following sections, I show you how to convert decimals to fractions when you have to work with mixed numbers and reduce the terms.

Getting mixed results

When you convert a decimal greater than 1 to a fraction, the result is a mixed number. Fortunately, this process is easy because the whole number part is unaffected by the conversion. So focusing only on the decimal part, follow the same steps I outline in the preceding section.

For example, suppose you want to change 4.51 to a fraction. The result will be a mixed number with a whole number part of 4. To find the fractional part, follow these steps:

1. **Draw a line (fraction bar) under the decimal and place a 1 underneath it.**

$$\frac{0.51}{1}$$

2. **Move the decimal point one place to the right and add a 0 after the 1.**

$$= \frac{5.1}{10}$$

3. **Repeat Step 2 until the decimal point moves all the way to the right so you can drop the decimal point entirely.**

   In this case, you have only one additional step:

$$= \frac{51}{100}$$

So the mixed-number equivalent of 4.51 is $4\frac{51}{100}$.

Examples

Q. Change the decimal 0.83 to a fraction.

A. $\frac{83}{100}$. Create a "fraction" with 0.83 in the numerator and 1.0 in the denominator:

$$\frac{0.83}{1.0}$$

Move the decimal point in 0.83 two places to the right to turn it into a whole number; move the decimal point in the denominator the same number of places. I do this one decimal place at a time:

$$\frac{0.83}{1.0} = \frac{8.3}{10.0} = \frac{83.0}{100.0}$$

At this point, you can drop the decimal points and trailing zeros in both the numerator and denominator.

Q. Change the decimal 0.0205 to a fraction.

A. $\frac{41}{2,000}$. Create a "fraction" with 0.0205 in the numerator and 1.0 in the denominator:

$$\frac{0.0205}{1.0}$$

Move the decimal point in the 0.0205 four places to the right to turn the numerator into a whole number; move the decimal point in the denominator the same number of places:

$$\frac{0.0205}{1.0}$$
$$= \frac{0.205}{10.0}$$
$$= \frac{02.05}{100.0}$$
$$= \frac{020.5}{1,000.0}$$
$$= \frac{0205.0}{10,000.0}$$

Drop the decimal points, plus any leading or trailing zeros in both the numerator and denominator.

$$= \frac{205}{10,000}$$

Both the numerator and denominator are divisible by 5, so reduce this fraction:

$$= \frac{41}{2,000}$$

Practice Questions

1. Change the decimal 0.27 to a fraction.

2. Convert the decimal 0.0315 to a fraction.

3. Write 45.12 as a mixed number.

4. Change 100.001 to a mixed number.

Practice Answers

1. $\frac{27}{100}$. Create a "fraction" with 0.27 in the numerator and 1.0 in the denominator. Then move the decimal points to the right until both the numerator and denominator are whole numbers:

$$\frac{0.27}{1.0} = \frac{2.7}{10.0} = \frac{27.0}{100.0}$$

At this point, you can drop the decimal points and trailing zeros.

2. $\frac{63}{2,000}$. Create a "fraction" with 0.0315 in the numerator and 1.0 in the denominator.

Then move the decimal points in both the numerator and denominator to the right one place at a time. Continue until both the numerator and denominator are whole numbers:

$$\frac{0.0315}{1.0} = \frac{0.315}{10.0} = \frac{3.15}{100.0} = \frac{31.5}{1,000.0} = \frac{315.0}{10,000.0}$$

Drop the decimal points and trailing zeros. The numerator and denominator are both divisible by 5, so reduce the fraction:

$$\frac{315}{10,000} = \frac{63}{2,000}$$

3. $45\frac{3}{25}$. Before you begin, separate out the whole-number portion of the decimal (45).

Create a "fraction" with 0.12 in the numerator and 1.0 in the denominator. Move the decimal points in both the numerator and denominator to the right until both are whole numbers:

$$\frac{0.12}{1.0} = \frac{1.2}{10.0} = \frac{12.0}{100.0}$$

Drop the decimal points and trailing zeros. As long as the numerator and denominator are both divisible by 2 (that is, even numbers), you can reduce this fraction:

$$\frac{12}{100} = \frac{6}{50} = \frac{3}{25}$$

To finish up, reattach the whole-number portion that you separated at the beginning.

4. $100\frac{1}{1,000}$. Separate out the whole-number portion of the decimal (100) and create a "fraction" with 0.001 in the numerator and 1.0 in the denominator. Move the decimal points in both the numerator and denominator to the right one place at a time until both are whole numbers:

$$\frac{0.001}{1.0} = \frac{0.01}{10.0} = \frac{0.1}{100.0} = \frac{1.0}{1,000.0}$$

Drop the decimal points and trailing zeros and reattach the whole-number portion of the number you started with:

$$100\frac{1}{1,000}$$

Changing fractions to decimals

Converting fractions to decimals isn't difficult, but to do it, you need to know about decimal division. If you need to get up to speed on this, check out "Dividing decimals," earlier in this chapter.

Remember: To convert a fraction to a decimal, follow these steps:

1. **Set up the fraction as a decimal division, dividing the numerator (top number) by the denominator (bottom number).**

2. **Attach enough trailing zeros to the numerator so that you can continue dividing until you find that the answer is either a *terminating decimal* or a *repeating decimal*.**

Don't worry, I explain terminating and repeating decimals next.

The last stop: Terminating decimals

Sometimes when you divide the numerator of a fraction by the denominator, the division eventually works out evenly. The result is a *terminating decimal*.

For example, suppose you want to change the fraction $\frac{2}{5}$ to a decimal. Here's your first step:

$$5\overline{)2}$$

One glance at this problem, and it looks like you're doomed from the start because 5 doesn't go into 2. But watch what happens when I add a few trailing zeros. Notice that I also place another decimal point in the answer just above the first decimal point. This step is important — you can read more about it in "Dividing decimals":

$$5\overline{)2.000}$$

Now you can divide because, although 5 doesn't go into 2, 5 does go into 20 four times:

$$
\begin{array}{r}
0.4 \\
5\overline{)2.000} \\
\underline{20} \\
0
\end{array}
$$

You're done! As it turns out, you needed only one trailing zero, so you can ignore the rest:

$$\frac{2}{5} = 0.4$$

Because the division worked out evenly, the answer is an example of a *terminating decimal*.

As another example, suppose you want to find out how to represent $\frac{7}{16}$ as a decimal. As earlier, I attach three trailing zeros:

$$
\begin{array}{r}
0.437 \\
16\overline{)7.000} \\
\underline{64} \\
60 \\
\underline{48} \\
120 \\
\underline{112} \\
8
\end{array}
$$

This time, three trailing zeros aren't enough to get my answer, so I attach a few more and continue:

$$
\begin{array}{r}
0.4375 \\
16\overline{)7.000000} \\
\underline{64} \\
60 \\
\underline{48} \\
120 \\
\underline{112} \\
80 \\
\underline{80} \\
0
\end{array}
$$

At last, the division works out evenly, so again the answer is a terminating decimal. Therefore, $\frac{7}{16} = 0.4375$.

The endless ride: Repeating decimals

Sometimes when you try to convert a fraction to a decimal, the division *never* works out evenly. The result is a *repeating decimal* — a decimal that cycles through the same number pattern forever.

You may recognize these pesky little critters from your calculator, when an apparently simple division problem produces a long string of numbers.

For example, to change $\frac{2}{3}$ to a decimal, begin by dividing 2 by 3. As in the last section, start by adding three trailing zeros, and see where it leads:

$$
\begin{array}{r}
0.666 \\
3\overline{)2.000} \\
\underline{18} \\
20 \\
\underline{18} \\
20 \\
\underline{18} \\
2
\end{array}
$$

At this point, you still haven't found an exact answer. But you may notice that a repeating pattern has developed in the division. No matter how many trailing zeros you attach to the number 2, the same pattern continues forever. This answer, 0.666 . . . , is an example of a repeating decimal. You can write $\frac{2}{3}$ as

$$\frac{2}{3} = 0.\overline{6}$$

The bar over the 6 means that, in this decimal, the number 6 repeats forever. You can represent many simple fractions as repeating decimals. In fact, *every* fraction can be represented either as a repeating decimal or as a terminating decimal — that is, as an ordinary decimal that ends.

Now suppose you want to find the decimal representation of $\frac{5}{11}$. Here's how this problem plays out:

$$
\begin{array}{r}
0.4545 \\
11\overline{)5.0000} \\
\underline{44} \\
60 \\
\underline{55} \\
50 \\
\underline{44} \\
60 \\
\underline{55} \\
5
\end{array}
$$

This time, the pattern repeats every other number — 4, then 5, then 4 again, and then 5 again, forever. Attaching more trailing zeros to the original decimal only strings out this pattern indefinitely. So you can write

$$\frac{5}{11} = 0.\overline{45}$$

This time, the bar is over both the 4 and the 5, telling you that these two numbers alternate forever.

Remember: Repeating decimals are an oddity, but they aren't hard to work with. In fact, as soon as you can show that a decimal division is repeating, you've found your answer. Just remember to place the bar only over the numbers that keep on repeating.

Some decimals never end and never repeat. You can't write them as fractions, so mathematicians have agreed on some shorter ways of naming them so that writing them out doesn't take, well, forever.

When you're asked to find the exact decimal value of a fraction, feel free to attach trailing zeros to the *dividend* (the number you're dividing) as you go along. Keep dividing until the division either works out evenly (so the quotient is a terminating decimal) or a repeating pattern develops (so it's a repeating decimal).

Examples

Q. Convert the fraction $\frac{9}{16}$ to an exact decimal value.

A. **0.5625.** Divide $9 \div 16$:

$$16\overline{)9.000}$$

Because 16 is too big to go into 9, I attached a decimal point and some trailing zeros to the 9. Now you can divide as I show you earlier in this chapter:

$$
\begin{array}{r}
0.5625 \\
16\overline{)9.0000} \\
-80 \\
\hline
100 \\
-96 \\
\hline
40 \\
-32 \\
\hline
80 \\
-80 \\
\hline
0
\end{array}
$$

Q. What is the exact decimal value of the fraction $\frac{5}{6}$?

A. Divide $5 \div 6$:

Because 6 is too big to go into 5, I attached a decimal point and some trailing zeros to the 5. Now divide:

$$
\begin{array}{r}
.8333 \\
6\overline{)5.0000} \\
\underline{-48} \\
20 \\
\underline{-18} \\
20 \\
\underline{-18} \\
20 \\
\underline{-18} \\
2
\end{array}
$$

As you can see, a pattern has developed. No matter how many trailing zeros you attach, the quotient will never come out evenly. Instead, the quotient is the repeating decimal $.08\overline{3}$. The bar over the 3 indicates that the number 3 repeats forever: 0.83333333 . . . .

Practice Questions

1. Change $\frac{13}{16}$ to an exact decimal value.

2. Express $\frac{7}{9}$ exactly as a decimal.

Practice Answers

1. $\frac{13}{16}$ = **0.8125.** Divide 13 ÷ 16, attaching plenty of trailing zeros to the 13:

$$
\begin{array}{r}
0.8125 \\
16\overline{)13.00000} \\
\underline{-128} \\
20 \\
\underline{-16} \\
40 \\
\underline{-32} \\
80 \\
\underline{-80} \\
0
\end{array}
$$

This division eventually ends, so the quotient is a terminating decimal.

2. $\frac{7}{9} = 0.\overline{7}$.

$$\frac{7}{9} = 0.\overline{7}$$

Divide $7 \div 9$, attaching plenty of trailing zeros to the 7:

$$
\begin{array}{r}
0.77 \\
9\overline{)7.000} \\
-63 \\
\hline
70 \\
-63 \\
\hline
70
\end{array}
$$

A pattern has developed in the subtraction: 70 - 63 = 7, so when you bring down the next 0, you'll get 70 again. Therefore, the quotient is a repeating decimal.

11

PLAYING WITH PERCENTS

Like whole numbers and decimals, percents are a way to talk about parts of a whole. The word *percent* means "out of 100." So if you have 50% of something, you have 50 out of 100. If you have 25% of it, you have 25 out of 100. Of course, if you have 100% of anything, you have all of it.

In this chapter, I show you how to work with percents. Because percents resemble decimals, I first show you how to convert numbers back and forth between percents and decimals. No worries — this switch is easy to do. Next, I show you how to convert back and forth between percents and fractions — also not too bad. When you understand how conversions work, I show you the three basic types of percent problems, plus a method that makes the problems simple.

MAKING SENSE OF PERCENTS

The word *percent* literally means "for 100," but in practice, it means closer to "out of 100." For example, suppose that a school has exactly 100 children — 50 girls and 50 boys. You can say that "50 out of 100" children are girls — or you can shorten it to simply "50 percent." Even shorter than that, you can use the symbol %, which means *percent*.

Saying that 50% of the students are girls is the same as saying that $\frac{1}{2}$ of them are girls. Or if you prefer decimals, it's the same thing as saying that 0.5 of all the students are girls. This example shows you that percents, like fractions and decimals, are just another way of talking about parts of the whole. In this case, the whole is the total number of children in the school.

You don't literally have to have 100 of something to use a percent. You probably won't ever really cut a cake into 100 pieces, but that doesn't matter. The values are the same. Whether you're talking about cake, a dollar, or a group of children, 50% is still half, 25% is still one-quarter, 75% is still three-quarters, and so on.

Any percentage smaller than 100% means less than the whole — the smaller the percentage, the less you have. You probably know this fact well from the school grading system. If you get 100%, you get a perfect score. And 90% is usually A work, 80% is a B, 70% is a C, and, well, you know the rest.

Of course, 0% means "0 out of 100" — any way you slice it, you have nothing.

DEALING WITH PERCENTS GREATER THAN 100%

100% means "100 out of 100" — in other words, everything. So when I say I have 100% confidence in you, I mean that I have complete confidence in you.

What about percentages more than 100%? Well, sometimes percentages like these don't make sense. For example, you can't spend more than 100% of your time playing basketball, no matter how much you love the sport; 100% is all the time you have, and there ain't no more.

But a lot of times, percentages larger than 100% are perfectly reasonable. For example, suppose I own a hot dog wagon and sell the following:

> 10 hot dogs in the morning
>
> 30 hot dogs in the afternoon

The number of hot dogs I sell in the afternoon is 300% of the number I sold in the morning. It's three times as many.

Here's another way of looking at this: I sell 20 more hot dogs in the afternoon than in the morning, so this is a *200% increase* in the afternoon — 20 is twice as many as 10.

Spend a little time thinking about this example until it makes sense. You visit some of these ideas again in Chapter 12, when I show you how to do word problems involving percents.

CONVERTING TO AND FROM PERCENTS, DECIMALS, AND FRACTIONS

To solve many percent problems, you need to change the percent to either a decimal or a fraction. Then you can apply what you know about solving decimal and fraction problems. For this reason, I show you how to convert to and from percents before I show you how to solve percent problems.

Percents and decimals are similar ways of expressing parts of a whole. This similarity makes converting percents to decimals, and vice versa, mostly a matter of moving the decimal point. It's so simple you can probably do it in your sleep (but you should probably stay awake when you first read about the concept).

Percents and fractions both express the same idea — parts of a whole — in different ways. So converting back and forth between percents and fractions isn't quite as simple as just moving the decimal point back and forth. In this section, I cover the ways to convert to and from percents, decimals, and fractions, starting with percents to decimals.

CONVERTING PERCENTS TO DECIMALS

Percents and decimals are very similar forms, so everything you know about decimals (see Chapter 10) carries over when you're working with percents. All you need to do is convert your percent to a decimal, and you're good to go.

To change a whole-number percent to a decimal, simply replace the percent sign with a decimal point and then move this decimal point two places to the left; after this, you can drop any trailing zeros. Here are a few common conversions between percents and decimals:

$$100\% = 1 \qquad 75\% = 0.75 \qquad 50\% = 0.5$$
$$25\% = 0.25 \qquad 20\% = 0.2 \qquad 10\% = 0.1$$

Sometimes a percent already has a decimal point. In this case, just drop the percent sign and move the decimal point two places to the left. For instance, 12.5% = 0.125.

Examples

Q. Change 80% to a decimal.

A. **0.8.** Replace the percent sign with a decimal point — changing 80% to 80. — and then move the decimal point two places to the left:

$$80\% = 0.80$$

At the end, you can drop the trailing zero to get 0.8.

Q. Change 37.5% to a decimal.

A. **0.375.** Drop the percent sign and move the decimal point two places to the left:

$$37.5\% = 0.375$$

Practice Questions

1. Change 90% to a decimal.

2. A common interest rate on an investment such as a bond is 4%. Convert 4% to a decimal.

3. Find the decimal equivalent of 99.44%.

4. What is 243.1% expressed as a decimal?

Practice Answers

1. **0.9.** Replace the percent sign with a decimal point and then move this decimal point two places to the left:

$$90\% = 0.90$$

At the end, drop the trailing zero to get 0.9.

2. **0.04.** Replace the percent sign with a decimal point and then move this decimal point two places to the left:

$$4\% = 0.04$$

3. **0.9944.** Drop the percent sign and move the decimal point two places to the left:

$$99.44\% = 0.9944$$

4. **2.431.** Drop the percent sign and move the decimal point two places to the left:

$$243.1\% = 2.431$$

CHANGING DECIMALS TO PERCENTS

Calculating with percents is easiest when you convert to decimals first. When you're done calculating, however, you often need to change your answer from a decimal back to a percent. This is especially true when you're working with interest rates, taxes, or the likelihood of a big snowfall the night before a big test. All these numbers are most commonly expressed as percents.

To change a decimal to a percent, move the decimal point two places to the right and attach a percent sign. If the result is a whole number, you can drop the decimal point.

Example

Q. Change 0.6 to a percent.

A. **60%.** Move the decimal point two places to the right and attach a percent sign:

$$0.6 = 60\%$$

Q **Practice Questions**

1.   Convert 0.57 to a percent.

2.   What is 0.3 expressed as a percent?

3.   Change 0.015 to a percent.

4.   Express 2.222 as a percent.

A **Practice Answers**

1.   **57%.** Move the decimal point two places to the right and attach a percent sign:

$$0.57 = 057\%$$

At the end, drop the leading zero to get 57%.

2.   **30%.** Move the decimal point two places to the right and attach a percent sign:

$$0.3 = 030\%$$

At the end, drop the leading zero to get 30%.

3.   **1.5%.** Move the decimal point two places to the right and attach a percent sign:

$$0.015 = 01.5\%$$

At the end, drop the leading zero to get 1.5%.

4.   **222.2%.** Move the decimal point two places to the right and attach a percent sign:

$$2.222 = 222.2\%$$

SWITCHING FROM PERCENTS TO FRACTIONS

Some percents are easy to convert to fractions. Here are a few quick conversions that are worth knowing:

$$1\% = \frac{1}{100} \quad 5\% = \frac{1}{20} \quad 10\% = \frac{1}{10} \quad 20\% = \frac{1}{5}$$

$$25\% = \frac{1}{4} \quad 50\% = \frac{1}{2} \quad 75\% = \frac{3}{4} \quad 100\% = 1$$

Beyond these simple conversions, changing a percent to a fraction isn't a skill you're likely to use much outside of a math class. Decimals are much easier to work with.

However, teachers often test you on this skill to make sure you understand the ins and outs of percents, so here's the scoop on converting percents to fractions: To change a percent to a fraction, use the percent without the percent sign as the *numerator* (top number) of the fraction and use 100 as the *denominator* (bottom number). When necessary, reduce this fraction to lowest terms or change it to a mixed number. (For a refresher on reducing fractions, see Chapter 8.)

Example

Q. Change 35% to a fraction.

A. $\frac{7}{20}$. Place 35 in the numerator and 100 in the denominator:

$$35\% = \frac{35}{100}$$

You can reduce this fraction because the numerator and denominator are both divisible by 5:

$$\frac{7}{20}$$

Practice Questions

1. Change 19% to a fraction.

2. A common interest rate on credit cards and other types of loans is 8%. What is 8% expressed as a fraction?

3. Switch 123% to a fraction.

4. Convert 375% to a fraction.

Practice Answers

1. $\frac{19}{100}$. Place 19 in the numerator and 100 in the denominator.

2. $\frac{2}{25}$. Place 8 in the numerator and 100 in the denominator:

$$\frac{8}{100}$$

You can reduce this fraction by 2 twice:

$$= \frac{4}{50} = \frac{2}{25}$$

3. $1\frac{23}{100}$. Place 123 in the numerator and 100 in the denominator:

$$\frac{123}{100}$$

You can change this improper fraction to a mixed number:

$$= 1\frac{23}{100}$$

4. $3\frac{3}{4}$. Place 375 in the numerator and 100 in the denominator:

$$\frac{375}{100}$$

Change the improper fraction to a mixed number:

$$= 3\frac{75}{100}$$

Reduce the fractional part of this mixed number, first by 5 and then by another 5:

$$= 3\frac{15}{20} = 3\frac{3}{4}$$

CONVERTING FRACTIONS TO PERCENTS

Knowing how to make a few simple conversions from fractions to percents is a useful real-world skill. Here are some of the most common conversions:

$$\frac{1}{100} = 1\% \quad \frac{1}{20} = 5\% \quad \frac{1}{10} = 10\% \quad \frac{1}{5} = 20\%$$

$$\frac{1}{4} = 25\% \quad \frac{1}{2} = 50\% \quad \frac{3}{4} = 75\% \quad 1 = 100\%$$

Beyond these, you're not all that likely to need to convert a fraction to a percent outside of a math class. But then, passing your math class is important, so in this section I show you how to make this type of conversion.

Converting a fraction to a percent is a two-step process:

1. **Convert the fraction to a decimal, as I show you in Chapter 10.**

   In some problems, the result of this step may be a repeating decimal. This is fine — in this case, the percent will also contain a repeating decimal.

2. **Convert this decimal to a percent.**

   Move the decimal point two places to the right and add a percent sign.

Example

Q. Change the fraction $\frac{1}{9}$ to a percent.

A. $11.\overline{1}\%$. First, change $\frac{1}{9}$ to a decimal:

$$
\begin{array}{r}
0.111 \\
9\overline{)1.000} \\
\underline{-9} \\
10 \\
\underline{-9} \\
10 \\
\underline{-9} \\
1
\end{array}
$$

The result is the repeating decimal $0.\overline{1}$. Now change this repeating decimal to a percent:

$$0.\overline{1} = 11.\overline{1}\%$$

Practice Questions

1. Express $\frac{2}{5}$ as a percent.

2. Change $\frac{3}{20}$ to a percent.

3. Convert $\frac{7}{8}$ to a percent.

4. Change $\frac{2}{11}$ to a percent.

Practice Answers

1. **40%**. First, change $\frac{2}{5}$ to a decimal:

$$2.0 \div 5 = 0.4$$

Now change 0.4 to a percent by moving the decimal point two places to the right and adding a percent sign:

$$0.4 = 40\%$$

2.   **15%.** First, change $\frac{3}{20}$ to a decimal:

$$3.00 \div 20 = 0.15$$

Then change 0.15 to a percent:

$$0.15 = 15\%$$

3.   **87.5%.** First, change $\frac{7}{8}$ to a decimal:

$$7.000 \div 8 = 0.875$$

Now change 0.875 to a percent:

$$0.875 = 87.5\%$$

4.   **18.$\overline{18}$%.** First, change $\frac{2}{11}$ to a decimal:

$$
\begin{array}{r}
0.1818 \\
11\overline{)2.0000} \\
-11 \\
\hline
90 \\
-88 \\
\hline
20 \\
-11 \\
\hline
90 \\
-88 \\
\hline
2
\end{array}
$$

The result is the repeating decimal $0.\overline{18}$. Now change this repeating decimal to a percent:

$$0.\overline{18} = 18.\overline{18}\%$$

SOLVING PERCENT PROBLEMS

When you know the connection between percents and fractions, which I discuss earlier in "Converting to and from Percents, Decimals, and Fractions," you can solve a lot of percent problems with a few simple tricks. Other problems, however, require a bit more work. In this section, I show you how to tell an easy percent problem from a tough one, and I give you the tools to solve all of them.

Figuring out simple percent problems

Tip: A lot of percent problems turn out to be easy when you give them a little thought. In many cases, just remember the connection between percents and fractions, and you're half-way home:

- **Finding 100% of a number:** Remember that 100% means the whole thing, so 100% of any number is simply the number itself:

  100% of 5 is 5.

  100% of 91 is 91.

  100% of 732 is 732.

- **Finding 50% of a number:** Remember that 50% means half, so to find 50% of a number, just divide it by 2:

  50% of 20 is 10.

  50% of 88 is 44.

  50% of 7 is $\frac{7}{2}$ (or $3\frac{1}{2}$ or 3.5).

- **Finding 25% of a number:** Remember that 25% equals $\frac{1}{4}$, so to find 25% of a number, divide it by 4:

$$25\% \text{ of } 40 = 10$$
$$25\% \text{ of } 88 = 22$$
$$25\% \text{ of } 15 = \frac{15}{4} = 3\frac{3}{4} = 3.75$$

- **Finding 20% of a number:** Finding 20% of a number is handy if you like the service you've received in a restaurant, because a good tip is 20% of the check. Because 20% equals $\frac{1}{5}$, you can find 20% of a number by dividing it by 5. But I can show you an easier way: Remember that 20% is 2 times 10%, so to find 20% of a number, move the decimal point one place to the left and double the result:

$$20\% \text{ of } 80 = 8 \times 2 = 16$$
$$20\% \text{ of } 300 = 30 \times 2 = 60$$
$$20\% \text{ of } 41 = 4.1 \times 2 = 8.2$$

- **Finding 10% of a number:** Finding 10% of any number is the same as finding $\frac{1}{10}$ of that number. To do this, just move the decimal point one place to the left:

$$10\% \text{ of } 30 = 3$$
$$10\% \text{ of } 41 = 4.1$$
$$10\% \text{ of } 7 = 0.7$$

- **Finding 200%, 300%, and so on of a number:** Working with percents that are multiples of 100 is easy. Just drop the two 0s and multiply by the number that's left:

$$200\% \text{ of } 7 = 2 \times 7 = 14$$
$$300\% \text{ of } 10 = 3 \times 10 = 30$$
$$1{,}000\% \text{ of } 45 = 10 \times 45 = 450$$

(See the earlier "Dealing with Percents Greater than 100%" section for details on what having more than 100% really means.)

Examples

Q. What is 10% of 64?

A. **6.4.** Move the decimal point one place to the left: 10% of 64 = 6.4.

Q. What is 20% of 52?

A. **10.4.** Move the decimal point one place to the left and double the result: 20% of 52 = 5.2 × 2 = 10.4.

Q. What is 400% of 11?

A. **44.** Multiply the number by 4: 400% of 11 = 11 × 4 = 44.

Practice Questions

1. What is 10% of 83?

2. What is 20% of 75?

3. What is 25% of 84?

4. What is 50% of 172?

5. What is 100% of 9.9?

6. What is 500% of 36?

A Practice Answers

1. **8.3.** Move the decimal point one place to the left: 10% of 83 = 8.3.

2. **15.** Move the decimal one place to the left and multiply by 2: 20% of 75 = 7.5 × 2 = 15.

3. **21.** Divide the number by 4: 25% of 84 = 84 ÷ 4 = 21.

4. **86.** Divide the number by 2: 50% of 172 = 172 ÷ 2 = 86.

5. **9.9.** 100% of any number is the number itself: 100% of 9.9 = 9.9.

6. **180.** Multiply the number by 5: 500% of 36 = 36 × 5 = 180.

Turning the problem around

Here's a trick that makes certain tough-looking percent problems so easy that you can do them in your head. Simply move the percent sign from one number to the other and flip the order of the numbers.

Suppose someone wants you to figure out the following:

$$88\% \text{ of } 50$$

Finding 88% of anything isn't an activity anybody looks forward to. But an easy way of solving the problem is to switch it around:

$$88\% \text{ of } 50 = 50\% \text{ of } 88$$

This move is perfectly valid, and it makes the problem a lot easier. It works because the word *of* really means multiplication, and you can multiply either backward or forward and get the same answer. As I discuss in the preceding section, "Figuring out simple percent problems," 50% of 88 is simply half of 88:

$$88\% \text{ of } 50 = 50\% \text{ of } 88 = 44$$

As another example, suppose you want to find

$$7\% \text{ of } 200$$

Again, finding 7% is tricky, but finding 200% is simple, so switch the problem around:

$$7\% \text{ of } 200 = 200\% \text{ of } 7$$

In the preceding section, I tell you that, to find 200% of any number, you just multiply that number by 2:

$$7\% \text{ of } 200 = 200\% \text{ of } 7 = 2 \times 7 = 14$$

E **Example**

Q. What is 18% of 10?

A. 1.8. Reverse the two numbers in the problem and solve: 18% of 10 = 10 % of 18 = 1.8.

Q **Practice Questions**

1. What is 48% of 25?

2. What is 61% of 300?

3. What is 98% of 50?

4. What is 132% of 20?

A **Practice Answers**

1. **12.** Reverse the two numbers in the problem and solve: 48% of 25 = 25% of 48 = 48 ÷ 4 = 12.

2. **183.** Reverse the two numbers in the problem and solve: 61% of 300 = 300% of 61 = 61 × 3 = 183.

3. **49.** Reverse the two numbers in the problem and solve: 98% of 50 = 50% of 98 = 98 ÷ 2 = 49.

4. **26.4.** Reverse the two numbers in the problem and solve: 132% of 20 = 20% of 132 = 13.2 × 2 = 26.4.

Deciphering more-difficult percent problems

You can solve a lot of percent problems using the tricks I show you earlier in this chapter. For more difficult problems, you may want to switch to a calculator. If you don't have a calculator at hand, solve percent problems by turning them into decimal multiplication, as follows:

1. **Change the word *of* to a multiplication sign and the percent to a decimal (as I show you earlier in this chapter).**

   Suppose you want to find 35% of 80. Here's how you start:

   $$35\% \text{ of } 80 = 0.35 \times 80$$

2. **Solve the problem using decimal multiplication (see Chapter 10).**

Here's what the example looks like:

$$
\begin{array}{r}
0.35 \\
\times\ 80 \\
\hline
28.00
\end{array}
$$

So 35% of 80 is 28.

Example

Q. What is 16% of 30?

A. **4.8.** Change the word *of* to a multiplication sign and the percent to a decimal, then solve using decimal multiplication: 16 % of 30 = 0.16 × 30 = 4.8.

Practice Questions

1. What is 21% of 70?

2. What is 35% of 45?

3. What is 61% of 127?

4. What is 89% of 351?

Practice Answers

1. **14.7.** Change the word *of* to a multiplication sign and the percent to a decimal, then solve using decimal multiplication: 21% of 70 = 0.21 × 70 = 14.7.

2. **15.75.** Change the word *of* to a multiplication sign and the percent to a decimal, then solve using decimal multiplication: 35% of 45 = 0.35 × 45 = 15.75.

3. **77.47.** Change the word *of* to a multiplication sign and the percent to a decimal, then solve using decimal multiplication: 61% of 127 = 0.61 × 127 = 77.47.

4. **312.39.** Change the word *of* to a multiplication sign and the percent to a decimal, then solve using decimal multiplication: 89% of 351 = 0.89 × 351 = 312.39.

PUTTING ALL THE PERCENT PROBLEMS TOGETHER

In the preceding section, "Solving Percent Problems," I give you a few ways to find any percent of any number. This type of percent problem is the most common, which is why it gets top billing.

But percents crop up in a wide range of business applications, such as banking, real estate, payroll, and taxes. (I show you some real-world applications when I discuss word problems in Chapter 12.) And depending on the situation, two other common types of percent problems may present themselves.

In this section, I show you these two additional types of percent problems and how they relate to the type you now know how to solve. I also give you a simple tool to make quick work of all three types.

Identifying the three types of percent problems

Earlier in this chapter, I show you how to solve problems that look like this:

50% of 2 is ?

The answer, of course, is 1. (See "Solving Percent Problems" for details on how to get this answer.) Given two pieces of information — the percent and the number to start with — you can figure out what number you end up with.

Now suppose instead that I leave out the percent but give you the starting and ending numbers:

?% of 2 is 1

You can still fill in the blank without too much trouble. Similarly, suppose that I leave out the starting number but give the percent and the ending number:

50% of ? is 1

Again, you can fill in the blank.

If you get this basic idea, you're ready to solve percent problems. When you boil them down, nearly all percent problems are like one of the three types I show in Table 11-1.

Table 11-1   The Three Main Types of Percent Problems

Problem Type	What to Find	Example
Type #1	The ending number	50% of 2 is *what*?
Type #2	The percentage	*What* percent of 2 is 1?
Type #3	The starting number	50% of *what* is 1?

In each case, the problem gives you two of the three pieces of information, and your job is to figure out the remaining piece. In the next section, I give you a simple tool to help you solve all three of these types of percent problems.

Solving percent problems with equations

Remember: Here's how to solve any percent problem:

1. **Change the word *of* to a multiplication sign and the percent to a decimal (as I show you earlier in this chapter).**

   This step is the same as for more straightforward percent problems. For example, consider this problem:

   60% of what is 75?

   Begin by changing as follows:

60%	of	what	is	75
0.6	×			75

2. **Turn the word *is* to an equals sign and the word *what* into the letter *n*.**

   Here's what this step looks like:

60%	of	what	is	75
0.6	×	n	=	75

   This equation looks more normal, as follows:

   $$0.6 \times n = 75$$

3. **Find the value of *n*.**

   Technically, the last step involves a little bit of algebra, but I know you can handle it. (For a complete explanation of algebra, see Part V of this book.) In the equation, *n* is being multiplied by 0.6. You want to "undo" this operation by *dividing* by 0.6 on both sides of the equation:

   $$0.6 \times n \div 0.6 = 75 \div 0.6$$

   Almost magically, the left side of the equation becomes a lot easier to work with because multiplication and division by the same number cancel each other out:

   $$n = 75 \div 0.6$$

   Remember that *n* is the answer to the problem. If your teacher lets you use a calculator, this last step is easy; if not, you can calculate it using some decimal division, as I show you in Chapter 10:

   $$n = 125$$

   Either way, the answer is 125 — so 60% of 125 is 75.

As another example, suppose you're faced with this percent problem:

What percent of 250 is 375?

To begin, change the *of* into a multiplication sign and the percent into a decimal.

What	percent	of	250	is	375
	× 0.01	×	250		375

Notice here that, because I don't know the percent, I change the word *percent* to × 0.01. Next, change *is* to an equals sign and *what* to the letter *n*:

What	percent	of	250	is	375
n	× 0.01	×	250	=	375

Consolidate the equation and then multiply:

$$n \times 2.5 = 375$$

Now divide both sides by 2.5:

$$n = 375 \div 2.5 = 150$$

Therefore, the answer is 150 — so 150% of 250 is 375.

Here's one more problem: 49 is what percent of 140? Begin, as always, by translating the problem into words:

49	is	what	percent	of	140
49	=	*n*	× 0.01	×	140

Simplify the equation:

$$49 = n \times 1.4$$

Now divide both sides by 1.4:

$$49 \div 1.4 = n \times 1.4 \div 1.4$$

Again, multiplication and division by the same number allows you to cancel on the left side of the equation and complete the problem:

$$49 \div 1.4 = n$$
$$35 = n$$

Therefore, the answer is 35, so 49 is 35% of 140.

In summary, here are the three pieces — and the kinds of questions that ask for each piece:

- **The percent:** The problem may give you the starting and ending numbers and ask you to find the percent. Here are some ways this problem can be asked:

  ?% of 4 is 1

  What percent of 4 is 1?

  1 is what percent of 4?

  The answer is 25%, because 25% × 4 = 1.

- **The starting number:** The problem may give you the percent and the ending number and ask you to find the starting number:

  10% of ? is 40

  10% of what number is 40?

  40 is 10% of what number?

  This time, the answer is 400, because 10% × 400 = 40.

- **The ending number:** The most common type of percent problem gives you a percentage and a starting number and asks you to figure out the ending number:

  50% of 6 is ?

  50% of 6 equals what number?

  Can you find 50% of 6?

  No matter how I phrase it, notice that the problem always includes *50% of 6*. The answer is 3, because 50% × 6 = 3.

Remember: Each type of percent problem gives you *two* pieces of information and asks you to find the third. Place the information into an equation using the following translations from words into symbols:

What (number) → *n*

is → =

percent → × 0.01

of → ×

Examples

Q. Place the statement *25% of 12 is 3* into an equation.

A. **25 × 0.01 × 12 = 3.**

This is a direct translation, as follows:

25	%	of	12	is	3
25	× 0.01	×	12	=	3

Q. What is 18% of 90?

A. **16.2.** Translate the problem into an equation:

What	is	18	%	of	90
n	=	18	× 0.01	×	90

Solve this equation:

$$n = 18 \times 0.01 \times 90 = 16.2$$

Practice Questions

1. Place the statement *20% of 350 is 70* into an equation. Check your work by simplifying the equation.

2. What percent of 150 is 25.5?

3. What is 79% of 11?

4. 30% of what number is 10?

Practice Answers

1. $20 \times 0.01 \times 350 = $ **70.**

   Turn the problem into an equation as follows:

20	percent	of	350	is	70
20	× 0.01	×	350	=	70

   Check this equation:

   $$20 \times 0.01 \times 350 = 70$$
   $$0.2 \times 350 = 70$$
   $$70 = 70$$

2. **17%.**

   Turn the problem into an equation:

What	percent	of	150	is	25.5
n	× 0.01	×	150	=	25.5

   Solve the equation for *n*:

   $$n \times 0.01 \times 150 = 25.5$$
   $$1.5n = 25.5$$
   $$\frac{1.5n}{1.5} = \frac{25.5}{1.5}$$
   $$n = 17$$

3. **8.69.**

   Turn the problem into an equation:

What	is	79	%	of	11
n	=	79	× 0.01	×	11

To find the answer, solve this equation for n:

$$n = 79 \times 0.01 \times 11 = 8.69$$

4.   **$33.\overline{3}$.**

Turn the problem into an equation:

30	%	of	what number	is	10
30	× 0.01	×	n	=	10

Solve for n:

$$30 \times 0.01 \times n = 10$$
$$0.3n = 10$$
$$\frac{0.3n}{0.3} = \frac{10}{0.3}$$
$$n = 33.\overline{3}$$

The answer is the repeating decimal $33.\overline{3}$.

12

WORD PROBLEMS WITH FRACTIONS, DECIMALS, AND PERCENTS

IN THIS CHAPTER

- Adding and subtracting fractions, decimals, and percents in word equations
- Translating the word *of* as multiplication
- Changing percents to decimals in word problems
- Tackling business problems involving percent increase and decrease

In Chapter 5, I show you how to solve word problems (also known as story problems) by setting up word equations that use the Big Four operations (adding, subtracting, multiplying, and dividing). In this chapter, I show you how to extend these skills to solve word problems with fractions, decimals, and percents.

First, I show you how to solve relatively easy problems, in which all you need to do is add or subtract fractions, decimals, or percents. Next, I show you how to solve problems that require you to multiply fractions. Such problems are easy to spot because they almost always contain the word *of*. After that, you discover how to solve percent problems by setting up a word equation and changing the percent to a decimal. Finally, I show you how to handle problems of percent increase and decrease. These problems are often practical money problems in which you figure out information about raises and salaries, costs and discounts, or amounts before and after taxes.

ADDING AND SUBTRACTING PARTS OF THE WHOLE IN WORD PROBLEMS

Certain word problems involving fractions, decimals, and percents are really just problems in adding and subtracting. You may add fractions, decimals, or percents in a variety of real-world settings that rely on weights and measures — such as cooking and carpentry. (In Chapter 14, I discuss these applications in depth.)

To solve these problems, you can use the skills that you pick up in Chapters 9 (for adding and subtracting fractions), 10 (for adding and subtracting decimals), and 11 (for adding and subtracting percents).

Sharing a pizza: Fractions

You may have to add or subtract fractions in problems that involve splitting up part of a whole. For example, consider the following:

Joan ate $\frac{1}{6}$ of a pizza, Tony ate $\frac{1}{4}$ and Sylvia ate $\frac{1}{3}$. What fraction of the pizza was left when they were finished?

In this problem, just jot down the information that's given as word equations:

$$\text{Joan} = \frac{1}{6} \quad \text{Tony} = \frac{1}{4} \quad \text{Sylvia} = \frac{1}{3}$$

These fractions are part of one total pizza. To solve the problem, you need to find out how much all three people ate, so form the following word equation:

$$\text{all three} = \text{Joan} + \text{Tony} + \text{Sylvia}$$

Now you can substitute as follows:

$$\text{all three} = \frac{1}{6} + \frac{1}{4} + \frac{1}{3}$$

Chapter 9 gives you several ways to add these fractions. Here's one way:

$$\text{all three} = \frac{2}{12} + \frac{3}{12} + \frac{4}{12} = \frac{9}{12} = \frac{3}{4}$$

However, the question asks what fraction of the pizza was left after they finished, so you have to subtract that amount from the whole:

$$1 - \frac{3}{4} = \frac{1}{4}$$

Thus, the three people left $\frac{1}{4}$ of a pizza.

Buying by the pound: Decimals

You frequently work with decimals when dealing with money, metric measurements (see Chapter 14), and food sold by the pound. The following problem requires you to add and subtract decimals, which I discuss in Chapter 10. Even though the decimals may look intimidating, this problem is fairly simple to set up:

Antonia bought 4.53 pounds of beef and 3.1 pounds of lamb. Lance bought 5.24 pounds of chicken and 0.7 pounds of pork. Which of them bought more meat, and how much more?

To solve this problem, you first find out how much each person bought:

$$\text{Antonia} = 4.53 + 3.1 = 7.63$$
$$\text{Lance} = 5.24 + 0.7 = 5.94$$

You can already see that Antonia bought more than Lance. To find how much more, subtract:

$$7.63 - 5.94 = 1.69$$

So Antonia bought 1.69 pounds more than Lance.

Splitting the vote: Percents

When percents represent answers in polls, votes in an election, or portions of a budget, the total often has to add up to 100%. In real life, you may see such info organized as a pie chart (which I discuss in Chapter 16). Solving problems about this kind of information often involves nothing more than adding and subtracting percents. Here's an example:

> In a recent mayoral election, five candidates were on the ballot. Faber won 39% of the vote, Gustafson won 31%, Ivanovich won 18%, Dixon won 7%, Obermayer won 3%, and the remaining votes went to write-in candidates. What percentage of voters wrote in their selection?

The candidates were in a single election, so all the votes have to total 100%. The first step here is just to add up the five percentages. Then subtract that value from 100%:

$$39\% + 31\% + 18\% + 7\% + 3\% = 98\%$$
$$100\% - 98\% = 2\%$$

Because 98% of voters voted for one of the five candidates, the remaining 2% wrote in their selections.

Examples

Q. Jeff is 1.13 meters tall and his younger brother, Ryan, is 0.25 meter shorter than he is. Their older sister, Trisha, is 0.4 meter shorter than the combined height of the two boys. How tall is Trisha?

A. **1.61 meters.** Jeff is 1.13 meters tall and Ryan is 0.25 meter less (1.13 – 0.25 = 0.88), so Ryan is 0.88 meter tall. Trisha is 0.4 meter shorter than their combined height (1.13 + 0.88 – 0.4 = 1.61).

Practice Questions

1. Anita pays 25% of her monthly income on rent, 20% on bills, and 14% paying off her student loans. What percentage of her income does this leave her for other things?

2. On a four-day car trip from Chicago to San Diego, Phyllis drove $\frac{3}{10}$ of the total distance the first day, $\frac{1}{3}$ the second day, and $\frac{1}{4}$ the third day. What fraction of the distance was left for her on the fourth day?

3.   Kevin always changes the oil in his car every 2,000 miles. Last month, he changed his oil, and then filled up his tank four times, driving the car 297.8 miles, 317.4 miles, 304.8 miles, and 315.6 miles on these four tanks of gasoline. How many more miles will he have to drive before it's time to change his oil again?

4.   A cake recipe calls for $3\frac{1}{2}$ cups of flour, $1\frac{1}{3}$ cup of white sugar, $\frac{3}{4}$ cup of brown sugar, and 1/8 cup of shaved chocolate. How many cups do these four ingredients total?

Practice Answers

1.   **41%.** Add up the three percentages (25% + 20% + 14% = 59%), and subtract the result from 100% (100% − 59% = 41%).

2. $\frac{7}{60}$. Begin by adding the three fractions:

$$\frac{3}{10} + \frac{1}{3} + \frac{1}{4} = \frac{18}{60} + \frac{20}{60} + \frac{15}{60} = \frac{53}{60}$$

Now, subtract the result from 1:

$$1 - \frac{53}{60} = \frac{60}{60} - \frac{53}{60} = \frac{7}{60}$$

3.   **764.4 miles.** Kevin drove the car a total of 1,235.6 miles (297.8 + 317.4 + 304.8 + 315.6 = 1,235.6). Subtract this amount from 2,000 to find the answer (2,000 − 1,235.6 = 764.4).

4. $5\frac{17}{24}$ **cups.** Add the two mixed numbers and the two fractions by changing to a common denominator:

$$3\frac{1}{2} + 1\frac{1}{3} + \frac{3}{4} + \frac{1}{8} = 3\frac{12}{24} + 1\frac{8}{24} + \frac{18}{24} + \frac{3}{24} = 5\frac{17}{24}$$

PROBLEMS ABOUT MULTIPLYING FRACTIONS

Remember: In word problems, the word *of* almost always means multiplication. So whenever you see the word *of* following a fraction, decimal, or percent, you can usually replace it with a times sign.

When you think about it, *of* means multiplication even when you're not talking about fractions. For example, when you point to an item in a store and say, "I'll take three of those," in a sense you're saying, "I'll take that one multiplied by three."

The following examples give you practice turning word problems that include the word *of* into multiplication problems that you can solve with fraction multiplication.

Renegade grocery shopping: Buying less than they tell you to

When you understand that the word *of* means multiplication, you have a powerful tool for solving word problems. For instance, you can figure out how much you'll spend if you don't buy food in the quantities listed on the signs. Here's an example:

If beef costs $4 a pound, how much does $\frac{5}{8}$ of a pound cost?

Here's what you get if you simply change the *of* to a multiplication sign:

$$\frac{5}{8} \times 1 \text{ pound of beef}$$

So you know how much beef you're buying. However, you want to know the cost. Because the problem tells you that 1 pound = $4, you can replace 1 pound of beef with $4:

$$= \frac{5}{8} \times \$4$$

Now you have an expression you can evaluate. Use the rules of multiplying fractions from Chapter 9 and solve:

$$= \frac{5 \times \$4}{8} = \$\frac{20}{8}$$

This fraction reduces to $\$\frac{5}{2}$. However, the answer looks weird because dollars are usually expressed in decimals, not fractions. So convert this fraction to a decimal using the rules I show you in Chapter 10:

$$\$\frac{5}{2} = \$2.5 = \$2.50$$

At this point, recognize that $2.5 is more commonly written as $2.50, and you have your answer.

Easy as pie: Working out what's left on your plate

Sometimes when you're sharing something such as a pie, not everyone gets to it at the same time. The eager pie-lovers snatch the first slice, not bothering to divide the pie into equal servings, and the people who were slower, more patient, or just not that hungry cut their own portions from what's left over. When someone takes a part of the leftovers, you can do a bit of multiplication to see how much of the whole pie that portion represents.

Consider the following example:

Jerry bought a pie and ate $\frac{1}{5}$ of it. Then his wife, Doreen, ate $\frac{1}{6}$ of what was left. How much of the total pie was left?

To solve this problem, begin by jotting down what the first sentence tells you:

$$\text{Jerry} = \frac{1}{5}$$

Doreen ate part of what was left, so write a word equation that tells you how much of the pie was left after Jerry was finished. He started with a whole pie, so subtract his portion from 1:

$$\text{pie left after Jerry} = 1 - \frac{1}{5} = \frac{4}{5}$$

Next, Doreen ate $\frac{1}{6}$ of this amount. Rewrite the word *of* as multiplication and solve as follows. This answer tells you how much of the whole pie Doreen ate:

$$\text{Doreen} = \frac{1}{6} \times \frac{4}{5} = \frac{4}{30}$$

To make the numbers a little smaller before you go on, notice that you can reduce the fraction:

$$\text{Doreen} = \frac{2}{15}$$

Now you know how much Jerry and Doreen both ate, so you can add these amounts together:

$$\text{Jerry} + \text{Doreen} = \frac{1}{5} + \frac{2}{15}$$

Solve this problem as I show you in Chapter 9:

$$= \frac{3}{15} + \frac{2}{15} = \frac{5}{15}$$

This fraction reduces to $\frac{1}{3}$. Now you know that Jerry and Doreen ate $\frac{1}{3}$ of the pie, but the problem asks you how much is left. So finish up with some subtraction and write the answer:

$$1 - \frac{1}{3} = \frac{2}{3}$$

The amount of pie left over was $\frac{2}{3}$.

Example

Q. A coffee canister holds 10 cups of coffee when filled in the morning. At the morning meeting, employees drink $\frac{2}{3}$ of the coffee in the canister. Then at lunch, another employee drinks $\frac{3}{4}$ of what's left. How many cups of coffee are in the canister after that?

A. $\frac{5}{6}$ **cup.** At the morning meeting, employees drink $\frac{2}{3}$ of the contents of the full canister, so $\frac{1}{3}$ of 10 cups is left:

$$10 \times \frac{1}{3} = \frac{10}{3}$$

At lunch, another employee drinks $\frac{3}{4}$ of what's left, leaving $\frac{1}{4}$ of $\frac{10}{3}$ of a cup, so calculate this amount and reduce:

$$\frac{1}{4} \times \frac{10}{3} = \frac{10}{12} = \frac{5}{6}$$

Q

Practice Questions

1. A recipe calls for $\frac{3}{4}$ of a cup of chocolate chips. How many cups do you need if you triple the recipe?

2. Brynn lives $\frac{5}{8}$ of a mile from school. If she walks to school and back every day for five days, how many miles does she walk?

3. A rectangular field is $1\frac{2}{5}$ of a mile long and $\frac{9}{10}$ of a mile wide. What is the total area of the field in square miles? (*Hint:* To find the area of a rectangle, multiply the length times the width.)

4. A king gives a parcel of land to his four daughters. He gives $\frac{1}{3}$ of the land to his eldest daughter, $\frac{2}{5}$ of what remains to his second daughter, and $\frac{3}{4}$ of what's left to his third daughter. What fraction of the land remains for the king's youngest daughter?

A

Practice Answers

1. $2\frac{1}{4}$ **cups.** Multiply $\frac{3}{4}$ by 3:

$$\frac{3}{4} \times 3 = \frac{9}{4}$$

Change this improper fraction to a mixed number:

$$= 2\frac{1}{4}$$

2. $6\frac{1}{4}$ **miles.** Multiply $\frac{5}{8}$ by 10:

$$\frac{5}{8} \times 10 = \frac{50}{8}$$

Change this improper fraction to a mixed number, then reduce it to lowest terms:

$$= 6\frac{2}{8} = 6\frac{1}{4}$$

3. $1\frac{13}{50}$ **square miles.** To find the area, multiply $1\frac{2}{5}$ by $\frac{9}{10}$. Begin by changing $1\frac{2}{5}$ to an improper fraction, then multiply:

$$1\frac{2}{5} \times \frac{9}{10} = \frac{7}{5} \times \frac{9}{10} = \frac{63}{50}$$

Change this improper fraction to a mixed number:

$$= 1\frac{13}{50}$$

4. $\frac{1}{10}$. The king gives $\frac{1}{3}$ of his land to his eldest daughter, so $\frac{2}{3}$ of the land remains. Of this, the king gives $\frac{2}{5}$ to his second daughter, which leaves $\frac{3}{5}$ of $\frac{2}{3}$:

$$\frac{3}{5} \times \frac{2}{3} = \frac{6}{15} = \frac{2}{5}$$

Of this remaining $\frac{2}{5}$ of the original parcel of land, $\frac{3}{4}$ goes to the third daughter, so $\frac{1}{4}$ of $\frac{2}{5}$ goes to the youngest daughter:

$$\frac{1}{4} \times \frac{2}{5} = \frac{2}{20} = \frac{1}{10}$$

MULTIPLYING DECIMALS AND PERCENTS IN WORD PROBLEMS

In the preceding section, "Problems about Multiplying Fractions," I show you how the word *of* in a fraction word problem usually means multiplication. This idea is also true in word problems involving decimals and percents. The method for solving these two types of problems is similar, so I lump them together in this section.

Tip: You can easily solve word problems involving percents by changing the percents into decimals (see Chapter 11 for details). Here are a few common percents and their decimal equivalents:

25% = 0.25   50% = 0.5   75% = 0.75   99% = 0.99

To the end: Figuring out how much money is left

One common type of problem gives you a starting amount — and a bunch of other information — and then asks you to figure out how much you end up with. Here's an example:

Maria's grandparents gave her $125 for her birthday. She put 40% of the money in the bank, spent 35% of what was left on a pair of shoes, and then spent the rest on a dress. How much did the dress cost?

Start at the beginning, forming a word equation to find out how much money Maria put in the bank:

money in bank = 40% of $125

To solve this word equation, change the percent to a decimal and the word *of* to a multiplication sign; then multiply:

money in bank = 0.4 × $125 = $50

Tip: Pay special attention to whether you're calculating how much of something was used up or how much of something is left over. If you need to work with the portion that remains, you may have to subtract the amount used from the amount you started with.

Because Maria started with $125, she had $75 left to spend:

$$\text{money left to spend}$$
$$= \text{money from grandparents} - \text{money in bank}$$
$$= \$125 - \$50$$
$$= \$75$$

The problem then says that she spent 35% of this amount on a pair of shoes. Again, change the percent to a decimal and the word *of* to a multiplication sign:

$$\text{shoes} = 35\% \text{ of } \$75 = 0.35 \times \$75 = \$26.25$$

She spent the rest of the money on a dress, so

$$\text{dress} = \$75 - \$26.25 = \$48.75$$

Therefore, Maria spent $48.75 on the dress.

Finding out how much you started with

Some problems give you the amount that you end up with and ask you to find out how much you started with. In general, these problems are harder because you're not used to thinking backward. Here's an example, and it's kind of a tough one, so fasten your seat belt:

> Maria received some birthday money from her aunt. She put her usual 40% in the bank and spent 75% of the rest on a purse. When she was done, she had $12 left to spend on dinner. How much did her aunt give her?

This problem is similar to the one in the preceding section, but you need to start at the end and work backward. Notice that the only dollar amount in the problem comes after the two percent amounts. The problem tells you that she ends up with $12 after two transactions — putting money in the bank and buying a purse — and asks you to find out how much she started with.

To solve this problem, set up two word equations to describe the two transactions:

$$\text{money from aunt} - \text{money for bank} = \text{money after bank}$$
$$\text{money after bank} - \text{money for purse} = \$12$$

Notice what these two word equations are saying. The first tells you that Maria took the money from her aunt, subtracted some money to put in the bank, and left the bank with a new amount of money, which I'm calling *money after bank*. The second word equation starts where the first leaves off. It tells you that Maria took the money left over from the bank, subtracted some money for a purse, and ended up with $12.

This second equation already has an amount of money filled in, so start here. To solve this problem, realize that Maria spent 75% of her money *at that time* on the purse — that is, 75% of the money she still had after the bank:

$$\text{money after bank} - 75\% \text{ of money after bank} = \$12$$

I'm going to make one small change to this equation so you can see what it's really saying:

100% of money after bank – 75% of money after bank = $12

Adding *100% of* doesn't change the equation because it really just means you're multiplying by 1. In fact, you can slip these two words in anywhere without changing what you mean, though you may sound ridiculous saying "Last night, I drove 100% of my car home from work, walked 100% of my dog, then took 100% of my wife to see 100% of a movie."

In this particular case, however, these words help you to make a connection because 100% – 75% = 25%; here's an even better way to write this equation:

25% of money after bank = $12

Before moving on, make sure you understand the steps that have brought you here.

You know now that 25% of money after bank is $12, so the total amount of money after bank is 4 times this amount — that is, $48. Therefore, you can plug this number into the first equation:

money from aunt – money for bank = $48

Now you can use the same type of thinking to solve this equation (and it goes a lot more quickly this time!). First, Maria placed 40% of the money from her aunt in the bank:

money from aunt – 40% of money from aunt = $48

Again, rewrite this equation to make what it's saying clearer:

100% of money from aunt – 40% of money from aunt = $48

Now, because 100% – 40% = 60%, rewrite it again:

60% of money from aunt = $48

Thus, $0.6 \times$ money from aunt = $48. Divide both sides of this equation by 0.6: money from aunt = $48 ÷ 0.6 = $80

So Maria's aunt gave her $80 for her birthday.

Example

Q. Kylie brings home a monthly paycheck totaling $3,400. She has a savings plan that automatically places 8% of this amount in a retirement account. Then, her monthly mortgage and other regular bills consume 65% of what's left. How much does this leave Kylie to spend on other things?

A. **$1,094.80.** Begin by finding 8% of $3,400 ($3,400 × 0.08 = $272) and subtracting this from $3,400 ($3,400 – $272 = $3,128). Next, find 65% of $3,128 ($3,128 × 0.65 = $2,033.20) and subtracting this from $3,128 ($3,128 – $2,033.20 = $1,094.80).

Practice Questions

1. If you bought a stock for $1,200 and sold it for $1,428, what percent profit did you make?

2. Jacob bought a television for $752.50, using a special coupon that gave him 14% off the retail price. What would he have paid without the coupon?

3. Antonietta received a $12,500 inheritance. She invested this in a mutual fund for one year and earned 7%, then rolled the money over into a CD and earned an additional 4%. How much money did she have after these two investments?

4. Jack won some money playing poker with his friends. He spent 70% of his winnings taking his girlfriend out for dinner, and then spent 40% of the remaining money on tickets to a basketball game. This left him with exactly $25.92. How much did he win originally at poker?

Practice Answers

1. **19%.** The profit on the investment is $228, so divide 228 by 1,200 and convert the answer from a decimal to a percent:

$$228 \div 1,200 = 0.19 = 19\%$$

2. **$875.** The cost of the television with a 14% discount is $752.50, so 86% of the cost of the television equals $752.50:

$$86\% \text{ of } n \text{ is } \$752.50$$

Change this to an equation:

$$0.86n = 752.5$$
$$n = \frac{752.5}{0.86}$$
$$n = 875$$

3. **$13,910.** Antonietta earned 7% on her first investment ($12,500 × 0.07 = $875), so she had $13,375 after the first year ($12,500 + $875 = $13,375). She earned 4% on this investment (13,375 × 0.04 = $535), so she ended up with $13,910 ($13,375 + $535 = $13,910).

4. **$144.** To begin, find out how much money Jack had before he bought the basketball tickets. He spent 40% of some this amount on the tickets and ended up with $25.92, so 60% of this amount equals $25.92:

$$0.6n = 25.92$$
$$n = \frac{25.92}{0.6}$$
$$n = 43.20$$

Thus, Jack had \$43.20 before buying the basketball tickets, but after buying dinner. He spent 70% of his poker winnings on this dinner, so \$43.20 represents the remaining 30% of his poker winnings:

$$0.3p = 43.20$$
$$p = \frac{43.20}{0.3}$$
$$n = 144$$

Therefore, Jack won \$144 at poker.

HANDLING PERCENT INCREASES AND DECREASES IN WORD PROBLEMS

Word problems that involve increasing or decreasing by a percentage add a final spin to percent problems. Typical percent-increase problems involve calculating the amount of a salary plus a raise, the cost of merchandise plus tax, or an amount of money plus interest or dividend. Typical percent decrease problems involve the amount of a salary minus taxes or the cost of merchandise minus a discount.

To tell you the truth, you may have already solved problems of this kind earlier in "Multiplying Decimals and Percents in Word Problems." But people often get thrown by the language of these problems — which, by the way, is the language of business — so I want to give you some practice in solving them.

Raking in the dough: Finding salary increases

A little street smarts should tell you that the words *salary increase* or *raise* mean more money, so get ready to do some addition. Here's an example:

Alison's salary was \$40,000 last year, and at the end of the year, she received a 5% raise. What will she earn this year?

To solve this problem, first realize that Alison got a raise. So whatever she makes this year, it will be more than she made last year. The key to setting up this type of problem is to think of percent increase as "100% of last year's salary plus 5% of last year's salary." Here's the word equation:

this year's salary = 100% of last year's salary + 5% of last year's salary

Now you can just add the percentages (see the nearby sidebar for why this works):

this year's salary = 105% of last year's salary

Change the percent to a decimal and the word *of* to a multiplication sign; then fill in the amount of last year's salary:

$$\text{this year's salary} = 1.05 \times \$40,000$$

Now you're ready to multiply:

$$\text{this year's salary} = \$42,000$$

So Alison's new salary is $42,000.

Earning interest on top of interest

The word *interest* means more money. When you receive interest from the bank, you get more money. And when you pay interest on a loan, you pay more money. Sometimes people earn interest on the interest they earned earlier, which makes the dollar amounts grow even faster. Here's an example:

Bethany placed $9,500 in a one-year CD that paid 4% interest. The next year, she rolled this over into a bond that paid 6% per year. How much did Bethany earn on her investment in those two years?

This problem involves interest, so it's another problem in percent increase — only this time, you have to deal with two transactions. Take them one at a time.

The first transaction is a percent increase of 4% on $9,500. The following word equation makes sense:

$$\text{money after first year} = 100\% \text{ of initial deposit} + 4\% \text{ of initial deposit}$$
$$= 104\% \text{ of initial deposit}$$

Now, substitute $9,500 for the initial deposit and calculate:

$$= 104\% \text{ of } \$9,500$$
$$= 1.04 \times \$9,500$$
$$= \$9,880$$

At this point, you're ready for the second transaction. This is a percent increase of 6% on $9,880:

$$\text{final amount} = 106\% \text{ of } \$9,880$$
$$= 1.06 \times \$9,880$$
$$= \$10,472.80$$

Then subtract the initial deposit from the final amount:

$$\text{earnings} = \text{final amount} - \text{initial deposit}$$
$$= \$10,472.80 - \$9,500$$
$$= \$972.80$$

So Bethany earned $972.80 on her investment.

Getting a deal: Calculating discounts

When you hear the words *discount* or *sale price,* think of subtraction. Here's an example:

> Greg has his eye on a television with a listed price of $2,100. The salesman offers him a 30% discount if he buys it today. What will the television cost with the discount?

In this problem, you need to realize that the discount lowers the price of the television, so you have to subtract:

$$\text{Sale price} = 100\% \text{ of regular price} - 30\% \text{ of regular price}$$
$$= 70\% \text{ of regular price}$$
$$= 0.7 \times \$2,100 = \$1,470$$

Thus, the television costs $1,470 with the discount.

Examples

Q. After lunch at a restaurant, Beth wants to leave her server a tip of at least 18% on a meal that cost $16.50. What is the minimum amount that Beth should pay, including the tip?

A. **$19.47.** Calculate a percent increase of 18% on $16.50:

$$\$16.50 \times 1.18 = \$19.47$$

Practice Questions

1. The Johnsons bought their home for $220,000, and sold it ten years later for an increase of 37%. What was the selling price of the house?

2. Finding that a car he wanted to buy had unusually high mileage, Myron negotiated a 7% discount on the Bluebook price of $11,000. What price did he pay for the car?

3. Andrea invested a total of $12,000 this year. She invested $7,500 on a tech stock, which gained 17%. But she also invested $4,500 on a friend's business, losing 24% of her investment. How much money did Andrea have after these investments?

4. Fred bought a lawnmower that normally sold for $640, using a coupon that saved him 15%. But he still had to pay 7.5% sales tax on the discounted cost of his purchase. How much did he end up paying for the lawnmower?

A

Practice Answers

1. **$301,400.** Find a percent increase of 37% on $220,000:

$$\$220,000 \times 1.37 = \$301,400$$

2. **$10,230.** Calculate a percent decrease of 7% on $11,000:

$$\$11,000 \times 0.93 = \$10,230$$

3. **$12,195.** First, find a 17% percent increase on $7,500:

$$\$7,500 \times 1.17 = \$8,775$$

Next, find a 24% decrease on $4,500:

$$\$4,500 \times 0.76 = \$3,420$$

To complete the problem, add the two numbers together:

$$\$8,775 + \$3,420 = \$12,195$$

4. **$584.80** To begin, find the cost of the lawn mower with a percent decrease of 15%:

$$\$640 \times 0.75 = \$544$$

Next, find a percent increase of 7.5% on this amount:

$$\$544 \times 1.085 = \$584.80$$

PART IV

picturing and measuring

MEASURES, STATS, AND SETS

For find out how to use probability to calculate the odds in dice games, go to www.dummies.com/extras/ucanbasicmathprealgebra.

IN THIS PART . . .

- Represent very large and very small numbers with scientific notation.

- Weigh and measure with both the English and metric systems.

- Understand basic geometry, including points, lines, and angles, plus basic shapes and solids.

- Answer real-world questions with statistics and probability.

- Get familiar with some basic set theory, including union and intersection.

13

A PERFECT TEN: CONDENSING NUMBERS WITH SCIENTIFIC NOTATION

IN THIS CHAPTER

- Knowing how to express powers of ten in exponential form
- Appreciating how and why scientific notation works
- Understanding order of magnitude
- Multiplying numbers in scientific notation

Scientists often work with very small or very large measurements — the distance to the next galaxy, the size of an atom, the mass of the Earth, or the number of bacteria cells growing in last week's leftover Chinese takeout. To save on time and space — and to make calculations easier — people developed a sort of shorthand called *scientific notation.*

Scientific notation uses a sequence of numbers known as the powers of ten, which I introduce in Chapter 2:

$$1 \quad 10 \quad 100 \quad 1{,}000 \quad 10{,}000 \quad 100{,}000 \quad 1{,}000{,}000 \quad 10{,}000{,}000\ldots$$

Each number in the sequence is 10 times more than the preceding number.

$$10^0 \quad 10^1 \quad 10^2 \quad 10^3 \quad 10^4 \quad 10^5 \quad 10^6 \quad 10^7\ldots$$

Powers of ten are easy to work with, especially when you're multiplying and dividing, because you can just add or drop zeros or move the decimal point.

Scientific notation is a handy system for writing very large and very small numbers without writing a bunch of 0s. It uses both decimals and exponents (so if you need a little brushing up on decimals, flip to Chapter 10). In this chapter, I introduce you to this powerful method of writing numbers. I also explain the order of magnitude of a number. Finally, I show you how to multiply numbers written in scientific notation.

FIRST THINGS FIRST: USING POWERS OF TEN AS EXPONENTS

Scientific notation uses powers of ten expressed as exponents, so you need a little background before you can jump in. In this section, I round out your knowledge of exponents, which I first introduce in Chapter 3.

Counting zeros and writing exponents

Numbers starting with a 1 and followed by only 0s (such 10, 100, 1,000, 10,000, and so forth) are called powers of ten, and they're easy to represent as exponents. Powers of ten are the result of multiplying 10 times itself any number of times.

Tip: To represent a number that's a power of 10 as an exponential number, count the zeros and raise 10 to that exponent. For example, 1,000 has three zeros, so $1,000 = 10^3$ (10^3 means to take 10 times itself three times, so it equals $10 \times 10 \times 10$). Table 13-1 shows a list of some powers of ten.

Table 13-1   Powers of Ten Expressed as Exponents

Standard Notation	Exponential Notation
1	10^0
10	10^1
100	10^2
1,000	10^3
10,000	10^4
100,000	10^5
1,000,000	10^6

When you know this trick, representing a lot of large numbers as powers of ten is easy — just count the 0s! For example, the number 1 trillion — 1,000,000,000,000 — is a 1 with twelve 0s after it, so

$$1,000,000,000,000 = 10^{12}$$

This trick may not seem like a big deal, but the higher the numbers get, the more space you save by using exponents. For example, a really big number is a googol, which is 1 followed by a hundred 0s. You can write this:

10,000,000,000,000,000,000,000,000,000,000,000,000,000,000,000,000,00

0,000,000,000,000,000,000,000,000,000,000,000,000,000,000

As you can see, a number of this size is practically unmanageable. You can save yourself some trouble and write 10^{100}.

Remember: A 10 raised to a negative number is also a power of ten.

You can also represent decimals using negative exponents. For example,

$$10^{-1} = 0.1 \quad 10^{-2} = 0.01 \quad 10^{-3} = 0.001 \quad 10^{-4} = 0.0001$$

Although the idea of negative exponents may seem strange, it makes sense when you think about it alongside what you know about positive exponents. For example, to find the value of 10^7, start with 1 and make it larger by moving the decimal point seven spaces to the right:

$$10^7 = 10,000,000$$

Similarly, to find the value of 10^{-7}, start with 1 and make it smaller by moving the decimal point seven spaces to the left:

$$10^{-7} = 0.0000001$$

Warning: Negative powers of 10 always have one fewer 0 between the 1 and the decimal point than the power indicates. In this example, notice that 10^{-7} has six 0s between them.

As with very large numbers, using exponents to represent very small decimals makes practical sense. For example,

$$10^{-23} = 0.00000000000000000000001$$

As you can see, this decimal is easy to work with in its exponential form but almost impossible to read otherwise.

Examples

Q. Write 10^6 in standard notation.

A. **1,000,000.** The exponent is 6, so the standard notation is a 1 with six 0s after it.

Q. Write 100,000 in exponential notation.

A. **10^5.** The number 100,000 has five 0s, so the exponential notation has 5 in the exponent.

Q. Write 10^{-5} in standard notation.

A. **0.00001.** The exponent is -5, so the standard notation is a decimal with five 0s (including the leading 0) followed by a 1.

Q. Write 0.0000001 in exponential notation.

A. **10^{-7}.** The decimal has seven 0s (including the leading 0), so the exponential notation has -7 in the exponent.

Q **Practice Questions**

1. Write each of the following powers of ten in standard notation:

   a. 10^4

   b. 10^7

   c. 10^{14}

   d. 10^{22}

2. Write each of the following powers of ten in exponential notation:

   a. 1,000,000,000

   b. 1,000,000,000,000

   c. 10,000,000,000,000,000

   d. 100,000,000,000,000,000,000,000,000,000,000

3. Write each of the following powers of ten in standard notation:

   a. 10^{-1}

   b. 10^{-5}

   c. 10^{-11}

   d. 10^{-16}

4. Write each of the following powers of ten in exponential notation:

   a. 0.01

   b. 0.000001

   c. 0.000000000001

   d. 0.00000000000000000001

A **Practice Answers**

1. In each case, write the digit 1 followed by the number of 0s indicated by the exponent:

   a. **10,000**

   b. **10,000,000**

   c. **100,000,000,000,000**

   d. **10,000,000,000,000,000,000,000**

2. In each case, count the number of 0s; then write a power of ten with this number as the exponent.

    a. 10^9

    b. 10^{12}

    c. 10^{16}

    d. 10^{32}

3. Write a decimal beginning with all 0s and ending in 1. The exponent indicates the number of 0s in this decimal (including the leading 0):

    a. 0.1

    b. 0.00001

    c. 0.00000000001

    d. 0.0000000000000001

4. In each case, count the number of 0s (including the leading 0); then write a power of ten using this number *negated (with a minus sign) as the exponent*:

    a. 10^{-2}

    b. 10^{-6}

    c. 10^{-12}

    d. 10^{-18}

EXPONENTIAL ARITHMETIC: MULTIPLYING AND DIVIDING POWERS OF TEN

Tip: Multiplying and dividing powers of ten in exponential notation is a snap because you don't have to do any multiplying or dividing at all — it's nothing more than simple addition and subtraction:

- **Multiplication:** To multiply two powers of ten in exponential notation, find the sum of the numbers' exponents; then write a power of ten using that sum as the exponent.

- **Division:** To divide one power of ten by another, subtract the second exponent from the first; then write a power of ten using this resulting sum as the exponent.

This rule works equally well when one or both exponents are negative — just use the rules for adding negative numbers, which I discuss in Chapter 3.

Examples

Q. Multiply 10^7 by 10^4.

A. $\mathbf{10^{11}}$. Add the exponents $7 + 4 = 11$, and use this as the exponent of your answer:

$$10^7 \times 10^4 = 10^{7+4} = 10^{11}$$

Q. Find $10^9 \div 10^6$.

A. $\mathbf{10^3}$. For division, you subtract. Subtract the exponents $9 - 6 = 3$ and use this as the exponent of your answer:

$$10^9 \div 10^6 = 10^{(9-6)} = 10^3$$

Practice Questions

1. Multiply each of the following powers of ten:

    a. $10^9 \times 10^2$

    b. $10^5 \times 10^5$

    c. $10^{13} \times 10^{-16}$

    d. $10^{100} \times 10^{21}$

    e. $10^{-15} \times 10^0$

    f. $10^{-10} \times 10^{-10}$

2. Divide each of the following powers of ten:

    a. $10^6 \div 10^4$

    b. $10^{12} \div 10^1$

    c. $10^{-7} \div 10^{-7}$

    d. $10^{18} \div 10^0$

    e. $10^{100} \div 10^{-19}$

    f. $10^{-50} \div 10^{50}$

Practice Answers

1. Add the exponents and use this sum as the exponent of the answer.

    a. $10^9 \times 10^2 = 10^{9+2} = \mathbf{10^{11}}$

    b. $10^5 \times 10^5 = 10^{5+5} = \mathbf{10^{10}}$

    c. $10^{13} \times 10^{-16} = 10^{13+-16} = \mathbf{10^{-3}}$

    d. $10^{100} \times 10^{21} = 10^{100+21} = \mathbf{10^{121}}$

    e. $10^{-15} \times 10^0 = 10^{-15+0} = \mathbf{10^{-15}}$

    f. $10^{-10} \times 10^{-10} = 10^{-10+-10} = \mathbf{10^{-20}}$

2. In each case, subtract the first exponent minus the second and use this result as the exponent of the answer.

    a. $10^6 \div 10^4 = 10^{6-4} = \mathbf{10^2}$

    b. $10^{12} \div 10^1 = 10^{12-1} = \mathbf{10^{11}}$

    c. $10^{-7} \div 10^{-7} = 10^{-7-(-7)} = 10^{-7+7} = \mathbf{10^0}$

    d. $10^{18} \div 10^0 = 10^{18-0} = \mathbf{10^{18}}$

    e. $10^{100} \div 10^{-19} = 10^{100-(-19)} = 10^{100+19} = \mathbf{10^{119}}$

    f. $10^{-50} \div 10^{50} = 10^{-50-50} = 10^{-100}$

WORKING WITH SCIENTIFIC NOTATION

Scientific notation is a system for writing very large and very small numbers that makes them easier to work with. Every number can be written in scientific notation as the product of two numbers (two numbers multiplied together):

- A decimal greater than or equal to 1 and less than 10 (see Chapter 10 for more on decimals)

- A power of ten written as an exponent (see the preceding section)

Writing in scientific notation

Remember: Here's how to write any number in scientific notation:

1. **Write the number as a decimal (if it isn't one already).**

   Suppose you want to change the number 360,000,000 to scientific notation. First, write it as a decimal:

   $$360,000,000.0$$

2. **Move the decimal point just enough places to change this number to a new number that's between 1 and 10.**

   Move the decimal point to the right or left so that only one nonzero digit comes before the decimal point. Drop any leading or trailing zeros as necessary.

   Using 360,000,000.0, only the 3 should come before the decimal point. So move the decimal point eight places to the left, drop the trailing zeros, and get 3.6:

$$360,000,000.0 \text{ becomes } 3.6.$$

3. **Multiply the new number by 10 raised to the number of places you moved the decimal point in Step 2.**

   You moved the decimal point eight places, so multiply the new number by 10^8:

$$3.6 \times 10^8$$

4. **If you moved the decimal point to the right in Step 2, put a minus sign on the exponent.**

   You moved the decimal point to the left, so you don't have to take any action here. Thus, 360,000,000 in scientific notation is 3.6×10^8.

Changing a decimal to scientific notation basically follows the same process. For example, suppose you want to change the number 0.00006113 to scientific notation:

1. **Write 0.00006113 as a decimal (this step's easy because it's already a decimal):**

$$0.00006113$$

2. **To change 0.00006113 to a new number between 1 and 10, move the decimal point five places to the right and drop the leading zeros:**

$$6.113$$

3. **Because you moved the decimal point five places to the right, multiply the new number by 10^{-5}:**

$$6.113 \times 10^{-5}$$

So 0.00006113 in scientific notation is 6.113×10^{-5}.

When you get used to writing numbers in scientific notation, you can do it all in one step. Here are a few examples:

$$17,400 = 1.74 \times 10^4$$
$$212.04 = 2.1204 \times 10^2$$
$$0.003002 = 3.002 \times 10^{-3}$$

Examples

Q. Change the number 70,000 to scientific notation.

A. 7.0×10^4. First, write the number as a decimal:

$$70,000.0$$

Move the decimal point just enough places to change this decimal to a new decimal that's between 1 and 10. In this case, move the decimal four places to the left. You can drop all but one trailing zero:

$$7.0$$

You moved the decimal point four places, so multiply the new number by 10^4:

$$7.0 \times 10^4$$

Because you moved the decimal point to the left (you started with a big number), the exponent is a positive number, so you're done.

Q. Change the decimal 0.000000439 to scientific notation.

A. 4.39×10^{-7}. You're starting with a decimal, so Step 1 — writing the number as a decimal — is already taken care of:

$$0.000000439$$

To change 0.000000439 to a decimal that's between 1 and 10, move the decimal point seven places to the right and drop the leading zeros:

$$4.39$$

Because you moved seven places to the right, multiply the new number by 10^{-7}:

$$4.39 \times 10^{-7}$$

Practice Questions

1. Change 2,591 to scientific notation.

2. Write the decimal 0.087 in scientific notation.

3. Write 1.00000783 in scientific notation.

4. Convert 20,002.00002 to scientific notation.

Practice Answers

1. **2.591×10^3.** Write 2,591 as a decimal:

$$2,591.0$$

   To change 2,591.0 to a decimal between 1 and 10, move the decimal point three places to the left and drop the trailing zero:

$$2.591$$

   Because you moved the decimal point three places, multiply the new decimal by 10^3:

$$2.591 \times 10^3$$

   You moved the decimal point to the left, so the exponent stays positive. The answer is 2.591×10^3.

2. **8.7×10^{-2}.** To change 0.087 to a decimal between 1 and 10, move the decimal point two places to the right and drop the leading zero:

$$8.7$$

   Because you moved the decimal point two places to the right, multiply the new decimal by 10^{-2}:

$$8.7 \times 10^{-2}$$

3. **1.00000783.** The decimal 1.00000783 is already between 1 and 10, so no change is needed.

4. **2.000200002×10^4.** The number 20,002.00002 is already a decimal. To change it to a decimal between 1 and 10, move the decimal point four places to the left:

$$2.000200002$$

   Because you moved the decimal point four places, multiply the new decimal by 10^4:

$$2.000200002 \times 10^4$$

   You moved the decimal point to the left, so the answer is 2.000200002×10^4.

Understanding order of magnitude

A good question to ask is why scientific notation always uses a decimal between 1 and 10. The answer has to do with order of magnitude. Order of magnitude is a simple way to keep track of roughly how large a number is so you can compare numbers more easily. The order of magnitude of a number is its exponent in scientific notation. For example,

$$703 = 7.03 \times 10^2 \text{ (order of magnitude is 2)}$$

$$600,000 = 6 \times 10^5 \text{ (order of magnitude is 5)}$$

$$0.00095 = 9.5 \times 10^{-4} \text{ (order of magnitude is } -4)$$

Every number starting with 10 but less than 100 has an order of magnitude of 1. Every number starting with 100 but less than 1,000 has an order of magnitude of 2.

Examples

Q. What is the order of magnitude of 5.6×10^7?

A. 7. The number is expressed in scientific notation and the exponent is 7, so its magnitude is 7.

Q. What is the order of magnitude of 893,441?

A. 5. The number is expressed in standard notation, so change it to scientific notation:

$$893,441 = 8.93441 \times 10^5$$

Therefore, its order of magnitude is 5.

Practice Questions

1. What is the order of magnitude of each of the following numbers?

   a. 8×10^4

   b. 6.02×10^{23}

   c. 7.77×10^{-8}

   d. 9×10^1

2. What is the order of magnitude of each of the following numbers?

   a. 0.8

   b. 538

   c. 6,000,000,000

   d. 0.000004321

Practice Answers

1. The exponent of a number that's expressed in scientific notation is the magnitude of that number.

   a. 4

   b. 23

   c. –8

   d. 1

2. To find the magnitude of a number that's expressed in standard notation, convert that number to scientific notation.

   a. -1. $0.8 = 8 \times 10^{-1}$.

   b. 2. $538 = 5.38 \times 10^2$.

   c. 9. $6,000,000,000 = 6 \times 10^9$.

   d. -6. $0.000004321 = 4.321 \times 10^{-6}$.

Multiplying and dividing with scientific notation

Multiplying numbers that are in scientific notation is fairly simple because multiplying powers of ten is easy, as you see earlier in this chapter in "Adding exponents to multiply." Here's how to multiply two numbers that are in scientific notation:

1. **Multiply the two decimal parts of the numbers.**

   Suppose you want to multiply the following:

   $$(6.02 \times 10^{23})(9 \times 10^{-28})$$

   Multiplication is commutative (see Chapter 3), so you can change the order of the numbers without changing the result. And because of the associative property, you can also change how you group the numbers. Therefore, you can rewrite this problem as

   $$(6.02 \times 9)(10^{23} \times 10^{-28})$$

   Multiply what's in the first set of parentheses — 6.02×9 — to find the decimal part of the solution:

   $$6.02 \times 9 = 54.18$$

2. **Multiply the two exponential parts by adding their exponents.**

   Now multiply $10^{23} \times 10^{-28}$:

   $$10^{23} \times 10^{-28} = 10^{-5}$$

3. **Write the answer as the product of the numbers you found in Steps 1 and 2.**

   $$54.18 \times 10^{-5}$$

4. **If the decimal part of the solution is 10 or greater, move the decimal point one place to the left and add 1 to the exponent.**

   $$5.418 \times 10^{-4}$$

This method works even when one or both of the exponents are negative numbers. For example, if you follow the preceding series of steps, you find that $(6.02 \times 10^{23})(9 \times 10^{-28}) = 5.418 \times 10^{-4}$.

MULTIPLYING AND DIVIDING WITH SCIENTIFIC NOTATION

Here's how to divide two numbers in scientific notation:

1. **Divide the decimal part of the first number by the decimal part of the second number to find the decimal part of the answer.**

2. **To find the power of ten in the answer, subtract the exponent on the second power of ten from the exponent on the first.**

   You're really just dividing the first power of ten by the second.

3. **If the decimal part of the result is less than 1, adjust the result by moving the decimal point one place to the right and subtracting 1 from the exponent.**

Examples

Q. Multiply 2.0×10^3 by 4.1×10^4.

A. **8.2×10^7.** Multiply the two decimal parts:

$$2.0 \times 4.1 = 8.2$$

Then multiply the powers of ten by adding the exponents:

$$10^3 \times 10^4 = 10^{3+4} = 10^7$$

In this case, no adjustment is necessary because the resulting decimal part is less than 10.

Q. Divide 3.4×10^4 by 2.0×10^9.

A. **1.7×10^{-5}.** Divide the first decimal part by the second:

$$3.4 \div 2.0 = 1.7$$

Then divide the first power of ten by the second by subtracting the exponents:

$$10^4 \div 10^9 = 10^{4-9} = 10^{-5}$$

In this case, no adjustment is necessary because the resulting decimal part isn't less than 1.

Practice Questions

1. Multiply 1.5×10^7 by 6.0×10^5.

2. Divide 6.6×10^8 by 1.1×10^3.

Practice Answers

1. **9.0×10^{12}.** Multiply the two decimal parts:

   $$1.5 \times 6.0 = 9.0$$

   Multiply the two powers of ten:

   $$10^7 \times 10^5 = 10^{7+5} = 10^{12}$$

   In this case, no adjustment is necessary because the decimal is less than 10.

2. **6.0×10^5.** Divide the first decimal part by the second:

   $$6.6 \div 1.1 = 6.0$$

   Divide the first power of ten by the second:

   $$10^8 \div 10^3 = 10^{8-3} = 10^5$$

   In this case, no adjustment is necessary because the decimal is greater than 1.

14

HOW MUCH HAVE YOU GOT? WEIGHTS AND MEASURES

IN THIS CHAPTER

- Using units for non-discrete measurement

- Discovering differences between the English and metric systems

- Estimating and calculating English and metric system conversions

In this chapter, I introduce you to *units,* which are items that can be counted, such as apples, coins, or hats. Apples, coins, and hats are easy to count because they're *discrete* — that is, you can easily see where one ends and the next one begins. But not everything is so easy. For example, how do you count water — by the drop? Even if you tried, exactly how big is a drop?

Units of measurement come in handy at this point. A *unit of measurement* allows you to count something that isn't discrete: an amount of a liquid or solid, the distance from one place to another, a length of time, the speed at which you're traveling, or the temperature of the air.

In this chapter, I discuss two important systems of measurement: English and metric. You're probably familiar with the English system already, and you may know more than you think about the metric system. Each of these measurement systems provides a different way to measure distance, volume, weight (or mass), time, and speed. Next, I show you how to estimate metric amounts in English units. Finally, I show how to convert from English units to metric and vice versa.

UNDERSTANDING UNITS

Anything that can be counted is a *unit.* That category is a pretty large one because almost anything that you can name can be counted. For now, just understand that all units can be counted, which means that you can apply the Big Four operations to units.

Adding and subtracting units

Adding and subtracting units isn't very different from adding and subtracting numbers. Just remember that you can add or subtract only when the units are the same. For example,

$$3 \text{ chairs} + 2 \text{ chairs} = 5 \text{ chairs}$$
$$4 \text{ oranges} - 1 \text{ orange} = 3 \text{ oranges}$$

What happens when you try to add or subtract different units? Here's an example:

$$3 \text{ chairs} + 2 \text{ tables} = ?$$

The only way you can complete this addition is to make the units the same:

$$3 \text{ pieces of furniture} + 2 \text{ pieces of furniture} = 5 \text{ pieces of furniture}$$

Multiplying and dividing units

You can always multiply and divide units by a *number*. For example, suppose you have four chairs but find that you need twice as many for a party. Here's how you represent this idea in math:

$$4 \text{ chairs} \times 2 = 8 \text{ chairs}$$

Similarly, suppose you have 20 cherries and want to split them among four people. Here's how you represent this idea:

$$20 \text{ cherries} \div 4 = 5 \text{ cherries}$$

But you have to be careful when multiplying or dividing units by units. For example:

$$2 \text{ apples} \times 3 \text{ apples} = ? \quad \text{WRONG!}$$
$$12 \text{ hats} \div 6 \text{ hats} = ? \quad \text{WRONG!}$$

Neither of these equations makes any sense. In these cases, multiplying or dividing by units is meaningless.

In many cases, however, multiplying and dividing units is okay. For example, multiplying *units of length* (such as inches, miles, or meters) results in *square units*. For example,

$$3 \text{ inches} \times 3 \text{ inches} = 9 \text{ square inches}$$
$$10 \text{ miles} \times 5 \text{ miles} = 50 \text{ square miles}$$
$$100 \text{ meters} \times 200 \text{ meters} = 20,000 \text{ square meters}$$

You find out more about units of length later in this chapter. Similarly, here are some examples of when dividing units makes sense:

$$12 \text{ slices of pizza} \div 4 \text{ people} = 3 \text{ slices of pizza/person}$$
$$140 \text{ miles} \div 2 \text{ hours} = 70 \text{ miles/hour}$$

In these cases, you read the fraction slash (/) as *per:* slices of pizza *per* person or miles *per* hour. You find out more about multiplying and dividing by units later in this chapter, when I show you how to convert from one unit of measurement to another.

EXAMINING DIFFERENCES BETWEEN THE ENGLISH AND METRIC SYSTEMS

The two most common measurement systems today are the *English system* and the *metric system*.

Most Americans learn the units of the English system — for example, pounds and ounces, feet and inches, and so forth — and use them every day. Unfortunately, the English system is awkward for use with math. English units such as inches and fluid ounces are often measured in fractions, which (as you may know from Chapters 8 and 9) can be difficult to work with.

The *metric system* was invented to simplify the application of math to measurement. Metric units are based on the number 10, which makes them much easier to work with. Parts of units are expressed as decimals, which (as Chapter 10 shows you) are much friendlier than fractions.

Yet despite these advantages, the metric system has been slow to catch on in the U.S. Many Americans feel comfortable with English units and are reluctant to part with them. For example, if I ask you to carry a 20-lb. bag for one-fourth of a mile, you know what to expect. However, if I ask you to carry a bag weighing 10 kilograms half a kilometer, you may not be sure.

In this section, I show you the basic units of measurement for both the English and metric systems.

If you want an example of the importance of converting carefully, you may want to look to NASA — they kind of lost a Mars orbiter in the late 1990s because an engineering team used English units and NASA used metric to navigate!

Looking at the English system

The *English system of measurement* is most commonly used in the United States (but, ironically, not in England). Although you're probably familiar with most of the English units of measurement, in the following list, I make sure you know the most important ones. I also show you some equivalent values that can help you do conversions from one type of unit to another.

- **Units of distance:** Distance — also called *length* — is measured in inches (in.), feet (ft.), yards (yd.), and miles (mi.):

  12 inches = 1 foot

  3 feet = 1 yard

  5,280 feet = 1 mile

- **Units of fluid volume:** Fluid volume (also called *capacity*) is the amount of space occupied by a liquid, such as water, milk, or wine. I discuss volume when I talk about geometry in Chapter 15. Volume is measured in fluid ounces (fl. oz.), cups (c.), pints (pt.), quarts (qt.), and gallons (gal.):

$$8 \text{ fluid ounces} = 1 \text{ cup}$$
$$2 \text{ cups} = 1 \text{ pint}$$
$$2 \text{ pints} = 1 \text{ quart}$$
$$4 \text{ quarts} = 1 \text{ gallon}$$

Units of fluid volume are typically used for measuring the volume of things that can be poured. The volume of solid objects is more commonly measured in cubic units of distance, such as cubic inches and cubic feet.

- **Units of weight:** Weight is the measurement of how strongly gravity pulls an object toward Earth. Weight is measured in ounces (oz.), pounds (lb.), and tons.

$$16 \text{ ounces} = 1 \text{ pound}$$
$$2,000 \text{ pounds} = 1 \text{ ton}$$

Warning: Don't confuse *fluid ounces,* which measure volume, with *ounces,* which measure weight. These units are two completely different types of measurements!

- **Units of time:** Time is hard to define, but everybody knows what it is. Time is measured in seconds, minutes, hours, days, weeks, and years:

$$60 \text{ seconds} = 1 \text{ minute}$$
$$60 \text{ minutes} = 1 \text{ hour}$$
$$24 \text{ hours} = 1 \text{ day}$$
$$7 \text{ days} = 1 \text{ week}$$
$$365 \text{ days} = 1 \text{ year}$$

The conversion from days to years is approximate because Earth's daily rotation on its axis and its yearly revolution around the sun aren't exactly synchronized. A year is closer to 365.25 days, which is why leap-years exist.

I left months out of the picture because the definition of a month is imprecise — it can vary from 28 to 31 days.

- **Unit of speed:** Speed is the measurement of how much time an object takes to move a given distance. The most common unit of speed is miles per hour (mph).

- **Unit of temperature:** Temperature measures how much heat an object contains. This object can be a glass of water, a turkey in the oven, or the air surrounding your house. Temperature is measured in degrees Fahrenheit ($°F$).

Table 14-1 summarizes the most commonly used units of English measurement. To use this chart, remember the following rules:

- When converting from a large unit to a smaller unit, always multiply. For example, 2 pints are in 1 quart, so to convert 10 quarts to pints, multiply by 2:

$$10 \text{ quarts} \times 2 = 20 \text{ pints}$$

- When converting from a small unit to a larger unit, always divide. For example, 3 feet are in 1 yard, so to convert 12 feet to yards, divide by 3:

$$12 \text{ feet} \div 3 = 4 \text{ yards}$$

Table 14-1   Commonly Used English Units of Measurement

Measure of . . .	English Units	Conversion Equations
Distance (length)	Inches (in.) Feet (ft.) Yards (yd.) Miles (mi.)	12 inches = 1 foot 3 feet = 1 yard 5,280 feet = 1 mile
Fluid volume (capacity)	Fluid ounces (fl. oz.) Cups (c.) Pints (pt.) Quarts (qt.) Gallons (gal.)	8 fluid ounces = 1 cup 2 cups = 1 pint 2 pints = 1 quart 4 quarts = 1 gallon
Weight	Ounces (oz.) Pounds (lb.) Tons	16 ounces = 1 pound 2,000 pounds = 1 ton
Time	Seconds Minutes Hours Days Weeks Years	60 seconds = 1 minute 60 minutes = 1 hour 24 hours = 1 day 7 days = 1 week 365 days = 1 year
Temperature	Degrees Fahrenheit (°F)	
Speed (rate)	Miles per hour (mph)	

When converting from large units to very small ones (for example, from tons to ounces), you may need to multiply more than once. Similarly, when converting from small units to much larger ones (for example, from minutes to days), you may need to divide more than once.

Tip: After doing a conversion, step back and apply a *reasonability test* to your answer — that is, think about whether your answer makes sense. For example, when you convert feet to inches, the number you end up with should be a lot bigger than the number you started with because there are lots of inches in a foot.

Examples

Q. How many minutes are in a day?

A. **1,440 minutes.** You're going from a larger unit (a day) to a smaller unit (minutes); 60 minutes are in an hour and 24 hours are in a day, so you simply multiply:

$$60 \times 24 = 1,440 \text{ minutes}$$

Therefore, 1,440 minutes are in a day.

Q. 5 pints = _____ fluid ounces.

A. **80 fluid ounces.** 8 fluid ounces are in a cup and 2 cups are in a pint, so

$$8 \times 2 = 16 \text{ fl. oz.}$$

16 fluid ounces are in a pint, so

$$5 \text{ pints} = 5 \times 16 \text{ fluid ounces} = 80 \text{ fl. oz.}$$

Q. If you have 32 fluid ounces, how many pints do you have?

A. **2 pints.** 8 fluid ounces are in a cup, so divide as follows:

$$32 \text{ fluid ounces} \div 8 = 4 \text{ cups}$$

And 2 cups are in a pint, so divide again:

$$4 \text{ cups} \div 2 = 2 \text{ pints}$$

Therefore, 32 fluid ounces equals 2 pints.

Q. 504 hours = _____ weeks.

A. **3 weeks.** 24 hours are in a day, so divide as follows:

$$504 \text{ hours} \div 24 = 21 \text{ days}$$

And 7 days are in a week, so divide again:

$$21 \text{ days} \div 7 = 3 \text{ weeks}$$

Therefore, 504 hours equals 3 weeks.

Practice Questions

1. Answer each of the following questions:

   a. How many inches are in a yard?

   b. How many hours are in a week?

   c. How many ounces are in a ton?

   d. How many cups are in a gallon?

2. Calculate each of the following:

   a. 7 quarts = _____ cups

   b. 5 miles = _____ inches

   c. 3 gallons = _____ fluid ounces

   d. 4 days = _____ seconds

3. Answer each of the following questions:

   a. If you have 420 minutes, how many hours do you have?

   b. If you have 144 inches, how many yards do you have?

   c. If you have 22,000 pounds, how many tons do you have?

   d. If you have 256 fluid ounces, how many gallons do you have?

4. Calculate each of the following:

   a. 168 inches = _____ feet

   b. 100 quarts = _____ gallons

   c. 288 ounces = _____ pounds

   d. 76 cups = _____ quarts

Practice Answers

1. All these questions ask you to convert a large unit to a smaller one, so use multiplication.

   a. How many inches are in a yard? **36 inches.** 12 inches are in a foot and 3 feet are in a yard, so

   $$3 \times 12 = 36 \text{ in.}$$

   b. How many hours are in a week? **168 hours.** 24 hours are in a day and 7 days are in a week, so

   $$24 \times 7 = 168 \text{ hours}$$

   c. How many ounces are in a ton? **32,000 ounces.** 16 ounces are in a pound and 2,000 pounds are in a ton, so

   $$16 \times 2,000 = 32,000 \text{ oz.}$$

   d. How many cups are in a gallon? **16 cups.** 2 cups are in a pint, 2 pints are in a quart, and 4 quarts are in a gallon, so

   $$2 \times 2 \times 4 = 16 \text{ c.}$$

2. All these problems ask you to convert a large unit to a smaller one, so use multiplication:

   a. 7 quarts = **28 cups.** 2 cups are in a pint and 2 pints are in a quart, so

   $$2 \times 2 = 4 \text{ c.}$$

   So 4 cups are in a quart; therefore,

   $$7 \text{ qt.} = 7 \times 4 \text{ c.} = 28 \text{ c.}$$

b. 5 miles = **316, 800 inches.** 12 inches are in a foot and 5,280 feet are in a mile, so

$$12 \times 5,280 = 63,360 \text{ in.}$$

63,360 inches are in a mile, so

$$5 \text{ mi.} = 5 \times 63,360 \text{ in.} = 316,800 \text{ in.}$$

c. 3 gallons = **384 fluid ounces.** 8 fluid ounces are in a cup, 2 cups are in a pint, 2 pints are in a quart, and 4 quarts are in a gallon, so

$$8 \times 2 \times 2 \times 4 = 128 \text{ fl. oz.}$$

Therefore, 128 fluid ounces are in a gallon, so

$$3 \text{ gal.} = 3 \times 128 \text{ fl. oz.} = 384 \text{ fl. oz.}$$

d. 4 days = **345,600 seconds.** 60 seconds are in a minute, 60 minutes are in an hour, and 24 hours are in a day, so

$$60 \times 60 \times 24 = 86,400 \text{ seconds}$$

86,400 seconds are in a day, so

$$4 \text{ days} = 4 \times 86,400 \text{ seconds} = 345,600 \text{ seconds}$$

3. All these problems ask you to convert a smaller unit to a larger one, so use division:

a. If you have 420 minutes, how many hours do you have? **7 hours.** There are 60 minutes in an hour, so divide by 60:

$$420 \text{ minutes} \div 60 = 7 \text{ hours}$$

b. If you have 144 inches, how many yards do you have? **4 yards.** There are 12 inches in a foot, so divide by 12:

$$144 \text{ in.} \div 12 = 12 \text{ ft.}$$

There are 3 feet in a yard, so divide by 3:

$$12 \text{ ft.} \div 3 = 4 \text{ yd.}$$

c. If you have 22,000 pounds, how many tons do you have? **11 tons.** There are 2,000 pounds in a ton, so divide by 2,000

$$22,000 \text{ lb} \div 2,000 = 11 \text{ tons}$$

d. If you have 256 fluid ounces, how many gallons do you have? **2 gallons.** There are 8 fluid ounces in a cup, so divide by 8:

$$256 \text{ ft. oz} \div 8 = 32 \text{ c.}$$

There are 2 cups in a pint, so divide by 2:

$$32 \text{ c.} \div 2 = 16 \text{ pt.}$$

There are 2 pints in a quart, so divide by 2:

$$16 \text{ pt.} \div 2 = 8 \text{ qt.}$$

There are 4 quarts in a gallon, so divide by 4:

$$8 \text{ qt.} \div 4 = 2 \text{ gal.}$$

4. All these problems ask you to convert a smaller unit to a larger one, so use division:

   a. 168 inches = **14 feet.** There are 12 inches in a foot, so divide by 12:

   $$168 \text{ in.} \div 12 = 14 \text{ ft.}$$

   b. 100 quarts = **25 gallons.** There are 4 quarts in a gallon, so divide by 4:

   $$100 \text{ qt.} \div 4 = 25 \text{ gal.}$$

   c. 288 ounces = **18 pounds.** There are 16 ounces in a pound, so divide by 16:

   $$288 \text{ oz.} \div 16 = 18 \text{ lbs.}$$

   d. 76 cups = **19 quarts.** There are 2 cups in a pint, so divide by 2:

   $$76 \text{ cups} \div 2 = 38 \text{ pints}$$

   There are 2 pints in a quart, so divide by 2:

   $$38 \text{ pints} \div 2 = 19 \text{ quarts}$$

Looking at the metric system

Like the English system, the metric system provides units of measurement for distance, volume, and so on. Unlike the English system, however, the metric system builds these units using a *basic unit* and a set of *prefixes*.

Table 14-2 shows five important basic units in the metric system.

Table 14-2   Five Basic Metric Units

Measure Of	Basic Metric Unit
Distance	Meter
Volume (capacity)	Liter
Mass (weight)	Gram
Time	Second
Temperature	Degrees Celsius (°C)

For scientific purposes, the metric system has been updated to the more rigorously defined *System of International Units (SI)*. Each basic SI unit correlates directly to a measurable scientific process that defines it. In SI, the kilogram (not the gram) is the basic unit of mass, the kelvin is the basic unit of temperature, and the liter is not considered a basic unit. For technical reasons, scientists tend to use the more rigidly defined SI, but most other people use the looser metric system. In everyday practice, you can think of the units in Table 14-2 as basic units.

Table 14-3 shows ten metric prefixes, with the three most commonly used in bold and italicized (see Chapter 13 for more information on powers of ten).

Table 14-3   Ten Metric Prefixes

Prefix	Meaning	Number	Power of Ten
Giga-	One billion	1,000,000,000	10^9
Mega-	One million	1,000,000	10^6
Kilo-	*One thousand*	*1,000*	10^3
Hecta-	One hundred	100	10^2
Deca-	Ten	10	10^1
(none)	One	1	10^0
Deci-	One tenth	0.1	10^{-1}
Centi-	*One hundredth*	*0.01*	10^{-2}
Milli-	*One thousandth*	*0.001*	10^{-3}
Micro-	One millionth	0.000001	10^{-6}
Nano-	One billionth	0.000000001	10^{-9}

Large and small metric units are formed by linking a basic unit with a prefix. For example, linking the prefix *kilo-* to the basic unit *meter* gives you the *kilometer,* which means 1,000 meters. Similarly, linking the prefix *milli-* to the basic unit *liter* gives you the *milliliter,* which means 0.001 (one thousandth) of a liter.

Here's a list giving you the basics:

- **Units of distance:** The basic metric unit of distance is the meter (m). Other common units are millimeters (mm), centimeters (cm), and kilometers (km):

$$1 \text{ kilometer} = 1,000 \text{ meters}$$
$$1 \text{ meter} = 100 \text{ centimeters}$$
$$1 \text{ meter} = 1,000 \text{ millimeters}$$

- **Units of fluid volume:** The basic metric unit of fluid volume (also called capacity) is the liter (L). Another common unit is the milliliter (mL):

$$1 \text{ liter} = 1,000 \text{ milliliters}$$

Note: One milliliter is equal to 1 cubic centimeter (cc).

- **Units of mass:** Technically, the metric system measures not weight, but mass. *Weight* is the measurement of how strongly gravity pulls an object toward Earth. *Mass,* however, is the measurement of the amount of matter an object has. If you traveled to the moon, your weight would change, so you would feel lighter. But your mass would remain the same, so all of you would still be there. Unless you're planning a trip into outer space or performing a scientific experiment, you probably don't need to know the difference between weight and mass. In this chapter, you can think of them as equivalent, and I use the word *weight* when referring to metric mass.

The basic unit of weight in the metric system is the gram (g). Even more commonly used, however, is the kilogram (kg):

$$1 \text{ kilogram} = 1{,}000 \text{ grams}$$

Note: 1 kilogram of water has a volume of 1 liter.

- **Units of time:** As in the English system, the basic metric unit of time is a second (s). For most purposes, people also use other English units, such as minutes and hours.

  For many scientific purposes, the second is the only unit used to measure time. Large numbers of seconds and small fractions of sections are represented with *scientific notation,* which I cover in Chapter 13.

- **Units of speed:** For most purposes, the most common metric unit of speed (also called velocity) is *kilometers per hour* (km/hr). Another common unit is *meters per second* (m/s).

- **Units of temperature (degrees Celsius or Centigrade):** The basic metric unit of temperature is the *Celsius degree* (°C), also called the *Centigrade degree.* The Celsius scale is set up so that, at sea level, water freezes at 0°C and boils at 100°C.

  Scientists often use another unit — the kelvin (K) — to talk about temperature. The degrees are the same size as in Celsius, but 0 K is set at *absolute zero,* the temperature at which atoms don't move at all. Absolute zero is approximately equal to –273.15°C.

Examples

Q. How many millimeters are in a meter?

A. **1,000.** The prefix *milli-* means one *thousandth,* so a millimeter is $\frac{1}{1,000}$ of a meter. Therefore, a meter contains 1,000 millimeters.

Q. A *dyne* is an old unit of force, the push or pull on an object. Using what you know about metric prefixes, how many dynes do you think are in 14 teradynes?

A. **14,000,000,000,000 (14 trillion).** The prefix *tera-* means *one trillion,* so 1,000,000,000,000 dynes are in a teradyne; therefore,

$$14 \text{ teradynes} = 14 \times 1{,}000{,}000{,}000{,}000 \text{ dynes} = 14{,}000{,}000{,}000{,}000 \text{ dynes}$$

Practice Questions

1. Give the basic metric unit for each type of measurement listed below:

   a. The amount of vegetable oil for a recipe

   b. The weight of an elephant

   c. How much water a swimming pool can hold

   d. How hot a swimming pool is

   e. How long you can hold your breath

   f. Your height

   g. Your weight

   h. How far you can run

2. Write down the number or decimal associated with each of the following metric prefixes:

   a. kilo-

   b. milli-

   c. centi-

   d. mega-

   e. micro-

   f. giga-

   g. nano-

   h. no prefix

3. Answer each of the following questions:

   a. How many centimeters are in a meter?

   b. How many milliliters are in a liter?

   c. How many milligrams are in a kilogram?

   d. How many centimeters are in a kilometer?

4. Using what you know about metric prefixes, calculate each of the following:

   a. 75 kilowatts = _____ watts

   b. 12 seconds = _____ microseconds

   c. 7 megatons = _____ tons

   d. 400 gigaHertz = _____ Hertz

Practice Answers

1. Note that you want only the base unit, without any prefixes.

   a. **liters**

   b. **grams**

   c. **liters**

   d. **degrees Celsius (Centigrade)**

  e. seconds

  f. meters

  g. grams

  h. meters

2. **a.** **1,000 (one thousand or 10^3)**

  b. **0.001 (one thousandth or 10^{-3})**

  c. **0.01 (one hundredth or 10^{-2})**

  d. **1,000,000 (one million or 10^6)**

  e. **0.000001 (one millionth or 10^{-6})**

  f. **1,000,000,000 (one billion or 10^9)**

  g. **0.000000001 (one billionth or 10^{-9})**

  h. **1 (one or 10^0)**

3. **a.** **100 centimeters**

  b. **1,000 milliliters**

  c. **1,000,000 milligrams.** 1,000 milligrams are in a gram and 1,000 grams are in a kilogram, so

$$1,000 \times 1,000 = 1,000,000 \text{ mg}$$

Therefore, 1,000,000 milligrams are in a kilogram.

  d. **100,000 centimeters.** 100 centimeters are in a meter and 1,000 meters are in a kilometer, so

$$100 \times 1,000 = 100,000 \text{ cm}$$

Therefore, 100,000 centimeters are in a kilometer.

4. **a.** **75,000 watts.** The prefix *kilo-* means *one thousand,* so 1,000 watts are in a kilowatt; therefore,

$$75 \text{ kilowatts} = 75 \times 1,000 \text{ watts} = 75,000 \text{ watts}$$

  b. **12,000,000 microseconds.** The prefix *micro-* means *one millionth,* so a microsecond is a millionth of a second. Therefore, 1,000,000 microseconds are in a second. Thus,

$$12 \text{ seconds} = 12 \times 1,000,000 \text{ microseconds} = 12,000,000 \text{ microseconds}$$

  c. **7,000,000 tons.** The prefix *mega-* means *one million,* so 1,000,000 tons are in a megaton; therefore,

$$7 \text{ megatons} = 7 \times 1,000,000 \text{ tons} = 7,000,000 \text{ tons}$$

  d. **400,000,000,000 Hertz.** The prefix *giga-* means *one billion,* so 1,000,000,000 Hertz are in a gigaHertz; thus,

$$400 \text{ gigaHertz} = 400 \times 1,000,000,000 \text{ Hertz} = 400,000,000,000 \text{ gigaHertz}$$

ESTIMATING AND CONVERTING BETWEEN THE ENGLISH AND METRIC SYSTEMS

Most Americans use the English system of measurement all the time and have only a passing acquaintance with the metric system. But metric units are being used more commonly as the units for tools, footraces, soft drinks, and many other things. Also, if you travel abroad, you need to know how far 100 kilometers is or how long you can drive on 10 liters of gasoline.

In this section, I show you how to make ballpark estimates of metric units in terms of English units, which can help you feel more comfortable with metric units. I also show you how to convert between English and metric units, which is a common type of math problem.

When I talk about *estimating,* I mean very loose ways of measuring metric amounts using the English units you are familiar with. In contrast, when I talk about *converting,* I mean using an equation to change from one system of units to the other. Neither method is exact, but converting provides a much closer approximation (and takes longer) than estimating.

Estimating in the metric system

One reason people sometimes feel uncomfortable using the metric system is that, when you're not familiar with it, estimating amounts in practical terms is hard. For example, if I tell you that we're going out to a beach that's $\frac{1}{4}$ mile away, you prepare yourself for a short walk. And if I tell you that it's 10 miles away, you head for the car. But what do you do with the information that the beach is 3 kilometers away?

Similarly, if I tell you that the temperature is 85°F, you'll probably wear a bathing suit or shorts. And if I tell you it's 40°F, you'll probably wear a coat. But what do you wear if I tell you that the temperature is 25°C?

In this section, I give you a few rules of thumb to estimate metric amounts. In each case, I show you how a common metric unit compares with an English unit that you already feel comfortable with.

Approximating short distances: 1 meter is about 1 yard (3 feet)

Tip: Here's how to convert meters to feet: 1 meter ≈ 3.28 feet. But for estimating, use the simple rule that 1 meter is about 1 yard (that is, about 3 feet).

By this estimate, a 6-foot man stands about 2 meters tall. A 15-foot room is 5 meters wide. And a football field that's 100 yards long is about 100 meters long. Similarly, a river with a depth of 4 meters is about 12 feet deep. A mountain that's 3,000 meters tall is about 9,000 feet. And a child who is only half a meter tall is about a foot and a half.

Estimating longer distances and speed

Tip: Here's how to convert kilometers to miles: 1 kilometer ≈ 0.62 miles. For a ballpark estimate, you can remember that 1 kilometer is about $\frac{1}{2}$ a mile. By the same token, 1 kilometer per hour is about $\frac{1}{2}$ mile per hour.

This guideline tells you that if you live 2 miles from the nearest supermarket, then you live about 4 kilometers from there. A marathon of 26 miles is about 52 kilometers. And if you run on a treadmill at 6 miles per hour, then you can run at about 12 kilometers per hour. By the same token, a 10-kilometer race is about 5 miles. If the Tour de France is about 4,000 kilometers, then it's about 2,000 miles. And if light travels about 300,000 kilometers per second, then it travels about 150,000 miles per second.

Approximating volume: 1 liter is about 1 quart (1/4 gallon)

Tip: Here's how to convert liters to gallons: 1 liter ≈ 0.26 gallons. A good estimate here is that 1 liter is about 1 quart (a gallon consists of about 4 liters).

Using this estimate, a gallon of milk is 4 quarts, so it's about 4 liters. If you put 10 gallons of gasoline in your tank, it's about 40 liters. In the other direction, if you buy a 2-liter bottle of cola, you have about 2 quarts. If you buy an aquarium with a 100-liter capacity, it holds about 25 gallons of water. And if a pool holds 8,000 liters of water, it holds about 2,000 gallons.

Estimating weight: 1 kilogram is about 2 pounds

Tip: Here's how to convert kilograms to pounds: 1 kilogram ≈ 2.20 pounds. For estimating, figure that 1 kilogram is equal to about 2 pounds.

By this estimate, a 5-kilogram bag of potatoes weighs about 10 pounds. If you can bench-press 70 kilograms, then you can bench-press about 140 pounds. And because a liter of water weighs exactly 1 kilogram, you know that a quart of water weighs about 2 pounds. Similarly, if a baby weighs 8 pounds at birth, he or she weighs about 4 kilograms. If you weigh 150 pounds, then you weigh about 75 kilograms. And if your New Year's resolution is to lose 20 pounds, then you want to lose about 10 kilograms.

Estimating temperature

The most common reason for estimating temperature in Celsius is in connection with the weather. The formula for converting from Celsius to Fahrenheit is kind of messy:

$$\text{Fahrenheit} = \text{Celsius} \times \frac{9}{5} + 32$$

Instead, use the handy chart in Table 14-4.

Any temperature below 0°C is cold, and any temperature over 30°C is hot. Most of the time, the temperature falls in this middling range. So now you know that when the temperature is 6°C, you want to wear a coat. When it's 14°C, you may want a sweater — or at least long sleeves. And when it's 25°C, head for the beach!

Table 14-4   Comparing Celsius and Fahrenheit Temperatures

Celsius (Centigrade)	Description	Fahrenheit
0°	Cold	32°
10°	Cool	50°
20°	Warm	68°
30°	Hot	86°

Example

Q.   While in training for lacrosse over the summer, Devin typically runs 3 miles and drinks 1 gallon of water per day. What are these approximate amounts in metric units?

A.   **6 kilometers; 4 liters.**

Practice Questions

1.   Elaine can comfortably carry a backpack of up to 30 kilograms a distance of 30 kilometers per day. What do this weight and this distance approximately equal in English units?

2.   A road is 17 miles long and 27 feet wide. What is its approximate length and width in metric units?

3.   An Olympic swimming pool is typically 50 meters long and holds 2,500,000 liters of water. What is its approximate length and capacity in English units?

4.   Driving home from college, Brayden filled his gas tank with 12.5 gallons of gasoline and then drove it 400 miles. What are these approximate amounts in metric units? How many kilometers per liter did he average on this trip?

5.   Do you think you would feel comfortable wearing a T-shirt with no jacket when the weather is 35°C? What about when the weather is 20°C? And what about 5°C?

Practice Answers

1.   **60 pounds; 15 miles.**

2.   **34 miles; 9 meters.**

3.   **150 feet; 625,000 gallons.**

4.   **50 liters; 800 kilometers; 16 kilometers per liter.** $800 \div 50 = 16$.

5.   **Yes; yes; no.** A temperature of 35°C is very hot (97°F); 20°C is warm (68°F); 5°C is between cool and cold (41°F).

Converting units of measurement

Many books give you one formula for converting from English to metric and another for converting from metric to English. People often find this conversion method confusing because they have trouble remembering which formula to use in which direction.

In this section, I show you a simple way to convert between English and metric units that uses only one formula for each type of conversion.

Here's a nice pair that's easy to remember: 16°C is about 61°F.

Understanding conversion factors

When you multiply any number by 1, that number stays the same. For example, $36 \times 1 = 36$. And when a fraction has the same numerator (top number) and denominator (bottom number), that fraction equals 1 (see Chapter 9 for details). So when you multiply a number by a fraction that equals 1, the number stays the same. For example:

$$36 \times \frac{5}{5} = 36$$

If you multiply a measurement by a special fraction that equals 1, you can switch from one unit of measurement to another without changing the value. People call such fractions *conversion factors*.

Take a look at some equations that show how metric and English units are related (all conversions between English and metric units are approximate):

- 1 meter ≈ 3.26 feet

- 1 kilometer ≈ 0.62 mile

- 1 liter ≈ 0.26 gallon

- 1 kilogram ≈ 2.20 pounds

Because the values on each side of the equations are equal, you can create

- $\dfrac{1 \text{ meter}}{3.26 \text{ feet}}$ or $\dfrac{3.26 \text{ feet}}{1 \text{ meter}}$

- $\dfrac{1 \text{ kilometer}}{0.62 \text{ mile}}$ or $\dfrac{0.62 \text{ mile}}{1 \text{ kilometer}}$

- $\dfrac{1 \text{ liter}}{0.26 \text{ gallon}}$ or $\dfrac{0.26 \text{ gallon}}{1 \text{ liter}}$

- $\dfrac{1 \text{ kilogram}}{2.2 \text{ pounds}}$ or $\dfrac{2.2 \text{ pounds}}{1 \text{ kilogram}}$

When you understand how units of measurement cancel (which I discuss in the next section), you can easily choose which fractions to use to switch between units of measurement.

Canceling units of measurement

When you're multiplying fractions, you can cancel any factor that appears in both the numerator and the denominator (see Chapter 9 for details). Just as with numbers, you can also cancel out units of measurement in fractions. For example, suppose you want to evaluate this fraction:

$$\frac{6 \text{ gallons}}{2 \text{ gallons}}$$

You already know that you can cancel out a factor of 2 in both the numerator and the denominator. But you can also cancel out the unit *gallons* in both the numerator and the denominator:

$$= \frac{6 \text{ gallons}}{2 \text{ gallons}}$$

So this fraction simplifies as follows:

$$= 3$$

Converting units

When you understand how to cancel out units in fractions and how to set up fractions equal to 1 (see the preceding sections), you have a foolproof system for converting units of measurement.

Suppose you want to convert 7 meters into feet. Using the equation 1 meter = 3.26 feet, you can make a fraction out of the two values, as follows:

$$\frac{1 \text{ meter}}{3.26 \text{ feet}} = 1 \quad \text{or} \quad \frac{3.26 \text{ feet}}{1 \text{ meter}} = 1$$

Both fractions equal 1 because the numerator and the denominator are equal. So you can multiply the quantity you're trying to convert (7 meters) by one of these fractions without changing it. Remember that you want the meters unit to cancel out. You already have the word *meters* in the numerator (to make this clear, place 1 in the denominator), so use the fraction that puts *1 meter* in the denominator:

$$\frac{7 \text{ meters}}{1} \times \frac{3.26 \text{ feet}}{1 \text{ meter}}$$

Now cancel out the unit that appears in both the numerator and the denominator:

$$= \frac{7 \ \cancel{\text{meters}}}{1} \times \frac{3.26 \text{ feet}}{1 \ \cancel{\text{meter}}}$$

At this point, the only value in the denominator is 1, so you can ignore it. And the only unit left is *feet,* so place it at the end of the expression:

$$= 7 \times 3.26 \text{ feet}$$

Now do the multiplication (Chapter 10 shows how to multiply decimals):

$$= 22.82 \text{ feet}$$

It may seem strange that the answer appears with the units already attached, but that's the beauty of this method: When you set up the right expression, the answer just appears.

CONVERTING BETWEEN ENGLISH AND METRIC UNITS

To convert between metric units and English units, use the four conversion equations shown in the first column of Table 14-5.

Table 14-5   Conversion Factors for English and Metric Units

Conversion Equation	English-to-Metric	Metric-to-English
1 meter ≈ 3.26 feet	(1 m)/(3.26 ft.)	(3.26 ft.)/(1 m)
1 kilometer ≈ 0.62 miles	(1 k)/(0.62 mi.)	(0.62 mi.)/(1 k)
1 liter ≈ 0.26 gallons	(1 L)/(0.26 gal.)	(0.26 gal.)/(1 L)
1 kilogram ≈ 2.20 pounds	(1 kg)/(2.20 lbs.)	(2.20 lbs.)/(1 kg)

The remaining columns in Table 14-5 show conversion factors (fractions) that you multiply by to convert from metric units to English or from English units to metric. To convert from one unit to another, multiply by the conversion factor and cancel out any unit that appears in both the numerator and denominator.

Tip: Always use the conversion factor that has units you're converting *from* in the *denominator.* For example, to convert from miles to kilometers, use the conversion factor that has miles in the denominator: that is, $\dfrac{1 \text{ km}}{0.62 \text{ mi.}}$.

Sometimes, you may want to convert between units for which no direct conversion factor exists. In these cases, set up a *conversion chain* to convert via one or more intermediate units. For instance, to convert centimeters into inches, you may go from centimeters to meters to feet to inches. When the conversion chain is set up correctly, every unit cancels out except for the unit that you're converting to. You can set up a conversion chain of any length to solve a problem.

Tip: With a long conversion chain, it's sometimes helpful to take an extra step and turn the whole chain into a single fraction. Place all the numerators above a single fraction bar and all the denominators below it, keeping multiplication signs between the numbers.

A chain can include conversion factors built from any conversion equation in this chapter. For example, you know from "The Basics of the English System" that 2 pints = 1 quart, so you can use the following two fractions:

$$\frac{2 \text{ pt.}}{1 \text{ qt.}} \quad \frac{1 \text{ qt.}}{2 \text{ pt.}}$$

Similarly, you know from "Going International with the Metric System" that 1 kilogram = 1,000 grams, so you can use these two fractions:

$$\frac{1\,\text{kg}}{1,000\,\text{g}} \qquad \frac{1,000\,\text{g}}{1\,\text{kg}}$$

Examples

Q. Convert 5 kilometers to miles.

A. **3.1 miles.** To convert *from* kilometers, multiply 5 kilometers by the conversion factor with kilometers in the denominator:

$$5\,\text{km} \times \frac{0.62\,\text{mi.}}{1\,\text{km}}$$

Now you can cancel the unit *kilometer* in both the numerator and denominator:

$$= 5\,\cancel{\text{km}} \times \frac{0.62\,\text{mi.}}{1\,\cancel{\text{km}}}$$

Calculate the result:

$$= 5 \times \frac{0.62\,\text{mi.}}{1} = 3.1\,\text{mi.}$$

Notice that when you set up the conversion correctly, you don't have to think about the unit — it changes from kilometers to miles automatically!

Q. Convert 21 grams to pounds.

A. **0.0462 pounds.** You don't have a conversion factor to convert grams to pounds directly, so set up a conversion chain that makes the following conversion:

$$\text{grams} \rightarrow \text{kilograms} \rightarrow \text{pounds}$$

To convert grams to kilograms, use the equation 1,000 g = 1 kg. Multiply by the fraction with kilograms in the numerator and grams in the denominator:

$$21\,\text{g} \times \frac{1\,\text{kg}}{1,000\,\text{g}}$$

To convert kilograms to pounds, use the equation 1 kg = 2.2 lb. and multiply by the fraction with pounds in the numerator and kilograms in the denominator:

$$= 21\,\text{g} \times \frac{1\,\text{kg}}{1,000\,\text{g}} \times \frac{2.2\,\text{lb.}}{1\,\text{kg}}$$

Cancel *grams* and *kilograms* in both the numerator and denominator:

$$= 21\,\cancel{\text{g}} \times \frac{1\,\cancel{\text{kg}}}{1,000\,\cancel{\text{g}}} \times \frac{2.2\,\text{lb.}}{1\,\cancel{\text{kg}}}$$

Finally, calculate the result:

$$= 21 \times \frac{1}{1,000} \times \frac{2.2 \text{ lb.}}{1}$$
$$= 0.021 \times 2.2 \text{ lbs} = 0.0462 \text{ lb.}$$

After doing the conversion, step back and apply a reasonability test to your answers. Does each answer make sense? For example, when you convert grams to kilograms, the number you end up with should be a lot smaller than the number you started with because each kilogram contains lots of grams.

Q. For a science project, Anita observed that an ant could walk a distance of 1 foot in 3.8 seconds. How many centimeters per second is that, to the nearest hundredth of a centimeter?

A. **8.07 centimeters per second.** Use the conversion equation 1 foot = 3.8 seconds. Then set up a conversion chain with centimeters in the numerator and seconds in the denominator.

$$\frac{1 \text{ ft.}}{3.8 \text{ sec}} \times \frac{1 \text{ m}}{3.26 \text{ ft.}} \times \frac{100 \text{ cm}}{1 \text{ m}} \approx 8.07 \frac{\text{cm}}{\text{sec.}}$$

Practice Questions

1. Convert 8 kilometers to miles.

2. If you weigh 72 kilograms, what's your weight in pounds to the nearest whole pound?

3. If you're 1.8 meters tall, what's your height in inches to the nearest whole inch?

4. Change 100 cups to liters, rounded to the nearest whole liter.

5. To the nearest whole gram, how many grams are in a ton?

6. Cathy has decided to cut back to drinking only 5 cups of coffee per day. How many liters of coffee is that, to the nearest tenth of a liter?

7. If 1 milliliter of water weighs exactly 1 gram, what is the weight of the water in an Olympic swimming pool that holds 660,000 gallons of water, to the nearest whole kilogram?

8. In 1973, Secretariat ran the 12-furlong Belmont Stakes in 144 seconds — a record that still stands to this day. If 1 furlong equals $\frac{1}{8}$ of a mile, what was Secretariat's average speed in kilometers per hour, to the nearest whole number of units? (**Hint:** Use the conversion equation 12 furlongs = 144 seconds. Then set up a conversion chain with miles in the numerator and seconds in the denominator.)

A Practice Answers

1. **4.96 miles.** To convert kilometers to miles, multiply by the conversion fraction with miles in the numerator and kilometers in the denominator:

   Cancel the unit *kilometer* in both the numerator and denominator:

   $$= 8 \; \cancel{\text{km}} \times \frac{0.62 \text{ mi.}}{1 \; \cancel{\text{km}}}$$

   Now calculate the result:

   $$= 8 \times 0.62 \text{ mi.} = 4.96 \text{ mi.}$$

2. **158 pounds.** To convert kilograms to pounds, multiply by the conversion factor with pounds in the numerator and kilograms in the denominator:

   $$72 \text{ kg} \times \frac{2.2 \text{ lb.}}{1 \text{ kg}}$$

   Cancel the unit *kilogram* in both the numerator and denominator:

   $$= 72 \; \cancel{\text{kg}} \times \frac{2.2 \text{ lb.}}{1 \; \cancel{\text{kg}}}$$
   $$= 72 \times 2.2 \text{ lb.}$$

   Multiply to find the answer:

   $$= 158.4 \text{ lb.}$$

   Round to the nearest whole pound:

   $$\approx 158 \text{ lb.}$$

3. **70 inches.** You don't have a conversion factor to change meters to inches directly, so set up a conversion chain as follows:

   $$\text{meters} \rightarrow \text{feet} \rightarrow \text{inches}$$

   Convert meters to feet with the fraction that puts meters in the denominator:

   $$1.8 \text{ m} \times \frac{3.26 \text{ ft.}}{1 \text{ m}}$$

   Convert the feet to inches with the conversion factor that has feet in the denominator:

   $$= 1.8 \text{ m} \times \frac{3.26 \text{ ft.}}{1 \text{ m}} \times \frac{12 \text{ in.}}{1 \text{ ft.}}$$

   Cancel the units *meters* and *feet* in both the numerator and denominator:

   $$= 1.8 \; \cancel{\text{m}} \times \frac{3.26 \; \cancel{\text{ft.}}}{1 \; \cancel{\text{m}}} \times \frac{12 \text{ in.}}{1 \; \cancel{\text{ft.}}}$$

Multiply to find the answer:

$$= 70.416 \text{ in.}$$

Round the answer to the nearest whole inch:

$$\approx 70 \text{ in.}$$

4. To the nearest liter, 100 cups = **24 liters.** You don't have a conversion factor to change cups to liters directly, so set up a conversion chain:

$$\text{cups} \rightarrow \text{pints} \rightarrow \text{quarts} \rightarrow \text{gallons} \rightarrow \text{liters}$$

Convert cups to pints:

$$100 \text{ c.} \times \frac{1 \text{ pt.}}{2 \text{ c.}}$$

Convert pints to quarts:

$$= 100 \text{ c.} \times \frac{1 \text{ pt.}}{2 \text{ c.}} \times \frac{1 \text{ qt.}}{2 \text{ pt.}}$$

Convert quarts to gallons:

$$= 100 \text{ c.} \times \frac{1 \text{ pt.}}{2 \text{ c.}} \times \frac{1 \text{ qt.}}{2 \text{ pt.}} \times \frac{1 \text{ gal.}}{4 \text{ qt.}}$$

Convert gallons to liters:

$$= 100 \text{ c.} \times \frac{1 \text{ pt.}}{2 \text{ c.}} \times \frac{1 \text{ qt.}}{2 \text{ pt.}} \times \frac{1 \text{ gal.}}{4 \text{ qt.}} \times \frac{1 \text{ L}}{0.26 \text{ gal.}}$$

Now all units *except* liters cancel out:

$$= 100 \text{ c.} \times \frac{1 \text{ pt.}}{2 \text{ c.}} \times \frac{1 \text{ qt.}}{2 \text{ pt.}} \times \frac{1 \text{ gal.}}{4 \text{ qt.}} \times \frac{1 \text{ L}}{0.26 \text{ gal.}}$$

$$= 100 \times \frac{1}{2} \times \frac{1}{2} \times \frac{1}{4} \times \frac{1 \text{ L}}{0.26}$$

To avoid confusion, set this chain up as a single fraction:

$$\frac{100 \times 1 \times 1 \times 1 \times 1 \text{ L}}{2 \times 2 \times 4 \times 0.26}$$

$$= \frac{100 \text{ L}}{4.16}$$

Use decimal division to find the answer to at least one decimal place:

$$\approx 24.04 \text{ L}$$

Round to the nearest whole liter:

$$\approx 24 \text{ L}$$

5. **909,091 grams.**

$$1\,\text{ton} \times \frac{2,000\,\text{lbs.}}{1\,\text{ton}} \times \frac{1\,\text{kg}}{2.2\,\text{lbs.}} \times \frac{1,000\,\text{g}}{1\,\text{kg}} \approx 909,091\,\text{g}$$

6. **1.2 liters.**

$$5\,\text{cu.} \times \frac{1\,\text{qt.}}{4\,\text{cu.}} \times \frac{1\,\text{gal.}}{4\,\text{qt.}} \times \frac{1\,\text{L}}{0.26\,\text{gal.}} \approx 1.2\,\text{L}$$

7. **2,538,462 kilograms.**

$$660,000\,\text{gal.} \times \frac{1\,\text{L}}{0.26\,\text{gal.}} \times \frac{1,000\,\text{mL}}{1\,\text{L}} \times \frac{1\,\text{g}}{1\,\text{mL}} \times \frac{1\,\text{kg}}{1,000\,\text{g}} \approx 2,538,462\,\text{kg}$$

8. **60 km per hour.**

$$\frac{1\,\text{km}}{0.62\,\text{mi.}} \times \frac{1\,\text{mi.}}{8\,\text{furlongs}} \times \frac{12\,\text{furlongs}}{144\,\text{sec.}} \times \frac{60\,\text{sec.}}{1\,\text{min.}} \times \frac{60\,\text{min.}}{1\,\text{hr.}} \approx 60\,\frac{\text{km}}{\text{hr.}}$$

15

PICTURE THIS: BASIC GEOMETRY

IN THIS CHAPTER

- Knowing the basic components of geometry: points, lines, angles, and shapes
- Examining two-dimensional shapes
- Looking at solid geometry
- Handling geometry word problems

Geometry is the mathematics of figures such as squares, circles, triangles, and lines. Because geometry is the math of physical space, it's one of the most useful areas of math. Geometry comes into play when measuring rooms or walls in your house, the area of a circular garden, the volume of water in a pool, or the shortest distance across a rectangular field.

Although geometry is usually a year-long course in high school, you may be surprised by how quickly you can pick up what you need to know about basic geometry. Much of what you discover in a geometry course is how to write geometric proofs, which you don't need for algebra — or trigonometry, or even calculus.

In this chapter, I give you a quick and practical overview of geometry. First, I show you four important concepts in plane geometry: points, lines, angles, and shapes. Then I give you the basics on geometric shapes, from flat circles to solid cubes. Finally, I discuss how to measure geometric shapes by finding the area and perimeter of two-dimensional forms and the volume and surface area of some geometric solids.

Of course, if you want to know more about geometry, the ideal place to look beyond this chapter is *Geometry For Dummies*, 2nd Edition, by Mark Ryan (published by Wiley)!

GETTING ON THE PLANE: POINTS, LINES, ANGLES, AND SHAPES

Plane geometry is the study of figures on a two-dimensional surface — that is, on a *plane*. You can think of the plane as a piece of paper with no thickness at all. Technically, a plane doesn't end at the edge of the paper — it continues forever.

In this section, I introduce you to four important concepts in plane geometry: points, lines, angles, and shapes (squares, circles, triangles, and so forth).

Making some points

A *point* is a location on a plane. It has no size or shape. Although in reality a point is too small to be seen, you can represent it visually in a drawing by using a dot.

When two lines intersect, as shown in this figure, they share a single point. Additionally, each corner of a polygon is a point. (Keep reading for more on lines and polygons.)

Knowing your lines

A *line* — also called a *straight line* — is pretty much what it sounds like; it marks the shortest distance between two points, but it extends infinitely in both directions. It has length but no width, making it a one-dimensional (1-D) figure.

Given any two points, you can draw exactly one line that passes through both of them. In other words, two points *determine* a line.

When two lines intersect, they share a single point. When two lines don't intersect, they are *parallel,* which means that they remain the same distance from each other everywhere. A good visual aid for parallel lines is a set of railroad tracks. In geometry, you draw a line with arrows at both ends. Arrows on either end of a line mean that the line goes on forever (as you can see in Chapter 1, where I discuss the number line).

A *line segment* is a piece of a line that has endpoints, as shown here.

A *ray* is a piece of a line that starts at a point and extends infinitely in one direction, kind of like a laser. It has one endpoint and one arrow.

Figuring the angles

An *angle* is formed when two rays extend from the same point.

Angles are typically used in carpentry to measure the corners of objects. They're also used in navigation to indicate a sudden change in direction. For example, when you're driving, it's common to distinguish when the angle of a turn is "sharp" or "not so sharp."

The sharpness of an angle is usually measured in *degrees*. The most common angle is the *right angle* — the angle at the corner of a square — which is a 90° (90-degree) angle:

Angles that have fewer than 90° — that is, angles that are sharper than a right angle — are called *acute angles*, like this one:

Angles that measure greater than 90° — that is, angles that aren't as sharp as a right angle — are called *obtuse angles*, as seen here:

When an angle is exactly 180°, it forms a straight line and is called a *straight angle*.

Shaping things up

A shape is any closed geometrical figure that has an inside and an outside. Circles, squares, triangles, and larger polygons are all examples of shapes.

Much of plane geometry focuses on different types of shapes. In the next section, I show you how to identify a variety of shapes. Later in this chapter, I show you how to measure these shapes.

GETTING IN SHAPE: POLYGON (AND NON-POLYGON) BASICS

You can divide shapes into two basic types: polygons and non-polygons. A *polygon* has all straight sides, and you can easily classify polygons by the number of sides they have:

Polygon	*Number of Sides*
Triangle	3
Quadrilateral	4
Pentagon	5
Hexagon	6
Heptagon	7
Octagon	8

Any shape that has at least one curved edge is a *non-polygon*. The most common non-polygon is the circle.

Remember: The area of a shape — the space inside — is usually measured in square units, such as square inches (in.^{2}), square meters (m^2), or square kilometers (km^2). If a problem mixes units of measurement, such as inches and feet, you have to convert to one or the other before doing the math (for more on conversions, see Chapter 14).

CLOSED ENCOUNTERS: SHAPING UP YOUR UNDERSTANDING OF 2-D SHAPES

Remember: A *shape* is any closed two-dimensional (2-D) geometrical figure that has an inside and an outside, separated by the *perimeter* (boundary) of the shape. The area of a shape is the measurement of the size inside that shape.

A few shapes that you're probably familiar with include the square, rectangle, and triangle. However, many shapes don't have names, as you can see in Figure 15-1.

Figure 15-1:
Unnamed shapes.

Measuring the perimeter and area of shapes is useful for a variety of applications, from land surveying (to get information about a parcel of land that you're measuring) to sewing (to figure out how much material you need for a project). In this section, I introduce you to a variety of geometric shapes. Later in the chapter, I show you how to find the perimeter and area of each, but for now, I just acquaint you with them.

Polygons

A *polygon* is any shape whose sides are all straight. Every polygon has three or more sides (if it had fewer than three, it wouldn't really be a shape at all). Following are a few of the most common polygons.

Triangles

The most basic shape with straight sides is the *triangle*, a three-sided polygon. You find out all about triangles when you study trigonometry (and what better place to begin than *Trigonometry For Dummies*, 2nd Edition, by Mary Jane Sterling [Wiley]?). Triangles are classified on the basis of their sides and angles. Take a look at the differences (and see Figure 15-2):

Figure 15-2:
Types of
triangles.

Equilateral   Isosceles   Scalene   Right

- **Equilateral:** An *equilateral triangle* has three sides that are all the same length and three angles that all measure 60°.

- **Isosceles:** An *isosceles triangle* has two sides that are the same length and two equal angles.

- **Scalene:** *Scalene triangles* have three sides that are all different lengths and three angles that are all unequal.

- **Right:** A *right triangle* has one right angle. It may be isosceles or scalene.

Quadrilaterals

A *quadrilateral* is any shape that has four straight sides. Quadrilaterals are one of the most common shapes you see in daily life. If you doubt this statement, look around and notice that most rooms, doors, windows, and tabletops are quadrilaterals. Here I introduce you to a few common quadrilaterals (Figure 15-3 shows you what they look like):

- **Square:** A *square* has four right angles and four sides of equal length; also, both pairs of opposite sides (sides directly across from each other) are parallel.

- **Rectangle:** Like a square, a *rectangle* has four right angles and two pairs of opposite sides that are parallel. Unlike the square, however, although opposite sides are equal in length, sides that share a corner — *adjacent* sides — may have different lengths.

- **Rhombus:** Imagine starting with a square and collapsing it as if its corners were hinges. This shape is called a *rhombus.* All four sides are equal in length, and both pairs of opposite sides are parallel.

- **Parallelogram:** Imagine starting with a rectangle and collapsing it as if the corners were hinges. This shape is a *parallelogram* — both pairs of opposite sides are equal in length, and both pairs of opposite sides are parallel.

- **Trapezoid:** The *trapezoid*'s only important feature is that at least two opposite sides are parallel.

- **Kite:** A kite is a quadrilateral with two pairs of adjacent sides that are the same length.

A quadrilateral can fit into more than one of these categories. For example, every parallelogram (with two sets of parallel sides) is also a trapezoid (with at least one set of parallel sides). Every rectangle and rhombus is also both a parallelogram and a trapezoid. And every square is also all five other types of quadrilaterals. In practice, however, it's common to identify a quadrilateral as descriptively as possible — that is, use the *first* word in the list that accurately describes it.

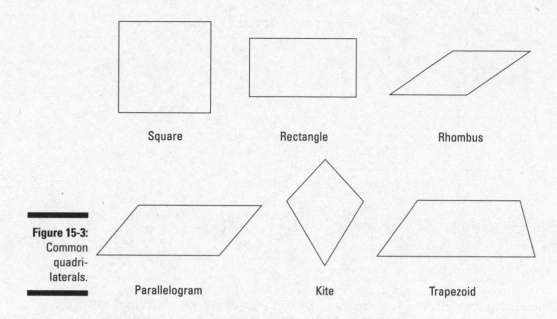

Figure 15-3:
Common
quadri-
laterals.

Square   Rectangle   Rhombus

Parallelogram   Kite   Trapezoid

Polygons on steroids — larger polygons

A polygon can have any number of sides. Polygons with more than four sides aren't as common as triangles and quadrilaterals, but they're still worth knowing about. Larger polygons come in two basic varieties: regular and irregular.

A *regular polygon* has equal sides and equal angles. The most common are regular pentagons (five sides), regular hexagons (six sides), and regular octagons (eight sides). See Figure 15-4.

Figure 15-4:
A pentagon,
a hexagon,
and an
octagon.

Every other polygon is an *irregular polygon* (see Figure 15-5).

Circles

A circle is the set of all points that are a constant distance from the circle's center. The distance from any point on the circle to its center is called the *radius* of the circle. The distance from any point on the circle straight through the center to the other side of the circle is called the *diameter* of the circle.

Figure 15-5:
Various
irregular
polygons.

Unlike polygons, a circle has no straight edges. The ancient Greeks — who invented much of geometry as we know it today — thought that the circle was the most perfect geometric shape.

SQUARING OFF WITH QUADRILATERALS

Any shape with four sides is a *quadrilateral*. Quadrilaterals include squares, rectangles, rhombuses, parallelograms, and trapezoids, plus a host of more irregular shapes. In this section, I show you how to find the area (*A*) and in some cases the perimeter (*P*) of these five basic types of quadrilaterals.

A *square* has four right angles and four equal sides. To find the area and perimeter of a square, use the following formulas, where *s* stands for the length of a side (see Figure 15-6):

$$A = s^2$$
$$P = 4 \times s$$

Square

side (*s*)

Figure 15-6:
The area
and perim-
eter of a
square
using the
length of a
side (*s*).

A *rectangle* has four right angles and opposite sides that are equal. The long side of a rectangle is called the *length,* and the short side is called the *width.* To find the area and perimeter of a rectangle, use the following formulas, where *l* stands for the length of a side and *w* stands for width (see Figure 15-7):

$$A = l \times w$$
$$P = 2 \times (l + w)$$

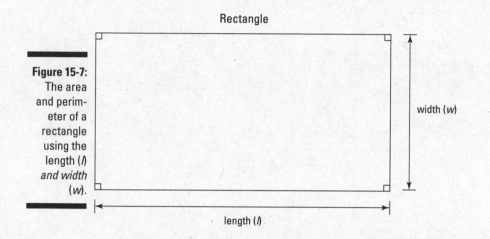

Figure 15-7:
The area and perimeter of a rectangle using the length (*l*) *and width* (*w*).

A *rhombus* resembles a collapsed square. It has four equal sides, but its four angles aren't necessarily right angles. Similarly, a *parallelogram* resembles a collapsed rectangle. Its opposite sides are equal, but its four angles aren't necessarily right angles. To find the area of a rhombus or a parallelogram, use the following formula, where *b* stands for the length of the base (either the bottom or top side) and *h* stands for the height (the shortest distance between the two bases); also see Figure 15-8:

$$A = b \times h$$

Figure 15-8:
The area of a rhombus or parallelogram using the base (*b*) *and height* (*h*).

A *trapezoid* is a quadrilateral whose only distinguishing feature is that it has two parallel bases (top side and bottom side). To find the area of a trapezoid, use the following formula, where b_1 and b_2 stand for the lengths of the two bases and h stands for the height (the shortest distance between the two bases); also see Figure 15-9:

$$A = \frac{1}{2} \times (b_1 + b_2) \times h$$

Trapezoid

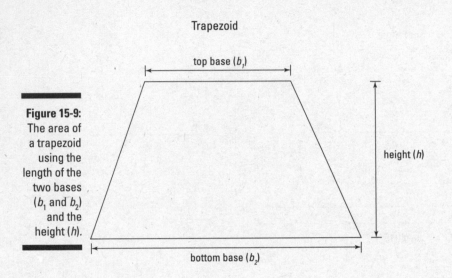

Figure 15-9: The area of a trapezoid using the length of the two bases (b_1 and b_2) and the height (h).

Examples

Q. Find the area and perimeter of a square with a side that measures 5 inches.

A. The area is 25 square inches, and the perimeter is 20 inches.

$$A = s^2 = (5 \text{ in.})^2 = 25 \text{ in.}^2$$
$$P = 4 \times s = 4 \times 5 \text{ in.} = 20 \text{ in.}$$

Q. Find the area and perimeter of a rectangle with a length of 9 centimeters and a width of 4 centimeters.

A. The area is 36 square centimeters, and the perimeter is 26 centimeters.

$$A = l \times w = 9 \text{ cm} \times 4 \text{ cm} = 36 \text{ cm}^2$$
$$P = 2 \times (l + w) = 2 \times (9 \text{ cm} + 4 \text{ cm}) = 2 \times 13 \text{ cm} = 26 \text{ cm}$$

Q. Find the area of a parallelogram with a base of 4 feet and a height of 3 feet.

A. The area is 12 square feet.

$$A = b \times h = 4 \text{ ft.} \times 3 \text{ ft.} = 12 \text{ ft.}^2$$

Q. Find the area of a trapezoid with bases of 3 meters and 5 meters and a height of 2 meters.

A. **The area is 8 square meters.**

$$A = \frac{1}{2} \times (b_1 + b_2) \times h$$
$$= \frac{1}{2} \times (3\,\text{m} + 5\,\text{m}) \times 2\,\text{m}$$
$$= \frac{1}{2} \times (8\,\text{m}) \times 2\,\text{m} = 8\,\text{m}^2$$

Practice Questions

1. What are the area and perimeter of a square with a side of 9 miles?

2. Find the area and perimeter of a square with a side of 31 centimeters.

3. Figure out the area and perimeter of a rectangle with a length of 10 inches and a width of 5 inches.

4. Determine the area and perimeter of a rectangle that has a length of 23 kilometers and a width of 19 kilometers.

5. What's the area of a rhombus with a base of 9 meters and a height of 6 meters?

6. Figure out the area of a parallelogram with a base of 17 yards and a height of 13 yards.

7. Write down the area of a trapezoid with bases of 6 feet and 8 feet and a height of 5 feet.

8. What's the area of a trapezoid that has bases of 15 millimeters and 35 millimeters and a height of 21 millimeters?

Practice Answers

1. **Area is 81 square miles; perimeter is 36 miles.** Use the formulas for the area and perimeter of a square:

$$A = s^2 = (9\,\text{mi.})^2 = 81\,\text{mi.}^2$$
$$P = 4 \times s = 4 \times 9\,\text{mi.} = 36\,\text{mi.}$$

2. **Area is 961 square centimeters; perimeter is 124 centimeters.** Plug in 31 cm for s in the formulas for the area and perimeter of a square.

$$A = s^2 = (31\,\text{cm})^2 = 961\,\text{cm}^2$$
$$P = 4 \times s = 4 \times 31\,\text{cm} = 124\,\text{cm}$$

3. **Area is 50 square inches; perimeter is 30 inches.** Plug the length and width into the area and perimeter formulas for a rectangle:

$$A = l \times w = 10 \text{ in.} \times 5 \text{ in.} = 50 \text{ in.}^2$$
$$P = 2 \times (l + w) = 2 \times (10 \text{ in.} + 5 \text{ in.}) = 30 \text{ in.}$$

4. **Area is 437 square kilometers; perimeter is 84 kilometers.** Use the rectangle area and perimeter formulas:

$$A = l \times w = 23 \text{ km} \times 19 \text{ km} = 437 \text{ km}^2$$
$$P = 2 \times (l + w) = 2 \times (23 \text{ km} + 19 \text{ km}) = 84 \text{ km}$$

5. **54 square meters.** Use the parallelogram/rhombus area formula:

$$A = b \times h = 9 \text{ m} \times 6 \text{ m} = 54 \text{ m}^2$$

6. **221 square yards.** Use the parallelogram/rhombus area formula:

$$A = b \times h = 17 \text{ yd.} \times 13 \text{ yd.} = 221 \text{ yd.}^2$$

7. **35 square feet.** Plug your numbers into the trapezoid area formula:

$$A = \frac{1}{2} \times (b_1 + b_2) \times h$$
$$= \frac{1}{2} \times (6 \text{ ft.} + 8 \text{ ft.}) \times 5 \text{ ft.}$$
$$= \frac{1}{2} \times 14 \text{ ft.} \times 5 \text{ ft.}$$
$$= 35 \text{ ft.}^2$$

8. **525 square millimeters.** Use the trapezoid area formula:

$$A = \frac{1}{2} \times (b_1 + b_2) \times h$$
$$= \frac{1}{2} \times (15 \text{ mm} + 35 \text{ mm}) \times 21 \text{ mm}$$
$$= \frac{1}{2} \times 50 \text{ mm} \times 21$$
$$= 525 \text{ mm}^2$$

MAKING A TRIPLE PLAY WITH TRIANGLES

Any shape with three straight sides is a *triangle*. To find the area of a triangle, use the following formula, in which b is the length of the base (one side of the triangle) and h is the height of the triangle (the shortest distance from the base to the opposite corner); also see Figure 15-10:

$$A = \frac{1}{2} \times b \times h$$

Triangle

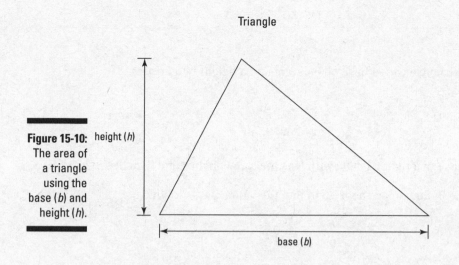

Figure 15-10:
The area of
a triangle
using the
base (*b*) and
height (*h*).

Any triangle that has a 90-degree angle is called a *right triangle*. The right triangle is one of the most important shapes in geometry. In fact, *trigonometry,* which is devoted entirely to the study of triangles, begins with a set of key insights and observations about right triangles.

The longest side of a right triangle (*c*) is called the *hypotenuse*, and the two short sides (*a* and *b*) are each called *legs*. The most important right triangle formula — called the *Pythagorean theorem* — allows you to find the length of the hypotenuse given only the length of the legs:

$$a^2 + b^2 = c^2$$

Figure 15-11 shows this theorem in action.

Right Triangle

Figure 15-11:
Using the
Pythagorean
theorem
to find the
hypotenuse
(*c*) of a right
triangle.

Examples

Q. Find the area of a triangle with a base of 5 meters and a height of 6 meters.

A. **15 square meters.**

$$A = \frac{1}{2} \times 5\,\text{m} \times 6\,\text{m} = 15\,\text{m}^2$$

Q. Find the hypotenuse of a right triangle with legs that are 6 inches and 8 inches in length.

A. **10 inches.** Use the Pythagorean theorem to find the value of c as follows:

$$a^2 + b^2 = c^2$$
$$6^2 + 8^2 = c^2$$
$$36 + 64 = c^2$$
$$100 = c^2$$

So when you multiply c by itself, the result is 100. Therefore, $c = 10$ in., because $10 \times 10 = 100$.

Practice Questions

1. What's the area of a triangle with a base of 7 centimeters and a height of 4 centimeters?

2. Find the area of a triangle with a base of 10 kilometers and a height of 17 kilometers.

3. Figure out the area of a triangle with a base of 2 feet and a height of 33 inches.

4. Discover the hypotenuse of a right triangle whose two legs measure 3 miles and 4 miles.

5. What's the hypotenuse of a right triangle with two legs measuring 5 millimeters and 12 millimeters?

6. Calculate the hypotenuse of a right triangle with two legs measuring 8 feet and 15 feet.

Practice Answers

1. **14 square centimeters.** Use the triangle area formula:

$$A = \frac{1}{2} \times b \times h = \frac{1}{2} \times 7\,\text{cm} \times 4\,\text{cm} = 14\,\text{cm}^2$$

2. **85 square kilometers.** Plug in the numbers for the base and height of the triangle:

$$A = \frac{1}{2} \times b \times h = \frac{1}{2} \times 10 \text{ km} \times 17 \text{ km} = 85 \text{ km}^2$$

3. **396 square inches.** First, convert feet to inches. Twelve inches are in 1 foot:

$$2 \text{ ft.} = 24 \text{ in.}$$

Now use the area formula for a triangle:

$$A = \frac{1}{2} \times b \times h$$
$$= \frac{1}{2} \times 24 \text{ in.} \times 33 \text{ in.}$$
$$= 396 \text{ in.}^2$$

Note: If you instead converted from inches to feet, the answer 2.75 square feet is also correct.

4. **5 miles.** Use the Pythagorean theorem to find the value of *c as follows:*

$$a^2 + b^2 = c^2$$
$$3^2 + 4^2 = c^2$$
$$9 + 16 = c^2$$
$$25 = c^2$$

When you multiply *c* by itself, the result is 25. Therefore,

$$c = 5 \text{ mi.}$$

5. **13 millimeters.** Use the Pythagorean theorem to find the value of *c:*

$$a^2 + b^2 = c^2$$
$$5^2 + 12^2 = c^2$$
$$25 + 144 = c^2$$
$$169 = c^2$$

When you multiply *c* by itself, the result is 169. The hypotenuse is longer than the longest leg, so *c* has to be greater than 12. Use trial and error, starting with 13:

$$13^2 = 169$$

Therefore, the hypotenuse is 13 mm.

6. **17 feet.** Use the Pythagorean theorem to find the value of *c:*

$$a^2 + b^2 = c^2$$
$$8^2 + 15^2 = c^2$$
$$64 + 225 = c^2$$
$$289 = c^2$$

When you multiply c by itself, the result is 289. The hypotenuse is longer than the longest leg, so c has to be greater than 15. Use trial and error, starting with 16:

$$16^2 = 256$$
$$17^2 = 289$$

Therefore, the hypotenuse is 17 ft.

GETTING AROUND WITH CIRCLE MEASUREMENTS

A *circle* is the set of all points that are a constant distance from a point inside it. Here are a few terms that are handy when talking about circles (see Figure 15-12):

- The *center (c)* of a circle is the point that's the same distance from any point on the circle itself.

- The *radius (r)* of a circle is the distance from the center to any point on the circle.

- The *diameter (d)* of a circle is the distance from any point on the circle through the center to the opposite point on the circle.

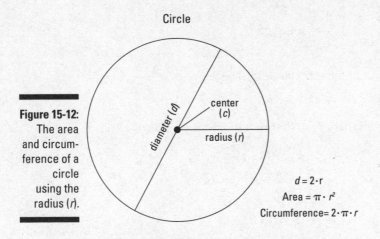

Circle

Figure 15-12:
The area
and circum-
ference of a
circle
using the
radius (*r*).

$d = 2 \cdot r$
Area $= \pi \cdot r^2$
Circumference $= 2 \cdot \pi \cdot r$

To find the area (*A*) of a circle, use the following formula:

$$A = \pi \times r^2$$

Remember: The symbol π is called *pi (pronounced pie)*. It's a decimal that goes on forever, so you can't get an exact value for π. However, the number 3.14 is a good approximation of π that you can use when solving problems that involve circles. (Note that when you use an approximation, the $\approx$ symbol replaces the = sign in problems.)

The perimeter of a circle has a special name: the *circumference (C)*. The formulas for the circumference of a circle also include π:

$$C = 2 \times \pi \times r$$
$$C = \pi \times d$$

These circumference formulas say the same thing because, as you can see in Figure 12-7, the diameter of a circle is always twice the radius of that circle. That gives you the following formula:

$$d = 2 \times r$$

Examples

Q. What's the diameter of a circle that has a radius of 3 inches?

A. **6 inches.**

$$d = 2 \times r = 2 \times 3 \text{ in.} = 6 \text{ in.}$$

Q. What's the approximate area of a circle that has a radius of 10 millimeters?

A. **314 square millimeters.**

$$A = \pi \times r^2$$
$$\approx 3.14 \times (10 \text{ mm})^2$$
$$\approx 3.14 \times 100 \text{ mm}^2$$
$$\approx 314 \text{ mm}^2$$

Q. What's the approximate circumference of a circle that has a radius of 4 feet?

A. **25.12 feet.**

$$C = 2\pi \times r$$
$$\approx 2 \times 3.14 \times 4 \text{ ft.}$$
$$\approx 25.12 \text{ ft.}$$

Practice Questions

1. What's the approximate area and circumference of a circle that has a radius of 3 kilometers?

2. Figure out the approximate area and circumference of a circle that has a radius of 12 yards.

3. Write down the approximate area and circumference of a circle with a diameter of 52 centimeters.

4. Find the approximate area and circumference of a circle that has a diameter of 86 inches.

Practice Answers

1. **Approximate area is 28.26 square kilometers; approximate circumference is 18.84 kilometers.** Use the area formula for a circle to find the area:

$$A = \pi \times r^2$$
$$\approx 3.14 \times (3 \text{ km})^2$$
$$= 3.14 \times 9 \text{ km}^2$$
$$= 28.26 \text{ km}^2$$

Use the circumference formula to find the circumference:

$$C = 2\pi \times r$$
$$\approx 2 \times 3.14 \times 3 \text{ km}$$
$$= 18.84 \text{ km}$$

2. **Approximate area is 452.16 square yards; approximate circumference is 75.36 yards.** Use the area formula for a circle to find the area:

$$A = \pi \times r^2$$
$$\approx 3.14 \times (12 \text{ yd.})^2$$
$$= 3.14 \times 144 \text{ yd.}^2$$
$$= 452.16 \text{ yd.}^2$$

Use the circumference formula to find the circumference:

$$C = 2\pi \times r$$
$$\approx 2 \times 3.14 \times 12 \text{ yd.}$$
$$= 75.36 \text{ yd.}$$

3. **Approximate area is 2,122.64 square centimeters; approximate circumference is 163.28 centimeters.** The diameter is 52 cm, so the radius is half of that, which is 26 cm. Use the area formula for a circle to find the area:

$$A = \pi \times r^2$$
$$\approx 3.14 \times (26 \text{ cm})^2$$
$$= 3.14 \times 676 \text{ cm}^2$$
$$= 2,122.64 \text{ cm}^2$$

Use the circumference formula to find the circumference:

$$C = 2\pi \times r$$
$$\approx 2 \times 3.14 \times 26 \text{ cm}$$
$$= 163.28 \text{ cm}$$

4. **Approximate area is 5,805.86 square inches; approximate circumference is 270.04 inches.** The diameter is 86 in., so the radius is half of that, which is 43 in. Use the area formula for a circle to find the area:

$$A = \pi \times r^2$$
$$\approx 3.14 \times (43 \text{ in.})^2$$
$$= 3.14 \times 1,849 \text{ in.}^2$$
$$= 5,805.86 \text{ in.}^2$$

Use the circumference formula to find the circumference:

$$C = 2\pi \times r$$
$$\approx 2 \times 3.14 \times 43 \text{ in.}$$
$$= 270.04 \text{ in.}$$

TAKING A TRIP TO ANOTHER DIMENSION: SOLID GEOMETRY

Solid geometry is the study of shapes in *space* — that is, the study of shapes in three dimensions. A *solid* is the spatial (three-dimensional, or 3-D) equivalent of a shape. Every solid has an *inside* and an *outside* separated by the surface of the solid. Here, I introduce you to a variety of solids.

The many faces of polyhedrons

A *polyhedron* is the three-dimensional equivalent of a polygon. As you may recall from earlier in the chapter, a polygon is a shape that has only straight sides. Similarly, a polyhedron is a solid that has only straight edges and flat faces (that is, faces that are polygons).

The most common polyhedron is the *cube* (see Figure 15-13). As you can see, a cube has 6 flat faces that are polygons — in this case, all the faces are square — and 12 straight edges. Additionally, a cube has eight *vertexes,* or *vertices* (corners). Later in this chapter, I show you how to measure the surface area and volume of a cube.

Figure 15-13: A typical cube.

Figure 15-14 shows a few common polyhedrons (or polyhedra).

Triangular prism   Hexagonal prism   Box

Figure 15-14:
Common
polyhedrons.

Cube   Pyramid

Later in this chapter, I show you how measure each of these polyhedrons to determine its volume — that is, the amount of space contained inside its surface.

One special set of polyhedrons is called the *five regular solids* (see Figure 15-15). Each regular solid has identical faces that are regular polygons. Notice that a cube is a type of regular solid. Similarly, the tetrahedron is a pyramid with four faces that are equilateral triangles.

3-D shapes with curves

Many solids aren't polyhedrons because they contain at least one curved surface. Here are a few of the most common of these types of solids (also see Figure 15-16):

- **Sphere:** A *sphere* is the solid, or three-dimensional, equivalent of a circle. A ball is a perfect visual aid for a sphere.

- **Cylinder:** A *cylinder* has a circular base and extends vertically from the plane. A good visual aid for a cylinder is a can of soup.

- **Cone:** A *cone* is a solid with a round base that extends vertically to a single point. A good visual aid for a cone is an ice-cream cone.

In the next section, I show you how to measure a sphere and a cylinder to determine their volume — that is, the amount of space contained within.

Figure 15-15:
The five
regular
solids.

Tetrahedron   Cube   Octahedron

Icosahedron   Dodecahedron

Figure 15-16:
Spheres,
cylinders,
and cones.

Sphere   Cylinder   Cone

BUILDING SOLID MEASUREMENT SKILLS

Solids take you into the real world, the third dimension. One of the simplest solids is the *cube*, a solid with six identical square faces. To find the volume of a cube, use the following formula, where *s* is the length of the side of any one face (check out Figure 15-17):

$$V = s^3$$

A *box* (also called a *rectangular solid*) is a solid with six rectangular faces. To find the volume of a box, use the following formula, where *l* is the length, *w* is the width, and *h* is the height (see Figure 15-18):

$$V = l \times w \times h$$

Cube

side (s)

Figure 15-17:
The volume
of a cube
using the
length of a
side (s).

Figure 15-18:
The volume
of a box
using the
length (*l*),
width (*w*),
and height
(*h*).

height (*h*)

width (*w*)

length (*l*)

A *prism* is a solid with two identical bases and a constant cross-section — that is, whenever you slice a prism parallel to the bases, the cross-section is the same size and shape as the bases. A *cylinder* is a solid with two identical circular bases and a constant cross-section. To find the volume of a prism or cylinder, use the following formula, where A_b is the area of the base and h is the height (see Figure 15-19). You can find the area formulas throughout this chapter:

$$V = A_b \times h$$

A *pyramid* is a solid that has a base that's a polygon (a shape with straight sides), with straight lines that extend from the sides of the base to meet at a single point. Similarly, a *cone* is a solid that has a base that's a circle, with straight lines extending from every point on the edge of the base to meet at a single point. The formula for the volume of a pyramid is the same as for the volume of a cone. In this formula, illustrated in Figure 15-20, A_b is the area of the base, and h is the height:

$$V = \frac{1}{3} \times A_b \times h$$

Remember: Volume measurements are usually in cubic units, such as cubic centimeters (cm^3) or cubic feet ($ft.^3$).

Figure 15-19:
The volume of a prism or cylinder using the area of the base (A_b) and the height (h).

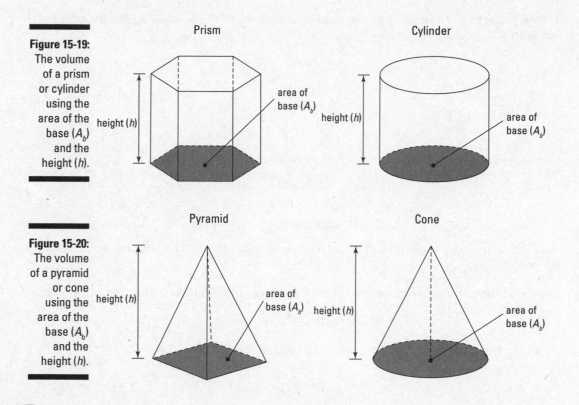

Prism

height (h)

area of base (A_b)

Cylinder

height (h)

area of base (A_b)

Figure 15-20:
The volume of a pyramid or cone using the area of the base (A_b) and the height (h).

Pyramid

height (h)

area of base (A_b)

Cone

height (h)

area of base (A_b)

Examples

Q. What's the volume of a cube with a side that measures 4 centimeters?

A. **64 cubic centimeters.**

$$V = s^3 = (4 \text{ cm})^3 = 4 \text{ cm} \times 4 \text{ cm} \times 4 \text{ cm} = 64 \text{ cm}^3$$

Q. Calculate the volume of a box with a length of 7 inches, a width of 4 inches, and a height of 2 inches.

A. **56 cubic inches.**

$$V = l \times w \times h = 7 \text{ in.} \times 4 \text{ in.} \times 2 \text{ in.} = 56 \text{ in.}^3$$

Q. Find the volume of a prism with a base that has an area of 6 square centimeters and a height of 3 centimeters.

A. **18 cubic centimeters.**

$$V = A_b \times h = 6 \text{ cm}^2 \times 3 \text{ cm} = 18 \text{ cm}^3$$

Q. Figure out the approximate volume of a cylinder with a base that has a radius of 2 feet and a height of 8 feet.

A. **100.48 cubic feet. To begin, find the approximate area of the base using the formula for the area of a circle:**

$$A_b = \pi \times r^2$$
$$\approx 3.14 \times (2 \text{ ft.})^2$$
$$\approx 3.14 \times 4 \text{ ft.}^2$$
$$\approx 12.56 \text{ ft.}^2$$

Now plug this result into the formula for the volume of a prism/cylinder:

$$V = A_b \times h$$
$$= 12.56 \text{ ft.}^2 \times 8 \text{ ft.}$$
$$= 100.48 \text{ ft.}^3$$

Q. Find the volume of a pyramid with a square base whose side is 10 inches and with a height of 6 inches.

A. **200 cubic inches.** First, find the area of the base using the formula for the area of a square:

$$A_b = s^2 = (10 \text{ in.})^2 = 100 \text{ in.}^2$$

Now plug this result into the formula for the volume of a pyramid/cone:

$$V = \frac{1}{3} \times A_b \times h$$
$$= \frac{1}{3} \times 100 \text{ in.}^2 \times 6 \text{ in.}$$
$$= 200 \text{ in.}^3$$

Q. Find the approximate volume of a cone with a base that has a radius of 2 meters and with a height of 3 meters.

A. **12.56 cubic meters.** First, find the approximate area of the base using the formula for the area of a circle:

$$A_b = \pi \times r^2$$
$$\approx 3.14 \times (2 \text{ m})^2$$
$$\approx 3.14 \times 4 \text{ m}^2$$
$$= 12.56 \text{ m}^2$$

Now plug this result into the formula for the volume of a pyramid/cone:

$$V = \frac{1}{3} \times A_b \times h$$
$$= \frac{1}{3} \times 12.56 \text{ m}^2 \times 3 \text{ m}$$
$$= 12.56 \text{ m}^3$$

Practice Questions

1. Find the volume of a cube that has a side of 19 meters.

2. Figure out the volume of a box with a length of 18 centimeters, a width of 14 centimeters, and a height of 10 centimeters.

3. Figure out the approximate volume of a cylinder whose base has a radius of 7 millimeters and whose height is 16 millimeters.

4. Find the approximate volume of a cone whose base has a radius of 3 inches and whose height is 8 inches.

Practice Answers

1. **6,859 cubic meters.** Substitute 19 m for s in the cube volume formula:

$$V = s^3 = (19 \text{ m})^3 = 6,859 \text{ m}^3$$

2. **2,520 cubic centimeters.** Use the box area formula:

$$V = l \times w \times h$$
$$= 18 \text{ cm} \times 14 \text{ cm} \times 10 \text{ cm} = 2,520 \text{ cm}^3$$

3. **Approximately 2,461.76 cubic millimeters.** First, use the area formula for a circle to find the area of the base:

$$A_b = \pi \times r^2$$
$$\approx 3.14 \times (7 \text{ mm})^2$$
$$= 3.14 \times 49 \text{ mm}^2$$
$$= 153.86 \text{ mm}^2$$

Plug this result into the formula for the volume of a prism/cylinder:

$$V = A_b \times h$$
$$= 153.86 \text{ mm}^2 \times 16 \text{ mm} = 2,461.76 \text{ mm}^3$$

4. **Approximately 75.36 cubic inches.** Use the area formula for a circle to find the area of the base:

$$A_b = \pi \times r^2$$
$$\approx 3.14 \times (3 \text{ in.})^2$$
$$= 3.14 \times 9 \text{ in.}^2$$
$$= 28.26 \text{ in.}^2$$

Plug this result into the formula for the volume of a pyramid/cone:

$$V = \frac{1}{3} \times A_b \times h = \frac{1}{3} \times 28.26 \text{ in.}^2 \times 8 \text{ in.} = 75.36 \text{ in.}^3$$

SOLVING GEOMETRY WORD PROBLEMS

Some geometry word problems present you with a picture. In other cases, you have to draw a picture yourself. Sketching figures is always a good idea because it can usually give you an idea of how to proceed. The following sections present you with both types of problems. (To solve these word problems, you need some of the geometry formulas I discuss in the previous section.)

Working from words and images

Sometimes you have to interpret a picture to solve a word problem. Read the problem carefully, recognize shapes in the drawing, pay attention to labels, and use whatever formulas you have to help you answer the question. In this problem, you get to work with a picture.

> Mr. Dennis is a farmer with two teenage sons. He gave them a rectangular piece of land with a creek running through it diagonally, as shown in Figure 15-21. The elder boy took the larger area, and the younger boy took the smaller. What is the area of each boy's land in square feet?

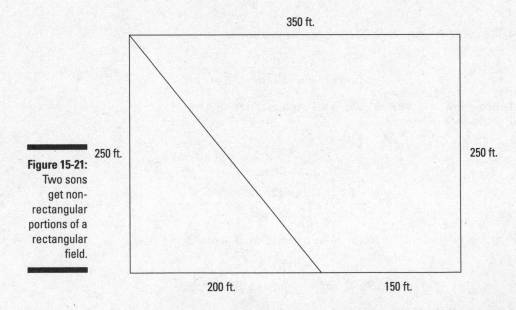

Figure 15-21: Two sons get non-rectangular portions of a rectangular field.

To find the area of the smaller triangular plot, use the formula for the area of a triangle, where A is the area, b is the base, and h is the height:

$$A = \frac{1}{2}(b \times h)$$

The whole piece of land is a rectangle, so you know that the corner the triangle shares with the rectangle is a right angle. Therefore, you know that the sides labeled 200 feet and

250 feet are the base and height. Find the area of this plot by plugging the base and height into the formula:

$$A = \frac{200 \text{ feet} \times 250 \text{ feet}}{2}$$

To make this calculation a little easier, notice that you can cancel a factor of 2 from the numerator and denominator:

$$A = \frac{\overset{100}{\cancel{200}} \text{ feet} \times 250 \text{ feet}}{\cancel{2}} = 25,000 \text{ square feeta}$$

The shape of the remaining area is a trapezoid. You can find its area by using the formula for a trapezoid, but there's an easier way. Because you know the area of the triangular plot, you can use this word equation to find the area of the trapezoid:

area of trapezoid = area of whole plot – area of triangle

To find the area of the whole plot, remember the formula for the area of a rectangle. Plug its length and width into the formula:

$$A = \text{length} \times \text{width}$$
$$A = 350 \text{ ft.} \times 250 \text{ ft.}$$
$$A = 87,500 \text{ square ft.}^2$$

Now just substitute the numbers that you know into the word equation you set up:

area of trapezoid = $87,500$ square feet $- 25,000$
$= 62,500$ square feet

So the area of the elder boy's land is 62,500 square feet, and the area of the younger boy's land is 25,000 square feet.

Breaking out those sketching skills

Geometry word problems may not make much sense until you draw some pictures. Here's an example of a geometry problem without a picture provided:

> In Elmwood Park, the flagpole is due south of the swing set and exactly 20 meters due west of the tree house. If the area of the triangle made by the flagpole, the swing set, and the tree house is 150 square meters, what is the distance from the swing set to the tree house?

This problem is bound to be confusing until you draw a picture of what it's telling you. Start with the first sentence, depicted in Figure 15-22. As you can see, I've drawn a right triangle whose corners are the swing set *(S)*, the flagpole *(F)*, and the tree house *(T)*. I've also labeled the distance from the flagpole to the tree house as 20 meters.

Figure 15-22:
A labeled
sketch
shows the
important
information
in a word
problem.

The next sentence tells you the area of this triangle:

$$A = 150\,\text{m}^2$$

Now you're out of information, so you need to remember anything you can from geometry. Because you know the area of the triangle, you may find the formula for the area of a triangle helpful:

$$A = \frac{1}{2}(b \times h)$$

Here b is the base and h is the height. In this case, you have a right triangle, so the base is the distance from F to T, and the height is the distance from S to F. So you already know the area of the triangle, and you also know the length of the base. Fill in the equation:

$$150 = \frac{1}{2}(20 \times h)$$

You can now solve this equation for h. Start by simplifying:

$$150 = 10 \times h$$
$$15 = h$$

Now you know that the height of the triangle is 15 meters, so you can add this information to your picture (see Figure 15-23).

Figure 15-23:
Update the
labels in
your sketch
as you work
through the
problem.

To solve the problem, though, you still need to find out the distance from S to T. Because this is a right triangle, you can use the Pythagorean theorem to figure out the distance:

$$a^2 + b^2 = c^2$$

Remember that a and b are the lengths of the short sides, and c is the length of the longest side, called the *hypotenuse*. (See the previous section for more on the Pythagorean theorem.) You can substitute numbers into this formula and solve, as follows:

$$15^2 + 20^2 = c^2$$
$$225 + 400 = c^2$$
$$625 = c^2$$
$$\sqrt{625} = \sqrt{c^2}$$
$$25 = c$$

So the distance from the swing set to the tree house is 25 meters.

Examples

Q. The long side of a rectangular field is 150 meters, and the diagonal length from one corner of the field to the opposite corner measures 170 meters. What is the area of the field?

A. **12,000 square meters.** Plug in the length of the field into the Pythagorean Theorem a and the diagonal length as c, and solve for b:

$$a^2 + b^2 = c^2$$
$$150^2 + b^2 = 170^2$$
$$22,500 + b^2 = 28,900$$
$$b^2 = 6,400$$
$$b = 80$$

Thus, the field has a width of 80 meters and a length of 150 meters:

$$A = 150 \times 80 = 12,000$$

Practice Questions

1. Becca and Sarah are sisters who share a room that measures 15 feet by 16 feet. Last year, they had a big fight and drew a diagonal line down the middle of the room, giving each girl the triangular half of the room containing her bed and desk. Can you find the area of each triangular half of the room?

2. A guard dog is kept in a circular area of approximately 1,017 square feet, limited by a chain attached to a stake in the ground. What is the length of the chain, to the nearest foot? (Use 3.14 to approximate the value of π.)

3. If a rectangular room has a width of 13 feet and a perimeter of 64 feet, what is its area?

4. The Great Pyramid at Giza has a height of 147 meters and a square base that measures 230 meters on each side. What is its volume?

5. A sign outside of a bowling alley includes a statue of a large bowling ball hitting ten pins. If the height of the bowling ball is 3.2 meters, what is its volume in cubic meters, to the nearest thousandth. (Use 3.14 to approximate the value of π.)

6. A small can of soup has a height of 4 inches and a radius of 1.5 inches, while a large can has a height of 6 inches and a radius of 2 inches. By volume, how much greater is the capacity of the large can than that of the small can? (Use 3.14 to approximate the value of π.)

7. A room has a height of 11 feet and a length of 40 feet. If the volume of the room measures 13,200 cubic feet, what is the distance from one corner of the floor to the opposite corner of the floor?

8. A circular room in a museum has diameter of 8 yards. At the center of the room is a large square pedestal whose edge measures 3 yards. To the nearest whole square yard, what is the area of the floor not covered by the pedestal? (Use 3.14 to approximate the value of π.)

Practice Answers

1. **120 square feet.** Plug the length and width of the room into the formula for the area of a triangle:

$$A = \frac{1}{2}bh = \frac{1}{2} \times 15 \times 16 = 120$$

2. **18 feet.** Plug 1,017 into the formula for the area of a circle, and solve for the radius:

$$A = \pi r^2$$
$$1,017 = 3.14r^2$$
$$323.88 \approx r^2$$
$$18 \approx r$$

3. **247 square feet.** Find the length of the room using the formula for the perimeter of a rectangle:

$$P = 2(l+w)$$
$$64 = 2(l+13)$$
$$64 = 2l + 26$$
$$38 = 2l$$
$$19 = l$$

So, the length of the room is 19 feet and the width is 13 feet. Plug these values into the formula for the area of a rectangle:

$$A = 19 \times 13 = 247$$

4. **2,592,100 cubic meters.** Plug the length of the side and the height of the pyramid into the formula for the volume of a pyramid:

$$V = \frac{1}{3} s^2 h = \frac{1}{3} 230^2 (147) = 2,592,100$$

5. **17.149 cubic meters.** The height of the bowling ball is 3.2 meters, so its radius is 1.6 meters. Plug this value into the formula for the volume of a sphere:

$$V = \frac{4}{3} \pi r^3 = \frac{4}{3} (3.14)(1.6)^3 \approx 17.149$$

6. **47.1 cubic inches.** Begin by finding the volume of the two cans, using the formula for the volume of a cylinder:

$$\text{Volume of large can} = \pi r^2 h = (3.14)(1.5^2)(4) = 28.26$$
$$\text{Volume of small can} = \pi r^2 h = (3.14)(2^2)(6) = 75.36$$

Now, subtract the volume of the large can minus that of the small can:

$$75.36 - 28.26 = 47.1$$

7. **50 feet.** To begin, plug the room's volume, height, and length into the formula for the volume of a box and calculate the width:

$$V = lwh$$
$$13,200 = (40)w(11)$$
$$13,200 = 440w$$
$$30 = w$$

Now, plug the width and the length into the Pythagorean Theorem as a and b and solve for c:

$$a^2 + b^2 = c^2$$
$$30^2 + 40^2 = c^2$$
$$900 + 1,600 = c^2$$
$$2,500 = c^2$$
$$50 = c$$

8. **41.24 square yards.** The room is a circle with a diameter of 8 yards, so its radius is 4:

$$\text{Area of room} = \pi r^2 = \pi(4^2) = 16\pi \approx 50.24$$

The pedestal is a square with a side of 3 yards:

$$\text{Area of pedestal} = s^2 = 3^2 = 9$$

Now, subtract the area of the room minus the area of the pedestal:

$$50.24 - 9 = 41.24$$

16

FIGURING YOUR CHANCES:
STATISTICS AND PROBABILITY

Statistics and probability are two of the most important and widely used applications of math. They're applicable to virtually every aspect of the real world — business, biology, city planning, politics, meteorology, and many more areas of study. Even physics, once thought to be devoid of uncertainty, now relies on probability.

In this chapter, I give you a basic understanding of these two mathematical ideas. First, I introduce you to statistics and the important distinction between qualitative and quantitative data. I show you how to work with both types of data to find meaningful answers. Then I give you the basics of probability. I show you how the probability that an event will occur is always a number from 0 to 1 — that is, usually a fraction, decimal, or percent. After that, I demonstrate how to build this number by counting both favorable outcomes and possible outcomes. Finally, I put these ideas to work by showing you how to calculate the probability of tossing coins.

GATHERING DATA MATHEMATICALLY: BASIC STATISTICS

Statistics is the science of gathering and drawing conclusions from data, which is information that's measured objectively in an unbiased, reproducible way.

An individual *statistic* is a conclusion drawn from this data. Here are some examples:

- The average working person drinks 3.7 cups of coffee every day.
- Only 52% of students who enter law school actually graduate.
- The cat is the most popular pet in the United States.
- In the last year, the cost of a high-definition TV dropped by an average of $575.

Statisticians do their work by identifying a population that they want to study: working people, law students, pet owners, buyers of electronics, whoever. Because most populations are far too large to work with, a statistician collects data from a smaller, randomly selected sample of this population. Much of statistics concerns itself with gathering data that's reliable and accurate. You can read all about this idea in *Statistics For Dummies,* 2nd Edition, by Deborah J. Rumsey (Wiley).

In this section, I give you a short introduction to the more mathematical aspects of statistics.

Understanding differences between qualitative and quantitative data

Data — the information used in statistics — can be either qualitative or quantitative. *Qualitative data* divides a data set (the pool of data that you've gathered) into discrete chunks based on a specific attribute. For example, in a class of students, qualitative data can include

- Each child's gender
- His or her favorite color
- Whether he or she owns at least one pet
- How he or she gets to and from school

Remember: You can identify qualitative data by noticing that it links an attribute — that is, a quality — to each member of the data set. For example, four attributes of Emma are that she's female, her favorite color is green, she owns a dog, and she walks to school.

On the other hand, *quantitative data* provides numerical information — that is, information about quantities, or amounts. For example, quantitative data on this same classroom of students can include the following:

- Each child's height in inches
- Each child's weight in pounds
- The number of siblings each child has
- The number of words each child spelled correctly on the most recent spelling test

Remember: You can identify quantitative data by noticing that it links a number to each member of the data set. For example, Carlos is 55 inches tall, weighs 68 pounds, has three siblings, and spelled 18 words correctly on his last spelling test.

Working with qualitative data

Qualitative data usually divides a sample into discrete chunks. As my sample — which is purely fictional — I use 25 children in Sister Elena's fifth-grade class. For example, suppose all 25 children in Sister Elena's class answer the three yes/no questions in Table 16-1.

Table 16-1 Sister Elena's Fifth-Grade Survey

Question	Yes	No
Are you an only child?	5	20
Do you own any pets?	14	11
Do you take the bus to school?	16	9

The students also answer the question "What is your favorite color?" with the results in Table 16-2.

Table 16-2 Favorite Colors in Sister Elena's Class

Color	Number of Students	Color	Number of Students
Blue	8	Orange	1
Red	6	Yellow	1
Green	5	Gold	1
Purple	3		

Even though the information that each child provided is non-numerical, you can handle it numerically by counting how many students made each response and working with these numbers.

Given this information, you can now make informed statements about the students in this class just by reading the charts. For instance,

- Exactly 20 children have at least one brother or sister.
- Nine children don't take the bus to school.
- Only one child's favorite color is yellow.

Playing the percentages

You can make more sophisticated statistical statements about qualitative data by finding out the percentage of the sample that has a specific attribute. Here's how you do so:

1. **Write a statement that includes the number of members who share that attribute and the total number in the sample.**

   Suppose you want to know what percentage of students in Sister Elena's class are only children. The chart tells you that 5 students have no siblings, and you know that 25 kids are in the class. So you can begin to answer this question as follows:

   Five out of 25 children are only children.

2. **Rewrite this statement, turning the numbers into a fraction:**

$$\frac{\text{number who share attribute}}{\text{number in sample}} = \frac{5}{25}$$

In the example, $\frac{5}{25}$ of the children are only children.

3. **Turn the fraction into a percent, using the method I show you in Chapter 11.**

You find that $\frac{5}{25} = \frac{1}{5} = 0.2$, so 20% of the children are only children.

Similarly, suppose you want to find out what percentage of children take the bus to school. This time, the chart tells you that 16 children take the bus, so you can write this statement:

Sixteen out of 25 children take the bus to school.

Now rewrite the statement as follows:

$\frac{16}{25}$ of the children take the bus to school.

Finally, turn this fraction into a percent: 16 ÷ 25 = 0.64, which equals 64%, so

64% of the children take the bus to school.

Getting into the mode

The *mode* tells you the most popular answer to a statistical question. For example, in the poll of Sister Elena's class (see Tables 16-1 and 16-2), the mode groups are children who

- Have at least one brother or sister (20 students)
- Own at least one pet (14 students)
- Take the bus to school (16 students)
- Chose blue as their favorite color (8 students)

Remember: When a question divides a data set into two parts (as with all yes/no questions), the mode group represents more than half of the data set. But when a question divides a data set into more than two parts, the mode doesn't necessarily represent more than half of the data set.

For example, 14 children own at least one pet, and the other 11 children don't own one. So the mode group — children who own a pet — is more than half the class. But 8 of the 25 children chose blue as their favorite color. So even though this is the mode group, fewer than half the class chose this color.

With a small sample, you may have more than one mode — for example, perhaps the number of students who like red is equal to the number who like blue. However, getting multiple modes isn't usually an issue with a larger sample because it becomes less likely that exactly the same number of people will have the same preference.

Examples

Q. When 25 children were asked, "Who usually helps you with your homework?", their answers were as follows:

Mother: 14

Father: 5

Older sibling: 3

Grandparent: 2

Aunt or uncle: 1

a. Which answer is the mode?

b. What percent of the children answered "Father."

c. What percent of the children named somebody other than a parent?

A. **a.** **Mother.** The most popular answer was "Mother."

b. **20%.** Five children out of 25 answered "Father," so divide 5 by 25 ($5 \div 25 = 0.2 = 20\%$).

c. **24%.** Six children out of 25 named somebody other than a parent ($3 + 2 + 1 = 6$), so divide 6 by 25 ($6 \div 25 = 0.24 = 24\%$)

Practice Questions

1. When 48 people were asked, "If you could go anywhere, where would you most like to take your next vacation?", their answers were as follows:

Europe: 16

North America: 14

Asia: 9

South America: 4

Someplace Else: 5

a. What is the mode answer?

b. What percentage of people chose Asia?

c. What percentage of people chose either North America or South America?

d. What percentage of people chose a place other than Europe?

2. When 80 people were asked their opinion of a politician, they answered as follows:

Very favorable: 16

Somewhat favorable: 30

Somewhat unfavorable: 22

Very unfavorable: 8

No opinion: 4

   a. What is the mode answer?

   b. What percentage of people answered either "Very favorable" or "Somewhat favorable?"

   c. What percentage of people answered either "Somewhat favorable" or "Somewhat unfavorable?"

   d. What percentage of people had no opinion?

A Practice Answers

1. **a.** **Europe.** The mode answer is the most popular, which is Europe.

   b. **18.75%.** Nine people out of 48 chose Asia, so divide 9 by 48 (9 ÷ 48 = 0.1875 = 18.75%).

   c. **37.5%.** Eighteen people (14 + 4 = 18) out of 48 chose either North America or South America, so divide 18 by 48 (18 ÷ 48 = 0.375 = 37.5%).

   d. **$66\frac{2}{3}$%.** Thirty-two people out of 48 chose a place other than Europe (48 – 16 = 32), so divide 32 by 48:

   $$\frac{32}{48} = \frac{2}{3} = 66\frac{2}{3}\%$$

2. **a.** **Somewhat favorable.** The mode answer is the most popular, which is "Somewhat favorable."

   b. **57.5%.** Forty-six people (30 + 16 = 46) answered either "Very favorable" or "Somewhat favorable," so divide 46 by 80 (46 ÷ 80 = 0.575 = 57.5%).

   c. **65%.** Fifty-two people (22 + 30 = 52) answered either "Somewhat favorable" or "Somewhat unfavorable," so divide 52 by 80 (52 ÷ 80 = 0.65 = 65%).

   d. **5%.** Four people answered "No opinion," so divide 4 by 80 (4 ÷ 80 = 0.05 = 5%).

Working with quantitative data

Quantitative data assigns a numerical value to each member of the sample. As my sample — again, fictional — I use five members of Sister Elena's basketball team. Suppose that the information in Table 16-3 has been gathered about each team member's height and most recent spelling test.

In this section, I show you how to use this information to find the mean and median for both sets of data. Both terms refer to ways to calculate the average value in a quantitative data set. An *average* gives you a general idea of where most individuals in a data set fall so you know what kinds of results are standard. For example, the average height of Sister Elena's fifth-grade class is probably less than the average height of the Los Angeles Lakers. As I show you in the sections that follow, an average can be misleading in some cases, so knowing when to use the mean versus the median is important.

Table 16-3   Height and Spelling Test Scores

Student	Height in Inches	Number of Words Spelled Correctly
Carlos	55	18
Dwight	60	20
Patrick	59	14
Tyler	58	17
William	63	18

Finding the mean

Remember: The mean is the most commonly used average. In fact, when most people use the word *average,* they're referring to the mean. Here's how you find the mean of a set of data:

1. **Add up all the numbers in that set.**

   For example, to find the average height of the five team members, first add up all their heights:

   $$55 + 60 + 59 + 58 + 63 = 295$$

2. **Divide this result by the total number of members in that set.**

   Divide 295 by 5 (that is, by the total number of boys on the team):

   $$295 \div 5 = 59$$

So the mean height of the boys on Sister Elena's team is 59 inches.

This procedure is summed up (so to speak) in simple formula:

$$\text{mean} = \frac{\text{sum of values}}{\text{number of values}}$$

You can use this formula to find the mean number of words that the boys spelled correctly. To do this, plug the number of words that each boy spelled correctly into the top part of the formula, and then plug the number of boys in the group into the bottom part:

$$\text{mean} = \frac{18 + 20 + 14 + 17 + 18}{5}$$

Now simplify to find the result:

$$= \frac{87}{5} = 17.4$$

As you can see, when you divide, you end up with a decimal in your answer. If you round to the nearest whole word, the mean number of words that the five boys spelled correctly is about 17 words. (For more information about rounding, see Chapter 2.)

Warning: The mean can be misleading when you have a strong skew in data — that is, when the data has many low values and a few very high ones, or vice versa.

For example, suppose that the president of a company tells you, "The average salary in my company is $200,000 a year!" But on your first day at work, you find out that the president's salary is $19,010,000 and each of his 99 employees earns $10,000. To find the mean, first plug the total salaries ($19,010,000 for the president plus $10,000 for each of 99 employees) into the top of the formula. Next, plug the number of employees (100) into the bottom:

$$\text{mean} = \frac{\$19,010,000 + (\$10,000 \times 99)}{100}$$

Now calculate:

$$= \frac{\$19,010,000 + \$990,000}{100} = \frac{\$20,000,000}{100} = \$200,000$$

So the president didn't lie. However, the skew in salaries resulted in a misleading mean.

Finding the median

When data values are skewed (when a few very high or very low numbers differ significantly from the rest of the data), the median can give you a more accurate picture of what's standard. Here's how to find the median of a set of data:

1. **Arrange the set from lowest to highest.**

   To find the median height of the boys in Table 16-3, arrange their five heights in order from lowest to highest.

   55   58   <u>59</u>   60   63

2. **Choose the middle number.**

   The middle value, 59 inches, is the median average height.

To find the median number of words that the boys spelled correctly (refer to Table 19-3), arrange their scores in order from lowest to highest:

$$14 \quad 17 \quad \underline{18} \quad 18 \quad 20$$

This time, the middle value is 18, so 18 is the median score.

Remember: If you have an even number of values in the data set, put the numbers in order and find the mean of the *two middle numbers* in the list (see the preceding section for details on the mean). For instance, consider the following:

$$2 \quad 3 \quad \underline{5} \quad \underline{7} \quad 9 \quad 11$$

The two center numbers are 5 and 7. Add them together to get 12, and then divide by 2 to get their mean. The median in this list is 6.

Now recall the company president who makes $19,010,000 a year and his 99 employees who each earn $10,000. Here's how this data looks:

$$10,000 \quad 10,000 \quad 10,000\ldots \quad 10,000 \quad 19,010,000$$

As you can see, if you wrote out all 100 salaries, the center numbers would obviously both be 10,000. The median salary is $10,000, and this result is much more reflective of what you'd probably earn if you worked at this company.

Examples

Q. Ten basketball players were each given five free throws. In order, the number of successful free throws were as follows:

$$2, 3, 4, 1, 4, 2, 5, 3, 1, \text{ and } 2.$$

a. What is the mean number of successful free throws, rounded to the nearest tenth?

b. What is the median number of free throws?

A. **a.** **2.7.** To find the mean, find the sum of successful free throws ($2 + 3 + 4 + 1 + 4 + 2 + 5 + 3 + 1 + 2 = 27$) and divide this number by 10 ($27 \div 10 = 2.7$).

b. **2.5.** To find the median, arrange the list in order from smallest to largest and find the two middle numbers:

$$1, 1, 2, 2, \mathbf{2, 3}, 3, 4, 4, 5$$

Take the average of the two middle numbers ($2 + 3 = 5$; $5 \div 2 = 2.5$).

Practice Questions

1. Thirteen children took a pop quiz scored on a scale from 0 to 10. Their scores, in order from least to greatest, were as follows

$$4, 6, 6, 7, 8, 8, 8, 8, 9, 9, 9, 10, 10$$

   a. What is the mean score on the quiz, rounded to the nearest tenth?

   b. What is the median score on the quiz?

2. Sixteen people were asked how many cups of coffee they drink each day. Their answers, in the order that they were asked, were as follows:

$$3, 0, 1, 2, 2, 3, 3, 2, 0, 5, 2, 2, 2, 1, 0, 1$$

   a. What is the mean number of cups of coffee, rounded to the nearest tenth?

   b. What is the median number of cups of coffee?

Practice Answers

1. **a.** **7.8.** To find the mean score, add the 13 scores ($4 + 6 + 6 + 7 + 8 + 8 + 8 + 8 + 9 + 9 + 9 + 10 + 10 = 102$) and divide their sum by 13 ($102 \div 13 \approx 7.2$)

   b. **8.** The median score is the middle number in the list:

$$4, 6, 6, 7, 8, 8, \mathbf{8}, 8, 9, 9, 9, 10, 10$$

2. **a.** **1.8.** To find the mean score, add the 16 numbers ($3 + 0 + 1 + 2 + 2 + 3 + + 3 + 2 + 0 + 5 + 2 + 2 + 2 + 1 + 0 + 1 = 29$) and divide their sum by 16 ($29 \div 16 = 1.8125 \approx 1.8$)

   b. **2.** To find the median, arrange the answers in order from smallest to largest and find the two middle numbers:

$$0, 0, 0, 1, 1, 1, 2, \mathbf{2, 2}, 2, 2, 2, 3, 3, 3, 5,$$

So the median is 2.

LOOKING AT LIKELIHOODS: BASIC PROBABILITY

Probability is the mathematics of deciding how likely an event is to occur. For example,

- What's the likelihood that the lottery ticket I bought will win?
- What's the likelihood that my new car will need repairs before the warranty runs out?
- What's the likelihood that more than 100 inches of snow will fall in Manchester, New Hampshire, this winter?

The Silver standard

Probability can be a powerful tool for predicting weather patterns, sports events, and election results. In his bestselling book *The Signal and the Noise*, Nate Silver discusses how statistical modeling, when done correctly, can permit mathematicians to peer into the future with spooky accuracy. He also discusses why a lot of apparently scientific predictions go wrong. Silver's work is cutting edge, and he does a good job of explaining what statisticians do without too much jargon or complicated equations. Check him out!

Probability has a wide variety of applications in insurance, weather prediction, biological sciences, and even physics. You can read all about the details of probability in *Probability For Dummies*, by Deborah J. Rumsey (Wiley). In this section, I give you a little taste of this fascinating subject.

Figuring the probability

The *probability* that an event will occur is a fraction whose numerator (top number) and denominator (bottom number) are as follows (for more on fractions, flip to Chapter 8):

$$\text{probability} = \frac{\text{target outcomes}}{\text{total outcomes}}$$

In this case, the number of *target outcomes* (or *successes*) is simply the number of outcomes in which the event you're examining does happen. In contrast, the number of *total outcomes* (or *sample space*) is the number of outcomes that *can* happen.

For example, suppose you want to know the probability that a tossed coin will land heads up. Notice that there are two total outcomes (heads or tails), but only one of these outcomes is the target — the outcome in which heads comes up. To find the probability of this event, make a fraction as follows:

$$\text{probability} = \frac{1}{2}$$

So the probability that the coin will land heads up is $\frac{1}{2}$.

So what's the probability that, when you roll a die, the number 3 will land face up? To figure this one out, notice that there are *six* total outcomes (1, 2, 3, 4, 5, and 6), but in only *one* of these does 3 land face up. To find the probability of this outcome, make a fraction as follows:

$$\text{probability} = \frac{1}{6}$$

So the probability that the number 3 will land face up is $\frac{1}{6}$.

And what's the probability that, if you pick a card at random from a deck, it'll be an ace? To figure this out, notice that there are *52* total outcomes (one for each card in the deck), but in only *4* of these do you pick an ace. So

$$probability = \frac{4}{52}$$

So the probability that you'll pick an ace is $\frac{4}{52}$, which reduces to $\frac{1}{13}$ (see Chapter 8 for more on reducing fractions).

Remember: Probability is always a number from 0 to 1. When the probability of an outcome is 0, the outcome is *impossible*. When the probability of an outcome is 1, the outcome is *certain*.

Oh, the possibilities! Counting outcomes with multiple coins

Although the basic probability formula isn't difficult, sometimes finding the numbers to plug into it can be tricky. One source of confusion is in counting the number of outcomes, both target and total. In this section, I focus on tossing coins.

When you flip a coin, you can generally get two total outcomes: heads or tails. When you flip two coins at the same time — say, a penny and a nickel — you can get four total outcomes:

Outcome	Penny	Nickel
#1	Heads	Heads
#2	Heads	Tails
#3	Tails	Heads
#4	Tails	Tails

When you flip three coins at the same time — say, a penny, a nickel, and a dime — eight outcomes are possible:

Outcome	Penny	Nickel	Dime
#1	Heads	Heads	Heads
#2	Heads	Heads	Tails
#3	Heads	Tails	Heads
#4	Heads	Tails	Tails
#5	Tails	Heads	Heads
#6	Tails	Heads	Tails
#7	Tails	Tails	Heads
#8	Tails	Tails	Tails

Notice the pattern: Every time you add a coin, the number of total outcomes doubles. So if you flip six coins, here's how many total outcomes you have:

$$2 \times 2 \times 2 \times 2 \times 2 \times 2 = 64$$

The number of total outcomes equals the number of outcomes per coin (2) raised to the number of coins (6): Mathematically, you have $2^6 = 64$.

Tip: Here's a handy formula for calculating the number of outcomes when you're flipping, shaking, or rolling multiple coins, dice, or other objects at the same time:

$$\text{total outcomes} = \text{number of outcomes per object}^{\text{number of objects}}$$

Suppose you want to find the probability that six tossed coins will all fall heads up. To do this, you want to build a fraction, and you already know that the denominator — the number of total outcomes — is 64. Only one outcome is the target outcome, so the numerator is 1:

$$\text{probability} = \frac{1}{64}$$

So the probability that six tossed coins will all fall heads up is $\frac{1}{64}$.

Here's a more subtle question: What's the probability that exactly five out of six tossed coins will all fall heads up? Again, you're building a fraction, and you already know that the denominator is 64. To find the numerator (target outcomes), think about it this way: If the first coin falls tails up, then all the rest must fall heads up. If the second coin falls tails up, then again all the rest must fall heads up. This is true of all six coins, so you have six target outcomes:

$$\text{probability} = \frac{6}{64}$$

Therefore, the probability that exactly five out of six coins will fall heads up is $\frac{6}{64}$, which reduces to $\frac{3}{32}$ (see Chapter 8 for more on reducing fractions).

Examples

Q. Suppose that you pick a random card from a deck of 52 cards:

   a. How many possible outcomes are there?

   b. What is the probability that the card is a heart? (*Hint:* There are 13 hearts in the deck.)

A. **a.** **52.** The deck has 52 cards, so there are 52 possible outcomes.

   b. $\frac{1}{4}$. There are a total of 52 possible outcomes, and 13 target outcomes:

$$\frac{\text{Target outcomes}}{\text{Total outcomes}} = \frac{13}{52} = \frac{1}{4}$$

Q Practice Questions

1. How many possible outcomes are there in each of the following cases?

   a. Flipping 4 coins

   b. Flipping 8 coins

   c. Rolling a pair of dice

   d. Rolling 4 dice

2. If you roll a six-sided die with the numbers 1 through 6 on it, what is the probability of each of the following outcomes?

   a. Rolling a 6

   b. Rolling either a 2 or a 3

   c. Rolling an odd number

   d. Rolling any number but a 1

3. If you flip three coins, what is the probability of each of the following outcomes?

   a. All three coins are heads.

   b. Exactly one coin is heads and exactly two are tails.

   c. At least one coin is heads and at least one coin is tails.

   d. At least two coins are heads and at least two coins are tails.

4. Suppose you pick one card at random from a deck of 52 cards. What is the probability of each of the following outcomes?

   a. The card is the ace of spades.

   b. The card is one of the four jacks.

   c. The card is either the jack of diamonds or the queen of spades.

   d. The card is not one of the 13 clubs.

A Practice Answers

1. a. **16.** There are 4 events, each with 2 possible outcomes ($2^4 = 16$).

   b. **256.** There are 8 events, each with 2 possible outcomes ($2^8 = 256$).

   c. **36.** There are 2 events, each with 6 possible outcomes ($6^2 = 36$).

   d. **1,296.** There are 4 events, each with 6 possible outcomes ($6^4 = 1,296$).

2. In each case, there are a total of 6 possible outcomes.

    a. $\frac{1}{6}$. There is 1 possible outcome:

$$\frac{\text{Target outcomes}}{\text{Total outcomes}} = \frac{1}{6}$$

    b. $\frac{1}{3}$. There are 2 possible outcomes:

$$\frac{\text{Target outcomes}}{\text{Total outcomes}} = \frac{2}{6} = \frac{1}{3}$$

    c. $\frac{1}{2}$. There are 3 possible outcomes:

$$\frac{\text{Targe outcomes}}{\text{Total outcomes}} = \frac{3}{6} = \frac{1}{2}$$

    d. $\frac{5}{6}$. There are 5 possible outcomes:

$$\frac{\text{Target outcomes}}{\text{Total outcomes}} = \frac{5}{6}$$

3. In each case, there are a total of 8 possible outcomes.

    a. $\frac{1}{8}$. There is one possible outcome:

$$\frac{\text{Target outcomes}}{\text{Total outcomes}} = \frac{1}{8}$$

    b. $\frac{3}{8}$. There are three possible outcomes (see the table earlier in this section):

$$\frac{\text{Target outcomes}}{\text{Total outcomes}} = \frac{3}{8}$$

    c. $\frac{3}{4}$. There are only 2 outcomes that are not target outcomes (3 heads and 3 tails), so there are 6 target outcomes:

$$\frac{\text{Target outcomes}}{\text{Total outcomes}} = \frac{6}{8} = \frac{3}{4}$$

    d. **0.** There are 0 target outcomes:

$$\frac{\text{Target outcomes}}{\text{Total outcomes}} = \frac{0}{8} = 0$$

4. In each case, the total number of possible outcomes is 52.

   a. $\frac{1}{52}$. There is only one target outcome:

   $$\frac{\text{Target outcomes}}{\text{Total outcomes}} = \frac{1}{52}$$

   b. $\frac{1}{13}$. There are four target outcomes:

   $$\frac{\text{Target outcomes}}{\text{Total outcomes}} = \frac{4}{52} = \frac{1}{13}$$

   c. $\frac{1}{26}$. There are two target outcomes:

   $$\frac{\text{Target outcomes}}{\text{Total outcomes}} = \frac{2}{52} = \frac{1}{26}$$

   d. $\frac{3}{4}$. There are 13 clubs and 39 other cards, so there are 39 target outcomes:

   $$\frac{\text{Target outcomes}}{\text{Total outcomes}} = \frac{39}{52} = \frac{3}{4}$$

17

SETTING THINGS UP WITH BASIC SET THEORY

IN THIS CHAPTER

- Defining a set and its elements

- Understanding subsets and the empty set

- Knowing the basic operations on sets, including union and intersection

A *set* is just a collection of things. But in their simplicity, sets are profound. At the deepest level, set theory is the foundation for everything in math.

Set theory provides a way to talk about collections of numbers, such as even numbers, prime numbers, or counting numbers, with ease and clarity. It also gives rules for performing calculations on sets that become useful in higher math. For these reasons, set theory becomes more important the higher up you go the math food chain — especially when you begin writing mathematical proofs. Studying sets can also be a nice break from the usual math stuff you work with.

In this chapter, I show you the basics of set theory. First, I show you how to define sets and their elements and how you can tell when two sets are equal. I also show you the simple idea of a set's cardinality. Next, I discuss subsets and the all-important empty set ($\varnothing$). After that, I discuss four operations on sets: union, intersection, relative complement, and complement.

UNDERSTANDING SETS

A *set* is a collection of things, in any order. These things can be buildings, earmuffs, lightning bugs, numbers, qualities of historical figures, names you call your little brother, whatever.

Remember: You can define a set in a few main ways:

- **Placing a list of the elements of the set in braces { }:** You can simply list everything that belongs in the set. When the set is too large, you use an ellipsis (. . .) to indicate elements of the set not mentioned. For example, to list the set of numbers from 1 to 100, you can write {1, 2, 3, . . . , 100}. To list the set of all the counting numbers, you can write {1, 2, 3, . . .}.

- **Using a verbal description:** If you use a verbal description of what the set includes, make sure the description is clear and unambiguous so you know exactly what's in the set and what isn't. For instance, the set of the four seasons is pretty clear-cut, but you may run into some debate on the set of words that describe my cooking skills because different people have different opinions.

- **Writing a mathematical rule (set-builder notation):** In later algebra, you can write an equation that tells people how to calculate the numbers that are part of a set. Check out *Algebra II For Dummies,* by Mary Jane Sterling (Wiley), for details.

Sets are usually identified with capital letters to keep them distinct from variables in algebra, which are usually small letters. (Chapter 18 talks about using variables.)

The best way to understand sets is to begin working with them. For example, here I define three sets:

A = {Empire State Building, Eiffel Tower, Roman Colosseum}

B = {Albert Einstein's intelligence, Marilyn Monroe's talent, Joe DiMaggio's athletic ability, Sen. Joseph McCarthy's ruthlessness}

C = the four seasons of the year

Set A contains three tangible objects: famous works of architecture. Set B contains four intangible objects: attributes of famous people. And set C also contains intangible objects: the four seasons. Set theory allows you to work with either tangible or intangible objects, provided that you define your set properly. In the following sections, I show you the basics of set theory.

Elementary, my dear: Considering what's inside sets

The things contained in a set are called *elements* (also known as *members*). Consider the first two sets I define in the section intro:

A = {Empire State Building, Eiffel Tower, Roman Colosseum}

B = {Albert Einstein's intelligence, Marilyn Monroe's talent, Joe DiMaggio's athletic ability, Sen. Joseph McCarthy's ruthlessness}

The Eiffel Tower is an element of A, and Marilyn Monroe's talent is an element of B. You can write these statements using the symbol $\in$, which means "is an element of":

$$\text{Eiffel Tower} \in A$$
$$\text{Marilyn Monroe's talent} \in B$$

However, the Eiffel Tower is not an element of B. You can write this statement using the symbol $\notin$, which means "is not an element of":

$$\text{Eiffel Tower} \notin B$$

These two symbols become more common as you move higher in your study of math. The following sections discuss what's inside those braces and how some sets relate to each other.

Cardinality of sets

The *cardinality* of a set is just a fancy word for the number of elements in that set.

When A is {Empire State Building, Eiffel Tower, Roman Colosseum}, it has three elements, so the cardinality of A is three. Set B, which is {Albert Einstein's intelligence, Marilyn Monroe's talent, Joe DiMaggio's athletic ability, Sen. Joseph McCarthy's ruthlessness}, has four elements, so the cardinality of B is four.

Equal sets

Remember: If two sets list or describe the exact same elements, the sets are equal (you can also say they're *identical* or *equivalent*). The order of elements in the sets doesn't matter. Similarly, an element may appear twice in one set, but only the distinct elements need to match.

Suppose I define some sets as follows:

C = the four seasons of the year

D = {spring, summer, fall, winter}

E = {fall, spring, summer, winter}

F = {summer, summer, summer, spring, fall, winter, winter, summer}

Set C gives a clear rule describing a set. Set D explicitly lists the four elements in C. Set E lists the four seasons in a different order. And set F lists the four seasons with some repetition. Thus, all four sets are equal. As with numbers, you can use the equals sign to show that sets are equal:

C = D = E = F

Subsets

When all the elements of one set are completely contained in a second set, the first set is a subset of the second. For example, consider these sets:

C = {spring, summer, fall, winter}

G = {spring, summer, fall}

As you can see, every element of G is also an element of C, so G is a subset of C. The symbol for subset is ⊂, so you can write the following:

G ⊂ C

Every set is a subset of itself. This idea may seem odd until you realize that all the elements of any set are obviously contained in that set.

Empty sets

The *empty set* — also called the *null set* — is a set that has no elements:

H = {}

As you can see, I define H by listing its elements, but I haven't listed any, so H is empty. The symbol ∅ is used to represent the empty set. So H = ∅.

You can also define an empty set by using a rule. For example,

I = types of roosters that lay eggs

Clearly, roosters are male and, therefore, can't lay eggs, so this set is empty.

Tip: You can think of ∅ as nothing. And because nothing is always nothing, there's only one empty set. All empty sets are equal to each other, so in this case, H = I.

Furthermore, ∅ is a subset of every other set (the preceding section discusses subsets), so the following statements are true:

∅ ⊃ A
∅ ⊃ B
∅ ⊃ C

This concept makes sense when you think about it. Remember that ∅ has no elements, so technically, every element in ∅ is in every other set.

Sets of numbers

One important use of sets is to define sets of numbers. As with all other sets, you can do so either by listing the elements or by verbally describing a rule that clearly tells you what's included in the set and what isn't. For example, consider the following sets:

J = {1, 2, 3, 4, 5}
K = {2, 4, 6, 8, 10,...}
L = the set of counting numbers

My definitions of J and K list their elements explicitly. Because K is infinitely large, you need to use an ellipsis (. . .) to show that this set goes on forever. The definition of L is a description of the set in words.

I discuss some especially significant sets of numbers in Chapter 22.

PERFORMING OPERATIONS ON SETS

In arithmetic, the Big Four operations (adding, subtracting, multiplying, and dividing) allow you to combine numbers in various ways (see Chapters 3 and 4 for more information). Set theory also has four important operations: union, intersection, relative complement, and complement. You'll see more of these operations as you move on in your study of math.

Here are definitions for three sets of numbers:

$$P = \{1, 7\}$$
$$Q = \{4, 5, 6\}$$
$$R = \{2, 4, 6, 8, 10\}$$

In this section, I use these three sets and a few others to discuss the four set operations and show you how they work. (***Note:*** Within equations, I relist the elements, replacing the names of the sets with their equivalent in braces. Therefore, you don't have to flip back and forth to look up what each set contains.)

Union: Combined elements

The union of two sets is the set of their *combined* elements. For example, the union of $\{1, 2\}$ and $\{3, 4\}$ is $\{1, 2, 3, 4\}$. The symbol for this operation is $\cup$, so

$$\{1, 2\} \cup \{3, 4\} = \{1, 2, 3, 4\}$$

Similarly, here's how to find the union of P and Q:

$$P \cup Q = \{1, 7\} \cup \{4, 5, 6\} = \{1, 4, 5, 6, 7\}$$

When two sets have one or more elements in common, these elements appear only once in their union set. For example, consider the union of Q and R. In this case, the elements 4 and 6 are in both sets, but each of these numbers appears once in their union:

$$Q \cup R = \{4, 5, 6\} \cup \{2, 4, 6, 8, 10\} = \{2, 4, 5, 6, 8, 10\}$$

The union of any set with itself is itself:

$$P \cup P = P$$

Similarly, the union of any set with $\varnothing$ (see the earlier "Empty sets" section) is itself:

$$P \cup \varnothing = P$$

Intersection: Elements in common

The intersection of two sets is the set of their common elements (the elements that appear in both sets). For example, the intersection of $\{1, 2, 3\}$ and $\{2, 3, 4\}$ is $\{2, 3\}$. The symbol for this operation is $\cap$. You can write the following:

$$\{1, 2, 3\} \cap \{2, 3, 4\} = \{2, 3\}$$

Similarly, here's how to write the intersection of Q and R:

$$Q \cap R = \{4, 5, 6\} \cap \{2, 4, 6, 8, 10\} = \{4, 6\}$$

When two sets have no elements in common, their intersection is the empty set ($\varnothing$):

$$P \cap Q = \{1, 7\} \cap \{4, 5, 6\} = \varnothing$$

The intersection of any set with itself is itself:

$$P \cap P = P$$

But the intersection of any set with $\varnothing$ is $\varnothing$:

$$P \cap \varnothing = \varnothing$$

Example

Q. Suppose L = the set of all even integers between 1 and 9, M = the set of all positive integers less than 7, and $N = \{3, 6, 9, 12\}$

   a. What elements are in L?

   b. What elements are in M?

   c. What is $(L \cup M) \cap N$?

A. **a.** **2, 4, 6, and 8.**

   b. **1, 2, 3, 4, 5, and 6.**

   c. **{3, 6}.** To begin, find the union of the two sets inside the parentheses:

$$L \cup M = \{1, 2, 3, 4, 5, 6, 8\}$$

Next, find the intersection of this set and N:

$$(L \cup M) \cap N = \{3, 6\}$$

Practice Questions

1. Suppose O = the set of all odd integers between 0 and 10, and P = the set of all integers greater than 6 and less than 14.

   a. What elements are in O?

   b. What elements are in P?

   c. What is $O \cup P$?

   d. What is $O \cap P$?

2. Let $R = \{1, 2, 3, 4, 5, 6, 7, 8, 9, 10\}$, $S = \{2, 4, 6, 8, 10\}$, and $T = \{2, 3, 5, 7\}$.

   a. What is $(R \cup S) \cap T$?

   b. What is $(R \cap S) \cup T$?

   c. What is $R \cup (S \cap T)$?

   d. What is $(R \cap S) \cup (S \cap T)$?

Practice Answers

1. **a.** 1, 3, 5, 7, and 9.

   b. 7, 8, 9, 10, 11, 12, and 13.

   c. {1, 3, 5, 7, 8, 9, 10, 11, 12, 13}. The union includes all elements from either set.

   d. {7, 9}. The intersection includes every element in both sets.

2. **a.** {2, 3, 5, 7}. Begin by solving the union inside the parentheses:

   $$R \cup S = \{1, 2, 3, 4, 5, 6, 7, 8, 9, 10\}$$

   Now, find the intersection of this set with T:

   $$(R \cup S) \cap T = \{2, 3, 5, 7\}$$

   b. {2, 3, 4, 5, 6, 7, 8, 10}. Begin by solving the intersection inside the parentheses:

   $$R \cap S = \{2, 4, 6, 8, 10\}$$

   Now, find the union of this set with T:

   $$(R \cup S) \cap T = \{2, 3, 4, 5, 6, 7, 8, 10\}$$

   c. {1, 2, 3, 4, 5, 6, 7, 8, 9, 10}. Begin by solving the intersection inside the parentheses:

   $$S \cap T = \{2\}$$

   Now, find the union of this set with R:

   $$(R \cup S) \cap T = \{1, 2, 3, 4, 5, 6, 7, 8, 9, 10\}$$

   d. {2, 4, 6, 8, 10}. Begin by solving the intersections inside both sets of parentheses:

   $$R \cap S = \{2, 4, 6, 8, 10\}$$
   $$S \cap T = \{2\}$$

   Now, find the union of these two sets:

   $$(R \cap S) \cup (S \cap T) = \{2, 4, 6, 8, 10\}$$

Relative complement: Subtraction (sorta)

The relative complement of two sets is an operation similar to subtraction. The symbol for this operation is the minus sign (–). Starting with the first set, you remove every element that appears in the second set to arrive at their relative complement. For example,

$$\{1, 2, 3, 4, 5\} - \{1, 2, 5\} = \{3, 4\}$$

Similarly, here's how to find the relative complement of R and Q. Both sets share a 4 and a 6, so you have to remove those elements from R:

$$R - Q = \{2, 4, 6, 8, 10\} - \{4, 5, 6\} = \{2, 8, 10\}$$

Note that the reversal of this operation gives you a different result. This time, you remove the shared 4 and 6 from Q:

$$Q - R = \{4, 5, 6\} - \{2, 4, 6, 8, 10\} = \{5\}$$

Remember: Like subtraction in arithmetic, the relative complement is not a commutative operation. In other words, order is important. (See Chapter 3 for more on commutative and non-commutative operations.)

Complement: Feeling left out

The complement of a set is everything that isn't in that set. Because "everything" is a difficult concept to work with, you first have to define what you mean by "everything" as the universal set (U). For example, suppose you define the universal set like this:

$$U = \{0, 1, 2, 3, 4, 5, 6, 7, 8, 9\}$$

Now, here are a couple of sets to work with:

$$M = \{1, 3, 5, 7, 9\}$$
$$N = \{6\}$$

The complement of each set is the set of every element in U that isn't in the original set:

$$U - M = \{0, 1, 2, 3, 4, 5, 6, 7, 8, 9\} - \{1, 3, 5, 7, 9\} = \{0, 2, 4, 6, 8\}$$
$$U - N = \{0, 1, 2, 3, 4, 5, 6, 7, 8, 9\} - \{6\} = \{0, 1, 2, 3, 4, 5, 7, 8, 9\}$$

The complement is closely related to the relative complement (see the preceding section). Both operations are similar to subtraction. The main difference is that the complement is *always* subtraction of a set from U, but the relative complement is subtraction of a set from any other set.

The symbol for the complement is ′, so you can write the following:

$$M' = \{0, 2, 4, 6, 8\}$$
$$N' = \{0, 1, 2, 3, 4, 5, 7, 8, 9\}$$

Example

Q. Let $X = \{1, 3, 5, 7\}$, $Y = \{2, 3, 4\}$, and $Z = \{4, 5, 6, 7, 8\}$

   a. What is $X - Y$?

   b. What is $Y - X$?

   c. What is $(Z - Y) \cup X$?

A.   **a.** {1, 5, 7}.

   b. {2, 4}.

   c. {1, 3, 5, 6, 7, 8}. To begin, find the relative complement of Z and Y:

$$Z - Y = \{5, 6, 7, 8\}$$

Next, find the union of this set and X:

$$(Z - Y) \cup X = \{1, 3, 5, 6, 7, 8\}$$

Practice Questions

  1. Suppose $F = \{1, 2, 3, 4, 6\}$ and $G = \{4, 5, 6, 7, 8\}$

    a. What is $F - G$?

    b. What is $G - F$?

  2. Let $H = \{1, 4, 7\}$, $J = \{2, 5, 7, 8, 9\}$, and $K = \{1, 2, 3, 4, 6, 10\}$

    a. What is $K - J$?

    b. What is $J - (K - H)$?

    c. What is $(H - J) \cup K$?

    d. What is $(H \cup K) - (J \cap H)$?

Practice Answers

  1. **a.** {1, 2, 3}.

    b. {5, 7, 8}.

  2. **a.** {1, 3, 4, 6, 3}.

    b. {5, 7, 8, 9}. To begin, find the relative complement of K and H:

$$K - H = \{2, 3, 6, 10\}$$

Next, find the relative complement of J and this set:

$$J - (K - H) = \{5, 7, 8, 9\}$$

c. **{1, 2, 3, 4, 6, 10}.** To begin, find the relative complement of H and J:

$$H - J = \{1, 4\}$$

Next, find the union of this set and K:

$$(H - J) \cup K = \{1, 2, 3, 4, 6, 10\}$$

d. **{1, 2, 3, 4, 6, 10}.** To begin, find the union of H and K and the intersection of J and H:

$$H \cup K = \{1, 2, 3, 4, 6, 7, 10\}$$
$$J \cap H = \{7\}$$

Next, find the relative complement of these two sets:

$$(H \cup K) - (J \cap H) = \{1, 2, 3, 4, 6, 10\}$$

PART V

the x-files

INTRODUCTION TO ALGEBRA

WEB EXTRA *For more info on simplifying algebraic expressions and to discover a trick that makes factoring quadratic polynomials easy, go to www.dummies.com/extras/ucanbasicmathprealgebra.*

IN THIS PART . . .

- Evaluate, simplify, and factor algebraic expressions.

- Keep algebraic equations balanced and solve them by isolating the variable.

- Use algebra to solve word problems too difficult to solve with just arithmetic.

18

ENTER MR. X: ALGEBRA AND ALGEBRAIC EXPRESSIONS

You never forget your first love, your first car, or your first *x*. Unfortunately for some folks, remembering their first *x* in algebra is similar to remembering their first love who stood them up at the prom or their first car that broke down someplace in Mexico.

The most well-known fact about algebra is that it uses letters — like *x* — to represent numbers. So if you have a traumatic *x*-related tale, all I can say is that the future will be brighter than the past.

What good is algebra? That question is a common one, and it deserves a decent answer. Algebra is used for solving problems that are just too difficult for ordinary arithmetic. And because number crunching is so much a part of the modern world, algebra is everywhere (even if you don't see it): architecture, engineering, medicine, statistics, computers, business, chemistry, physics, biology, and, of course, higher math. Anywhere that numbers are useful, algebra is there. That fact is why virtually every college and university insists that you leave (or enter) with at least a passing familiarity with algebra.

In this chapter, I introduce (or reintroduce) you to that elusive little fellow, Mr. *X,* in a way that's bound to make him seem a little friendlier. Then I show you how *algebraic expressions* are similar to and different from the arithmetic expressions that you're used to working with. (For a refresher on arithmetic expressions, see Chapter 4.)

SEEING HOW X MARKS THE SPOT

In math, *x* stands for a number — any number. Any letter that you use to stand for a number is a *variable,* which means that its value can *vary* — that is, its value is uncertain. In contrast, a number in algebra is often called a *constant* because its value is *fixed*.

Sometimes you have enough information to find out the identity of x. For example, consider the following:

$$2 + 2 = x$$

Obviously, in this equation, x stands for the number 4. But other times, what the number x stands for stays shrouded in mystery. For example:

$$x > 5$$

In this inequality, x stands for some number greater than 5 — maybe 6, maybe $7\frac{1}{2}$, maybe 542.002.

EXPRESSING YOURSELF WITH ALGEBRAIC EXPRESSIONS

In Chapter 4, I introduce you to arithmetic expressions: strings of numbers and operators that can be evaluated or placed on one side of an equation. For example:

$$2 + 3$$
$$7 \times 1.5 - 2$$
$$2^4 - |-4| - \sqrt{400}$$

In this chapter, I introduce you to another type of mathematical expression: the algebraic expression. An *algebraic expression* is any string of mathematical symbols that can be placed on one side of an equation and that includes at least one variable.

Here are a few examples of algebraic expressions:

$$5x$$
$$-5x + 2$$
$$x^2 y - xy^2 + \frac{z}{3} - xyz + 1$$

Remember: As you can see, the difference between arithmetic and algebraic expressions is simply that an algebraic expression includes at least one variable.

In this section, I show you how to work with algebraic expressions. First, I demonstrate how to evaluate an algebraic expression by substituting the values of its variables. Then I show you how to separate an algebraic expression into one or more terms, and I walk through how to identify the coefficient and the variable part of each term.

EVALUATING ALGEBRAIC EXPRESSIONS

Remember: To evaluate an algebraic expression, you need to know the numerical value of every variable. For each variable in the expression, substitute, or plug in, the number that it stands for and then evaluate the expression.

In Chapter 4, I show you how to evaluate an arithmetic expression. Briefly, this means finding the value of that expression as a single number (flip to Chapter 4 for more on evaluating).

Knowing how to evaluate arithmetic expressions comes in handy for evaluating algebraic expressions. For example, suppose you want to evaluate the following expression:

$$4x - 7$$

Note that this expression contains the variable x, which is unknown, so the value of the whole expression is also unknown.

An algebraic expression can have any number of variables, but you usually don't work with expressions that have more than two or maybe three, at the most. You can use any letter as a variable, but x, y, and z tend to get a lot of mileage.

Suppose in this case that $x = 2$. To evaluate the expression, substitute 2 for x everywhere it appears in the expression:

$$4(2) - 7$$

After you make the substitution, you're left with an arithmetic expression, so you can finish your calculations to evaluate the expression:

$$= 8 - 7 = 1$$

So given $x = 2$, the algebraic expression $4x - 7 = 1$.

Now suppose you want to evaluate the following expression, where $x = 4$:

$$2x^2 - 5x - 15$$

Again, the first step is to substitute 4 for x everywhere this variable appears in the expression:

$$= 2(4^2) - 5(4) - 15$$

Now evaluate according to the order of operations explained in Chapter 4. You do powers first, so begin by evaluating the exponent 4^2, which equals 4×4:

$$= 2(16) - 5(4) - 15$$

Now proceed to the multiplication, moving from left to right:

$$= 32 - 5(4) - 15$$
$$= 32 - 20 - 15$$

Then evaluate the subtraction, again from left to right:

$$= 12 - 15$$
$$= -3$$

So given $x = 4$, the algebraic expression $2x^2 - 5x - 15 = -3$.

You aren't limited to expressions of only one variable when using substitution. As long as you know the value of every variable in the expression, you can evaluate algebraic expressions with any number of variables. For example, suppose you want to evaluate this expression:

$$3x^2 + 2xy - xyz$$

To evaluate it, you need the values of all three variables:

$$x = 3$$
$$y = -2$$
$$z = 5$$

The first step is to substitute the equivalent value for each of the three variables wherever you find them:

$$3(3^2) + 2(3)(-2) - (3)(-2)(5)$$

Now use the rules for order of operations from Chapter 4. Begin by evaluating the exponent 3^2:

$$= 3(9) + 2(3)(-2) - (3)(-2)(5)$$

Next, evaluate the multiplication from left to right (if you need to know more about the rules for multiplying negative numbers, check out Chapter 3):

$$= 27 + (-12) - (-30)$$

Now all that's left is addition and subtraction. Evaluate from left to right, remembering the rules for adding and subtracting negative numbers in Chapter 3:

$$= 15 - (-30) = 15 + 30 = 45$$

So given the three values for x, y, and z, the algebraic expression $3x^2 + 2xy - xyz = 45$.

Tip: For practice, copy this expression and the three values on a separate piece of paper, close the book, and see whether you can substitute and evaluate on your own to get the same answer.

Examples

Q. Evaluate the algebraic expression $x^2 - 2x + 5$ when $x = 3$.

A. **8.** First, substitute 3 for x in the expression:

$$x^2 - 2x + 5 = 3^2 - 2(3) + 5$$

Evaluate the expression using the order of operations. Start by evaluating the power:

$$= 9 - 2(3) + 5$$

Next, evaluate the multiplication:

$$= 9 - 6 + 5$$

Finally, evaluate the addition and subtraction from left to right:

$$= 3 + 5 = 8$$

Q. Evaluate the algebraic expression $3x^2 + 2xy^4 - y^3$ when $x = 5$ and $y = 2$.

A. **227.** Plug in 5 for x and 3 for y in the expression:

$$3x^2 + 2xy^4 - y^3 = 3(5^2) + 2(5)(2^4) - 2^3$$

Evaluate the expression using the order of operations. Start by evaluating the three powers:

$$= 3(25) + 2(5)(16) - 8$$

Next, evaluate the multiplication:

$$= 75 + 160 - 8$$

Finally, evaluate the addition and subtraction from left to right:

$$= 235 - 8 = 227$$

Practice Questions

1. Evaluate the expression $x^2 + 5x + 4$ when $x = 3$.

2. Find the value of $5x^4 + x^3 - x^2 + 10x + 8$ when $x = -2$.

3. Evaluate the expression $x(x^2 - 6)(x - 7)$ when $x = 4$.

4. Evaluate $\dfrac{(x-9)^4}{(x+4)^3}$ when $x = 6$.

5. Find the value of $3x^2 + 5xy + 4y^2$ when $x = 5$ and $y = 7$.

6. Evaluate the expression $x^6y - 5xy^2$ when $x = -1$ and $y = 9$.

A Practice Answers

1. **28.** Substitute 3 for x in the expression:

$$x^2 + 5x + 4 = 3^2 + 5(3) + 4$$

Evaluate the expression using the order of operations. Start with the power:

$$= 9 + 5(3) + 4$$

Continue by evaluating the multiplication:

$$= 9 + 15 + 4$$

Finish up by evaluating the addition from left to right:

$$= 24 + 4 = 28$$

2. **56.** Plug in –2 for every x in the expression:

$$5x^4 + x^3 - x^2 + 10x + 8 = 5(-2)^4 + (-2)^3 - (-2)^2 + 10(-2) + 8$$

Evaluate the expression using the order of operations. Start with the powers:

$$= 5(16) + -8 - 4 + 10(-2) + 8$$

Do the multiplication:

$$= 80 + -8 - 4 - 20 + 8$$

Finish up by evaluating the addition and subtraction from left to right:

$$= 72 - 4 + -20 + 8 = 68 - 20 + 8 = 48 + 8 = 56$$

3. **–120.** Substitute 4 for x in the expression:

$$x(x^2 - 6)(x - 7) = 4 (4^2 - 6)(4 - 7)$$

Follow the order of operations as you evaluate the expression. Starting inside the first set of parentheses, evaluate the power and then the subtraction:

$$= 4(16 - 6)(4 - 7) = 4(10)(4 - 7)$$

Find the contents of the remaining set of parentheses:

$$= 4(10)(-3)$$

Evaluate the multiplication from left to right:

$$= 40(-3) = -120$$

4. $\dfrac{(x-9)^4}{(x+4)^3} = \dfrac{81}{1,000}$ **(or 0.081).** Replace every x in the expression with a 6:

$$\frac{(x-9)^4}{(x+4)^3} = \frac{(6-9)^4}{(6+4)^3}$$

Follow the order of operations. Evaluate the contents of the set of parentheses in the numerator and then in the denominator:

$$= \frac{(-3)^4}{(6+4)^3} = \frac{(-3)^4}{10^3}$$

$$= \frac{81}{10^3} = \frac{81}{1,000}$$

Continue by evaluating the powers from top to bottom:

You can also express this fraction as the decimal 0.081.

5. **446.** Substitute 5 for x and 7 for y in the expression:

$$3x^2 + 5xy + 4y^2 = 3(5)^2 + 5(5)(7) + 4(7)^2$$

Evaluate using the order of operations. Start with the two powers:

$$= 3(25) + 5(5)(7) + 4(49)$$

Evaluate the multiplication from left to right:

$$= 75 + 175 + 196$$

Finally, do the addition from left to right:

$$= 250 + 196 = 446$$

6. **414.** Plug in –1 for each x and 9 for each y in the expression:

$$x^6 y - 5xy^2 = (-1)^6(9) - 5(-1)(9)^2$$

Follow the order of operations. Start by evaluating the two powers:

$$= 1(9) - 5(-1)(81)$$

Continue by evaluating the multiplication from left to right:

$$= 9 - (-5)(81) = 9 - (-405)$$

Finally, do the subtraction:

$$= 414$$

KNOWING THE TERMS OF SEPARATION

Algebraic expressions begin to make more sense when you understand how they're put together, and the best way to understand this is to take them apart and know what each part is called. Every algebraic expression can be broken into one or more terms. A *term* is any chunk of symbols set off from the rest of the expression by either an addition or subtraction sign. For example,

- The expression $-7x + 2xy$ has two terms: $-7x$ and $2xy$.

- The expression $x^4 - \dfrac{x^2}{5} - 2x + 2$ has four terms: x^4, $\dfrac{-x^2}{5}$, $-2x$, and 2.

- The expression $8x^2y^3z^4$ has only one term: $8x^2y^3z^4$.

Remember: When you separate an algebraic expression into terms, group the plus or minus sign with the term that immediately follows it. Then you can rearrange terms in any order you like. After the terms are separated, you can drop the plus signs.

When a term doesn't have a variable, it's called a *constant.* (*Constant* is just a fancy word for number when you're talking about terms in an expression.) When a term has a variable, it's called an *algebraic term.* Every algebraic term has two parts:

- The *coefficient* is the signed numerical part of a term — that is, the number and the sign (+ or –) that go with that term. Typically, the coefficients 1 and –1 are dropped from a term, so when a term appears to have no coefficient, its coefficient is either 1 or –1, depending on the sign of that term. The coefficient of a constant is the constant itself.

- The *variable part* is everything else other than the coefficient.

Remember: When two terms have the same variable part, they're called *like terms or similar terms.* For two terms to be like, both the letters and their exponents have to be exact matches. For example, the terms $2x$, $-12x$, and $834x$ are all like terms because their variable parts are all x. Similarly, the terms $17xy$, $1,000xy$, and $0.625xy$ are all like terms because their variable parts are all xy.

E Examples

Q. In the expression $3x^2 - 2x - 9$, identify which terms are algebraic and which are constants.

A. **$3x^2$ and $-2x$ are both algebraic terms, and -9 is a constant.**

Q. Identify the coefficient of every term in the expression $2x^4 - 5x^3 + x^2 - x - 9$.

A. **2, –5, 1, –1, and –9.** The expression has five terms: $2x^4$, $-5x^3$, x^2, $-x$, -9. The coefficient of $2x^4$ is 2, and the coefficient of $-5x^3$ is –5. The term x^2 appears to have no coefficient, so its coefficient is 1. The term $-x$ also appears to have no coefficient, so its coefficient it –1. The term -9 is a constant, so its coefficient is –9.

Practice Questions

1. Write down all the terms in the expression $7x^2yz - 10xy^2z + 4xyz^2 + y - z + 2$. Which are algebraic terms, and which are constants?

2. Identify the coefficient of every term in $-2x^5 + 6x^4 + x^3 - x^2 - 8x + 17$.

3. Name every coefficient in the expression $-x^3y^3z^3 - x^2y^2z^2 + xyz - x$.

4. In the expression $12x^3 + 7x^2 - 2x - x^2 - 8x^4 + 99x + 99$, identify any sets of like terms.

Practice Answers

1. **The algebraic terms are $7x^2yz$, $-10xy^2z$, $4xyz^2$, y, and $-z$; the constant is 2.**

2. **The six coefficients, in order, are -2, 6, 1, -1, -8, and 17.**

3. **The four coefficients, in order, are -1, -1, 1, and -1.**

4. **$7x^2$ and $-x^2$ are like terms (the variable part is x^2); $-2x$ and $99x$ are also like terms (the variable part is x).**

Making the commute: Rearranging your terms

When you understand how to separate an algebraic expression into terms, you can go one step further by rearranging the terms in any order you like. Each term moves as a unit, kind of like a group of people carpooling to work together — everyone in the car stays together for the whole ride.

For example, suppose you begin with the expression $-5x + 2$. You can rearrange the two terms of this expression without changing its value. Notice that each term's sign stays with that term, although dropping the plus sign at the beginning of an expression is customary:

$$= 2 - 5x$$

Rearranging terms in this way doesn't affect the value of the expression because addition is *commutative* — that is, you can rearrange things that you're adding without changing the answer. (See Chapter 3 for more on the commutative property of addition.)

For example, suppose $x = 3$. Then the original expression and its rearrangement evaluate as follows (using the rules that I outline earlier in "Evaluating algebraic expressions"):

$$
\begin{array}{ll}
-5x + 2 & 2 - 5x \\
= -5(3) + 2 & = 2 - 5(3) \\
= -15 + 2 & = 2 - 15 \\
= -13 & = -13
\end{array}
$$

Rearranging expressions in this way becomes handy later in this chapter, when you simplify algebraic expressions. As another example, suppose you have this expression:

$$4x - y + 6$$

You can rearrange it in a variety of ways:

$$= 6 + 4x - y$$
$$= -y + 4x + 6$$

Because the term $4x$ has no sign, it's positive, so you can write in a plus sign as needed when rearranging terms.

Remember: As long as each term's sign stays with that term, rearranging the terms in an expression has no effect on its value.

For example, suppose that $x = 2$ and $y = 3$. Here's how to evaluate the original expression and the two rearrangements:

$$
\begin{array}{lll}
4x - y + 6 & 6 + 4x - y & -y + 4x + 6 \\
= 4(2) - 3 + 6 & = 6 + 4(2) - 3 & = -3 + 4(2) + 6 \\
= 8 - 3 + 6 & = 6 + 8 - 3 & = -3 + 8 + 6 \\
= 5 + 6 & = 14 - 3 & = 5 + 6 \\
= 11 & = 11 & = 11
\end{array}
$$

Example

Q. Rearrange the terms of the expression $-2b + 5a - 10 + c$ so that variable terms are all in alphabetical order and the constant is at the end.

A. **$5a - 2b + c - 10$.** Remember that each minus sign stays with the term it precedes.

Practice Questions

1. Rearrange the terms of the expression $-7k + 1 - j$ so that variable terms are both in alphabetical order and the constant is at the end.

2. Rearrange the terms of the expression $5z + 4y - 9 - x$ so that variable terms are all in alphabetical order and the constant is at the end.

3. Rearrange the terms of the expression $-75 + 15v - 100u - t$ so that variable terms are all in alphabetical order and the constant is at the end.

4. Rearrange the terms of the expression $q + 3r - 10 + s - 2p$ so that variable terms are all in alphabetical order and the constant is at the end.

Practice Answers

1. $-j - 7k + 1$

2. $-x + 4y + 5z - 9$

3. $-t - 100u + 15v - 75$

4. $-2p + q + 3r + s - 10$

Identifying the coefficient and variable

Every term in an algebraic expression has a coefficient. The *coefficient* is the signed numerical part of a term in an algebraic expression — that is, the number and the sign (+ or –) that goes with that term. For example, suppose you're working with the following algebraic expression:

$$-4x^3 + x^2 - x - 7$$

The following table shows the four terms of this expression, with each term's coefficient:

Term	Coefficient	Variable
$-4x^3$	-4	x^3
x^2	1	x^2
$-x$	-1	x
-7	-7	none

Notice that the sign associated with the term is part of the coefficient. So the coefficient of $-4x^3$ is –4.

Remember: When a term appears to have no coefficient, the coefficient is actually 1. So the coefficient of x^2 is 1, and the coefficient of $-x$ is –1. And when a term is a constant (just a number), that number with its associated sign is the coefficient. So the coefficient of the term –7 is simply –7.

By the way, when the coefficient of any algebraic term is 0, the expression equals 0 no matter what the variable part looks like:

$$0x = 0 \qquad 0xyz = 0 \qquad 0x^3y^4z^{10} = 0$$

In contrast, the *variable part* of an expression is everything except the coefficient. The previous table shows the four terms of the same expression, with each term's variable part.

Example

Q. What are the coefficients of the terms in the expression $5x - y + z - 3$?

A. 5, –1, 1, and –3.

Practice Questions

1. What are the coefficients of the terms in the expression $h + 4k - 3$?

2. What are the coefficients of the terms in the expression $3x - 2xy - z$?

3. What are the coefficients of the terms in the expression $-25a^3 - 60b + 10.5c - d + 300$?

4. What are the coefficients of the terms in the expression $-x^3 + 7x^2 - 9x + 1$?

Practice Answers

1. 1, 4, and –3.

2. 3, –2, and –1.

3. –25, – 60, 10.5, –1, and 300.

4. –1, 7, –9, and 1.

ADDING AND SUBTRACTING LIKE TERMS

In algebra, you can add and subtract only *like terms* – in other words, terms in which the variable parts match exactly. In this section, I show you how to identify like terms and then you discover how to add and subtract them.

Identifying like terms

Like terms (or *similar terms*) are any two algebraic terms that have the same variable part — that is, both the letters and their exponents have to be exact matches. Here are some examples:

Variable Part	Examples of Like Terms		
x	$4x$	$12x$	$99.9x$
x^2	$6x^2$	$-20x^2$	$\dfrac{8}{3}x^2$
y	y	$1,000y$	πy
xy	$-7xy$	$800xy$	$\dfrac{22}{7}xy$
x^3y^3	$3x^3y^3$	$-111x^3y^3$	$3.14x^3y^3$

As you can see, in each example, the variable part in all three like terms is the same. Only the coefficient changes, and it can be any real number: positive or negative, whole number, fraction, or decimal — or even an irrational number such as π. (For more on real numbers, see Chapter 22.)

Adding and subtracting terms

Remember: Add like terms by adding their coefficients and keeping the same variable part.

For example, suppose you have the expression $2x + 3x$. Remember that $2x$ is just shorthand for $x + x$, and $3x$ means simply $x + x + x$. So when you add them up, you get the following:

$$= x + x + x + x + x = 5x$$

As you can see, when the variable parts of two terms are the same, you add these terms by adding their coefficients: $2x + 3x = (2 + 3)x$. The idea here is roughly similar to the idea that 2 apples + 3 apples = 5 apples.

Warning: You *cannot* add nonlike terms. Here are some cases in which the variables or their exponents are different:

$$2x + 3y$$
$$2yz + 3y$$
$$2x^2 + 3x$$

In these cases, you can't simplify the expression. You're faced with a situation that's similar to 2 apples + 3 oranges. Because the units (apples and oranges) are different, you can't combine terms. (See Chapter 14 for more on how to work with units.)

Remember: Subtraction works much the same as addition. Subtract like terms by finding the difference between their coefficients and keeping the same variable part.

For example, suppose you have $3x - x$. Recall that $3x$ is simply shorthand for $x + x + x$. So doing this subtraction gives you the following:

$$x + x + x - x = 2x$$

No big surprises here. You simply find $(3 - 1)x$. This time, the idea roughly parallels the idea that $\$3 - \$1 = \$2$.

Here's another example:

$$2x - 5x$$

Again, no problem, as long as you know how to work with negative numbers (see Chapter 3 if you need details). Just find the difference between the coefficients:

$$= (2 - 5)x = -3x$$

In this case, recall that $\$2 - \$5 = -\$3$ (that is, a debt of $\$3$).

Warning: You *cannot* subtract nonlike terms. For example, you can't subtract either of the following:

$$7x - 4y$$
$$7x^2y - 4xy^2$$

As with addition, you can't do subtraction with different variables. Think of this as trying to figure out $\$7 - 4$ pesos. Because the units in this case (dollars versus pesos) are different, you're stuck. (See Chapter 14 for more on working with units.)

Examples

Q. What is $3x + 5x$?

A. **8x.** The variable part of both terms is x, so you can add:

$$3x + 5x = (3 + 5)x = 8x$$

Q. What is $24x^3 - 7x^3$?

A. **17x³.** Subtract the coefficients:

$$24x^3 - 7x^3 = (24 - 7)x^3 = 17x^3$$

Practice Questions

1. Add $4x^2y + (-9x^2y)$

2. Add $x^3y^2 + 20x^3y^2$.

3. Subtract $-2xy^4 - 20xy^4$.

4. Subtract $-xyz - (-xyz)$.

Practice Answers

1. $4x^2y + -(9x^2y) = (4-9)x^2y = $ **$-5x^2y$**

2. $x^3y^2 + 20x^3y^2 = (1+20)x^3y^2 = $ **$21x^3y^2$**.

3. $-2xy^4 - 20xy^4 = (-2-20)xy^4 = $ **$-22xy^4$**.

4. $-xyz - (-xyz) = [-1-(-1)]xyz = (-1+1)xyz = $ **0**.

MULTIPLYING AND DIVIDING TERMS

Remember: Unlike adding and subtracting, you can multiply non-like terms. Multiply *any* two terms by multiplying their coefficients and combining — that is, by collecting or gathering up — all the variables in each term into a single term, as I show you next.

For example, suppose you want to multiply $5x(3y)$. To get the coefficient, multiply 5×3. To get the algebraic part, combine the variables x and y:

$$= 5(3)xy = 15xy$$

Now suppose you want to multiply $2x(7x)$. Again, multiply the coefficients and collect the variables into a single term:

$$= 7(2)xx = 14xx$$

Remember that x^2 is shorthand for xx, so you can write the answer more efficiently:

$$= 14x^2$$

Here's another example. Multiply all three coefficients together and gather up the variables:

$$2x^2(3y)(4xy)$$
$$= 2(3)(4)x^2xyy$$
$$= 24x^3y^2$$

As you can see, the exponent 3 that's associated with x is just the count of how many x's appear in the problem. The same is true of the exponent 2 associated with y.

Tip: A fast way to multiply variables with exponents is to add the exponents together. For example:

$$(x^4y^3)(x^2y^5)(x^6y) = x^{12}y^9$$

In this example, I added the exponents of the x's ($4 + 2 + 6 = 12$) to get the exponent of x in the expression. Similarly, I added the exponents of the y's ($3 + 5 + 1 = 9$ — don't forget that $y = y^1$!) to get the exponent of y in the expression.

It's customary to represent division of algebraic expressions as a fraction instead of using the division sign ($\div$). So division of algebraic terms really looks like reducing a fraction to lowest terms (see Chapter 8 for more on reducing).

To divide one algebraic term by another, follow these steps:

1. **Make a fraction of the two terms.**

   Suppose you want to divide $3xy$ by $12x^2$. Begin by turning the problem into a fraction:

   $$\frac{3xy}{12x^2}$$

2. **Cancel out factors in coefficients that are in both the numerator and the denominator.**

   In this case, you can cancel out a 3. Notice that when the coefficient in xy becomes 1, you can drop it:

   $$= \frac{xy}{4x^2}$$

3. **Cancel out any variable that's in both the numerator and the denominator.**

   You can break x^2 out as xx:

   $$= \frac{xy}{4xx}$$

   Now you can clearly cancel an x in both the numerator and the denominator:

   $$= \frac{y}{4x}$$

   As you can see, the resulting fraction is really a reduced form of the original.

As another example, suppose you want to divide $-6x^2yz^3$ by $-8x^2y^2z$. Begin by writing the division as a fraction:

$$\frac{-6x^2yz^3}{-8x^2y^2z}$$

First, reduce the coefficients. Notice that, because both coefficients were originally negative, you can cancel out both minus signs as well:

$$= \frac{3x^2yz^3}{4x^2y^2z}$$

Now you can begin canceling variables. I do this in two steps, as before:

$$= \frac{3xxyzzz}{4xxyyz}$$

At this point, just cross out any occurrence of a variable that appears in both the numerator and the denominator:

$$= \frac{3zz}{4y}$$

$$= \frac{3z^2}{4y}$$

Warning: You can't cancel out variables or coefficients if either the numerator or the denominator has more than one term in it. This is a very common mistake in algebra, so don't let it happen to you!

Tip: A fast way to divide is to subtract the exponents of identical variables. For each variable, subtract the exponent in the numerator minus the exponent in the denominator. When a resulting exponent is a negative number, move that variable to the denominator and remove the minus sign.

Examples

Q. What is $2x(6y)$?

A. **12xy.** To get the final coefficient, multiply the coefficients $2 \times 6 = 12$. To get the variable part of the answer, combine the variables x and y:

$$2x(6y) = 2(6)xy = 12xy$$

Q. What is $4x(-9x)$?

A. **$-36x^2$.** To get the final coefficient, multiply the coefficients $4 \times -9 = -36$. To get the variable part of the answer, collect the variables into a single term:

$$4x(-9x) = 4(-9)xx$$

Remember that x^2 is shorthand for xx, so rewrite the answer as follows:

$$-36x^2$$

Q. What is $2x(4xy)(5y^2)$?

A. $40x^2y^3$. Multiply all three coefficients together and gather up the variables:

$$2x(4xy)(5y^2) = 2(4)(5)xxyy^2 = 40x^2y^3$$

In this answer, the exponent 2 that's associated with x is just a count of how many x's appear in the problem; the same is true of the exponent 3 that's associated with y.

Q. What is $(x^2y^3)(xy^5)(x^4)$?

A. x^7y^8. Add exponents of the three x's ($2 + 1 + 4 = 7$) to get the exponent of x in the answer (remember that x means x^1). Add the two y exponents ($3 + 5 = 8$) to get the exponent of y in the answer.

Q. What is $\dfrac{6x^3y^2}{3xy^2}$?

A. $2x^2$. Exponents mean repeated multiplication, so

$$\frac{6x^3y^2}{3xy^2} = \frac{6xxxyy}{3xyy}$$

Both the numerator and the denominator are divisible by 2, so reduce the coefficients just as you'd reduce a fraction:

$$= \frac{2xxxyy}{xyy}$$

Now cancel out any variables repeated in both the numerator and denominator — that is, one x and two y's both above and below the fraction bar:

$$= \frac{2xx\not{x}\not{y}\not{y}}{\not{x}\not{y}\not{y}} = 2xx$$

Rewrite this result using an exponent.

$$2x^2$$

Practice Questions

1. Multiply $4x(7x^2)$.

2. Multiply $-xy^3z^4(10x^2y^2z)(-2xz)$.

3. Divide $\dfrac{6x^4y^5}{8x^4y^4}$.

4. Divide $\dfrac{7x^2y}{21xy^3}$.

Practice Answers

1. **$28x^3$.** Multiply the two coefficients to get the coefficient of the answer; then gather the variables into one term:

    $$4x(7x^2) = 4(7)xx^2 = 28x^3$$

2. **$20x^4y^5z^6$.** Multiply the coefficients ($-1 \times 10 \times -2 = 20$) to get the coefficient of the answer. Add the x exponents ($1 + 2 + 1 = 4$) to get the exponent of x in the answer. Add the y exponents ($3 + 2 = 5$) to get the exponent of y in the answer. And add the z exponents ($4 + 1 + 1 = 6$) to get the exponent of z in the answer.

3. $\dfrac{6x^4y^5}{8x^4y^4} = \dfrac{\mathbf{3y}}{\mathbf{4}}$. Reduce the coefficients of the numerator and denominator just as you'd reduce a fraction:

    $$\frac{6x^4y^5}{8x^4y^4} = \frac{3x^4y^5}{4x^4y^4}$$

    To get the x exponent of the answer, take the x exponent in the numerator minus the x exponent in the denominator: $4 - 4 = 0$, so the x's cancel:

    $$= \frac{3y^5}{4y^4}$$

    To get the y exponent of the answer, take the y exponent in the numerator minus the y exponent in the denominator: $5 - 4 = 1$, so you have only y^1, or y, in the numerator:

    $$= \frac{3y}{4}$$

4. $\dfrac{7x^2y}{21xy^3} = \dfrac{\mathbf{x}}{\mathbf{3y^2}}$. Reduce the coefficients of the numerator and denominator just as you'd reduce a fraction:

    $$\frac{7x^2y}{21xy^3} = \frac{x^2y}{3xy^3}$$

    Take the x exponent in the numerator minus the x exponent in the denominator ($2 - 1 = 1$) to get the x exponent of the answer:

    $$= \frac{xy}{3y^3}$$

To get the y exponent, take the y exponent in the numerator minus the y exponent in the denominator $(1 - 3 = -2)$:

$$= \frac{xy^{-2}}{3}$$

To finish up, remove the minus sign from the y exponent and move this variable to the denominator:

$$= \frac{x}{3y^2}$$

SIMPLIFYING EXPRESSIONS BY COMBINING LIKE TERMS

Although an algebraic expression may have any number of terms, you can sometimes make it smaller and easier to work with. This process is called *simplifying* an expression. To simplify, you *combine like terms,* which means you add or subtract any terms that have identical variable parts. (To understand how to identify like algebraic terms, see "Knowing the Terms of Separation," earlier in this chapter. "Adding and Subtracting Like Terms" shows you how to do the math.)

In some expressions, like terms may not all be next to each other. In this case, you may want to rearrange the terms to place all like terms together before combining them.

Warning: When rearranging terms, you *must* keep the sign (+ or −) with the term that follows it.

To finish, you usually rearrange the answer alphabetically, from highest exponent to lowest, and place any constant last. In other words, if your answer is $-5 + 4y^3 + x^2 + 2x - 3xy$, you'd rearrange it to read $x^2 + 2x - 3xy + 4y^3 - 5$. (This step doesn't change the answer, but it kind of cleans it up, so teachers love it.)

For example, suppose you're working with the following expression:

$$4x - 3y + 2x + y - x + 2y$$

As it stands, this expression has six terms. But three terms have the variable x and the other three have the variable y. Begin by rearranging the expression so that all like terms are grouped together:

$$= 4x + 2x - x - 3y + y + 2y$$

Now you can add and subtract like terms. I do this in two steps, first for the x terms and then for the y terms:

$$= 5x - 3y + y + 2y$$
$$= 5x + 0y$$
$$= 5x$$

Notice that the x terms simplify to $5x$, and the y terms simplify to $0y$, which is 0, so the y terms drop out of the expression altogether.

Here's a somewhat more complicated example that has variables with exponents:

$$12x - xy - 3x^2 + 8y + 10xy + 3x^2 - 7x$$

This time, you have four different types of terms. As a first step, you can rearrange these terms so that groups of like terms are all together (I underline these four groups so you can see them clearly):

$$= \underline{12x - 7x} \; \underline{-xy + 10xy} \; \underline{-3x^2 + 3x^2} \; \underline{+8y}$$

Now combine each set of like terms:

$$= 5x + 9xy + 0x^2 + 8y$$

This time, the x^2 terms add up to 0, so they drop out of the expression altogether:

$$= 5x + 9xy + 8y$$

Examples

Q. Simplify the expression $x^2 + 2x - 7x + 1$.

A. $x^2 - 5x + 1$. The terms $2x$ and $-7x$ are like, so you can combine them:

$$x^2 + \underline{2x - 7x} + 1 = x^2 + (2 - 7)x + 1 = x^2 - 5x + 1$$

Q. Simplify the expression $4x^2 - 3x + 2 + x - 7x^2$.

A. $-3x^2 - 2x + 2$. I've begun by underlining the five terms of the expression:

$$\underline{4x^2} \; \underline{-3x} \; \underline{+2} \; \underline{+x} \; \underline{-7x^2}$$

Rearrange these terms as needed so that all like terms are together:

$$\underline{4x^2 - 7x^2} \; \underline{-3x + x} \; +2$$

At this point, combine the two underlined pairs of like terms:

$$(4 - 7)x^2 + (-3 + 1)x + 2 = -3x^2 - 2x + 2$$

Practice Questions

1. Simplify the expression $3x^2 + 5x^2 + 2x - 8x - 1$.

2. Simplify the expression $6x^3 - x^2 + 2 - 5x^2 - 1 + x$.

3. Simplify the expression $2x^4 - 2x^3 + 2x^2 - x + 9 + x + 7x^2$.

4. Simplify the expression $x^5 - x^3 + xy - 5x^3 - 1 + x^3 - xy + x$.

Practice Answers

1. $\mathbf{8x^2 - 6x - 1}$. Combine the following underlined like terms:

$$\underline{3x^2 + 5x^2} + \underline{2x - 8x} - 1$$
$$= 8x^2 - 6x - 1$$

2. $\mathbf{6x^3 - 6x^2 + x + 1}$. Rearrange the terms so like terms are next to each other:

$$6x^3 - x^2 + 2 - 5x^2 - 1 + x$$
$$= 6x^3 \underline{- x^2 - 5x^2} + x + \underline{2 - 1}$$

Now combine the underlined like terms:

$$= 6x^3 - 6x^2 + x + 1$$

3. $\mathbf{2x^4 - 2x^3 + 9x^2 + 9}$. Put like terms next to each other:

$$2x^4 - 2x^3 + 2x^2 - x + 9 + x + 7x^2$$
$$= 2x^4 - 2x^3 + \underline{2x^2 + 7x^2} - \underline{x + x} + 9$$

Now combine the underlined like terms:

$$= 2x^4 - 2x^3 + 9x^2 + 9$$

Note that the two x terms cancel each other out.

4. $\mathbf{x^5 - 5x^3 + x - 1}$. Rearrange the terms so like terms are next to each other:

$$x^5 - x^3 + xy - 5x^3 - 1 + x^3 - xy + x$$
$$= x^5 \underline{- x^3 - 5x^3 + x^3} + \underline{xy - xy} + x - 1$$

Now combine the underlined like terms:

$$= x^5 - 5x^3 + x - 1$$

Note that the two xy terms cancel each other out.

REMOVING PARENTHESES FROM AN ALGEBRAIC EXPRESSION

Parentheses keep parts of an expression together as a single unit. In Chapter 4, I show you how to handle parentheses in an arithmetic expression. This skill is also useful with algebraic expressions. As you find when you begin solving algebraic equations in Chapter 19, getting rid of parentheses is often the first step toward solving a problem. In this section, I show how to handle the Big Four operations with ease.

Drop everything: Parentheses with a plus sign

When an expression contains parentheses that come right after a plus sign (+), you can just remove the parentheses. Here's an example:

$$2x + (3x - y) + 5y$$
$$= 2x + 3x - y + 5y$$

Now you can simplify the expression by combining like terms:

$$= 5x + 4y$$

When the first term inside the parentheses is negative, when you drop the parentheses, the minus sign replaces the plus sign. For example:

$$6x + (-2x + y) - 4y$$
$$= 6x - 2x + y - 4y$$
$$= 4x - 3x$$

Sign turnabout: Parentheses with a minus sign

Remember: Sometimes an expression contains parentheses that come right after a minus sign (–). In this case, change the sign of every term inside the parentheses to the opposite sign; then remove the parentheses.

Consider this example:

$$6x - (2xy - 3y) + 5xy$$

A minus sign is in front of the parentheses, so you need to change the signs of both terms in the parentheses and remove the parentheses. Notice that the term $2xy$ appears to have no sign because it's the first term inside the parentheses. This expression really means the following:

$$= 6x - (+2xy - 3y) + 5xy$$

You can see how to change the signs:

$$= 6x - 2xy + 3y + 5xy$$

At this point, you can combine the two xy terms:

$$= 6x + 3xy + 3y$$

Distribution: Parentheses with no sign

When you see nothing between a number and a set of parentheses, it means multiplication. For example,

$$2(3) = 6 \qquad 4(4) = 16 \qquad 10(15) = 150$$

This notation becomes much more common with algebraic expressions, replacing the multiplication sign ($\times$) to avoid confusion with the variable x:

$$3(4x) = 12x \qquad 4x(2x) = 8x^2 \qquad 3x(7y) = 21xy$$

Remember: To remove parentheses without a sign, multiply the term outside the parentheses by every term inside the parentheses; then remove the parentheses. When you follow those steps, you're using the *distributive property*.

Here's an example:

$$2(3x - 5y + 4)$$

In this case, multiply 2 by each of the three terms inside the parentheses:

$$= 2(3x) + 2(-5y) + 2(4)$$

For the moment, this expression looks more complex than the original one, but now you can get rid of all three sets of parentheses by multiplying:

$$= 6x - 10y + 8$$

Multiplying by every term inside the parentheses is simply distribution of multiplication over addition — also called the *distributive property* — which I discuss in Chapter 3.

As another example, suppose you have the following expression:

$$-2x(-3x + y + 6) + 2xy - 5x^2$$

Begin by multiplying $-2x$ by the three terms inside the parentheses:

$$= -2x(-3x) - 2x(y) - 2x(6) + 2xy - 5x^2$$

The expression looks worse than when you started, but you can get rid of all the parentheses by multiplying:

$$= 6x^2 - 2xy - 12x + 2xy - 5x^2$$

Now you can combine like terms:

$$= x^2 - 12x$$

In summary, here are the four possible cases:

- **Parentheses preceded by a plus sign (+):** Just remove the parentheses. After that, you may be able to simplify the expression further by combining like terms, as I show you in the preceding section.

- **Parentheses preceded by a minus sign (–):** Change every term inside the parentheses to the opposite sign; then remove the parentheses. After the parentheses are gone, combine like terms.

- **Parentheses preceded by no sign (a term directly next to a set of parentheses):** Multiply the term next to the parentheses by every term inside the parentheses (make sure you include the plus or minus signs in your terms); then drop the parentheses. Simplify by combining like terms.

    Remember: To multiply identical variables, simply add the exponents. For instance, $x(x^2) = x^{1+2} = x^3$. Likewise, $-x^3(x^4) = -(x^{3+4}) = -x^7$.

- **Two sets of adjacent parentheses:** I discuss this case in the next section.

E

Examples

Q. Simplify the expression $7x + (x^2 - 6x + 4) - 5$.

A. $x^2 + x - 1$. Because a plus sign precedes this set of parentheses, you can drop the parentheses.

$$7x + (x^2 - 6x + 4) - 5$$
$$= 7x + x^2 - 6x + 4 - 5$$

Now combine like terms. I do this in two steps:

$$= \underline{7x - 6x} + x^2 + \underline{4 - 5}$$
$$= x + x^2 - 1$$

Finally, rearrange the answer from highest exponent to lowest:

$$x^2 + x - 1$$

Q. Simplify the expression $x - 3x(x^3 - 4x^2 + 2) + 8x^4$.

A. $5x^4 + 12x^3 - 5x$. The term $-3x$ precedes this set of parentheses without a sign in between, so multiply every term inside the parentheses by $-3x$ and then drop the parentheses:

$$x - 3x(x^3 - 4x^2 + 2) + 8x^4$$
$$= x + -3x(x^3) - 3x(-4x^2) - 3x(2) + 8x^4$$
$$= x - 3x^4 + 12x^3 - 6x + 8x^4$$

Now combine like terms. I do this in two steps:

$$= \underline{x - 6x} \ \underline{- 3x^4 + 8x^4} + 12x^3$$
$$= -5x + 5x^4 + 12x^3$$

As a final step, arrange the terms in order from highest to lowest exponent:

$$= 5x^4 + 12x^3 - 5x$$

Practice Questions

1. Simplify the expression $3x^3 + (12x^3 - 6x) + (5 - x)$.

2. Simplify the expression $2x^4 - (-9x^2 + x) + (x + 10)$.

3. Simplify the expression $x - (x^3 - x - 5) + 3(x^2 - x)$.

4. Simplify the expression $-x^3(x^2 + x) - (x^5 - x^4)$.

Practice Answers

1. $15x^3 - 7x + 5$. A plus sign precedes both sets of parentheses, so you can drop both sets:

    $$3x^3 + (12x^3 - 6x) + (5 - x)$$
    $$= 3x^3 + 12x^3 - 6x + 5 - x$$

    Now combine like terms:

    $$= \underline{3x^3 + 12x^3} \ \underline{- 6x - x} + 5$$
    $$= 15x^3 - 7x + 5$$

2. $2x^4 + 9x^2 + 10$. A minus sign precedes the first set of parentheses, so change the sign of every term inside this set and then drop these parentheses:

    $$2x^4 - (-9x^2 + x) + (x + 10)$$
    $$= 2x^4 + 9x^2 - x + (x + 10)$$

A plus sign precedes the second set of parentheses, so just drop this set:

$$= 2x^4 + 9x^2 - x + x + 10$$

Now combine like terms:

$$= 2x^4 + 9x^2 + 10$$

3.   **$-x^3 + 3x^2 - x + 5$.** A minus sign precedes the first set of parentheses, so change the sign of every term inside this set and then drop these parentheses:

$$x - (x^3 - x - 5) + 3(x^2 - x)$$
$$= x - x^3 + x + 5 + 3(x^2 - x)$$

You have no sign between the term 3 and the second set of parentheses, so multiply every term inside these parentheses by 3 and then drop the parentheses:

$$= x - x^3 + x + 5 + 3x^2 - 3x$$

Now combine like terms:

$$= \underline{x + x - 3x} - x^3 + 5 + 3x^2$$
$$= -x - x^3 + 5 + 3x^2$$

Rearrange the answer so that the exponents are in descending order:

$$= -x^3 + 3x^2 - x + 5$$

4.   **$-2x^5$.** You have no sign between the term $-x^3$ and the first set of parentheses, so multiply $-x^3$ by every term inside this set and then drop these parentheses:

$$-x^3(x^2 + x) - (x^5 - x^4)$$
$$= -x^5 - x^4 - (x^5 - x^4)$$

A minus sign precedes the second set of parentheses, so change every term inside this set and then drop the parentheses:

$$= -x^5 - x^4 - x^5 + x^4$$

Combine like terms, noting that the x^4 terms cancel out:

$$= \underline{-x^5 - x^5} \, \underline{-x^4 + x^4}$$
$$= -2x^5$$

FOILING: DEALING WITH TWO SETS OF PARENTHESES

Sometimes, expressions have two sets of parentheses next to each other without a sign between them. In that case, you need to multiply *every term* inside the first set by *every term* inside the second.

Tip: When you have two terms inside each set of parentheses, you can use a process called FOILing. This is really just the distributive property, as I show you below. The word *FOIL* is an acronym to help you make sure you multiply the correct terms. It stands for *F*irst, *O*utside, *I*nside, and *L*ast.

Here's how the process works. In this example, you're simplifying the expression $(2x - 2)(3x - 6)$:

1. **Start out by multiplying the two *First* terms in the parentheses.**

   The first term in the first set of parentheses is $2x$, and $3x$ is the first term in the second set of parentheses: $(2x - 2)(3x - 6)$.

   $$\text{F: Multiply the first terms: } 2x(3x) = 6x^2$$

2. **Multiply the two *Outside* terms.**

   The two outside terms, $2x$ and -6, are on the ends: $(2x - 2)(3x - 6)$

   $$\text{O: Multiply the outside terms: } 2x(-6) = -12x$$

3. **Multiply the two *Inside* terms.**

   The two terms in the middle are -2 and $3x$: $(2x - 2)(3x - 6)$

   $$\text{I: Multiply the middle terms: } -2(3x) = -6x$$

4. **Multiply the two *Last* terms.**

   The last term in the first set of parentheses is -2, and -6 is the last term in the second set: $(2x - 2)(3x - 6)$

   $$\text{L: Multiply the last terms: } -2(-6) = 12$$

Add these four results together to get the simplified expression:

$$(2x - 2)(3x - 6) = 6x^2 - 12x - 6x + 12$$

In this case, you can simplify this expression still further by combining the like terms $-12x$ and $-6x$:

$$= 6x^2 - 18x + 12$$

Notice that, during this process, you multiply every term inside one set of parentheses by every term inside the other set. FOILing just helps you keep track and make sure you've multiplied everything.

FOILing is really just an application of the distributive property, which I discuss in the section preceding this one. In other words, $(2x - 2)(3x - 6)$ is really the same as $2x(3x - 6) - 2(3x - 6)$ when distributed. Then distributing again gives you $6x^2 - 12x - 6x + 12$.

Examples

Q. Simplify the expression $(x + 4)(x - 3)$.

A. $x^2 + x - 12$. Start by multiplying the two first terms:

$$(\underline{x}+4)(\underline{x}-3) \quad x \times x = x^2$$

Next, multiply the two outside terms:

$$(\underline{x}+4)(x\,\underline{-3}) \quad x \times -3 = -3x$$

Now multiply the two inside terms:

$$(x+\underline{4})(\underline{x}-3) \quad 4 \times x = 4x$$

Finally, multiply the two last terms:

$$(x+\underline{4})(x\,\underline{-3}) \quad 4 \times -3 = -12$$

Add all four results together and simplify by combining like terms:

$$(x+4)(x-3) = x^2\underline{-3x+4x}-12 = x^2+x-12$$

Q. Simplify the expression $x^2 - (-2x + 5)(3x - 1) + 9$.

A. $7x^2 - 17x + 14$. Begin by FOILing the parentheses. Start by multiplying the two first terms:

$$(\underline{-2x}+5)(\underline{3x}-1)-2x \times 3x = -6x^2$$

Multiply the two outside terms:

$$(\underline{-2x}+5)(3x\,\underline{-1})-2x \times -1 = 2x$$

Multiply the two inside terms:

$$(-2x\,\underline{+5})(\underline{3x}-1)5 \times 3x = 15x$$

Finally, multiply the two last terms:

$$(-2x+\underline{5})(3x\,\underline{-1})5 \times -1 = -5$$

Add these four products together and put the result inside one set of parentheses, replacing the two sets of parentheses that were originally there:

$$x^2-(-2x+5)(3x-1)+9$$
$$= x^2-(-6x^2+2x+15x-5)+9$$

The remaining set of parentheses is preceded by a minus sign, so change the sign of every term in there to its opposite and drop the parentheses:

$$= x^2 + 6x^2 - 2x - 15x + 5 + 9$$

At this point, you can simplify the expression by combining like terms:

$$= x^2 + 6x^2 - 2x - 15x + 5 + 9$$
$$= 7x^2 - 17x + 14$$

Practice Questions

1. Simplify the expression $(x + 7)(x - 2)$.

2. Simplify the expression $(x - 1)(-x - 9)$.

3. Simplify the expression $6x - (x - 2)(x - 4) + 7x^2$.

4. Simplify the expression $3 - 4x(x^2 + 1)(x - 5) + 2x^3$.

Practice Answers

1. **$x^2 + 5x - 14$.** Begin by FOILing the parentheses. Start by multiplying the two first terms:

    $$(\underline{x} + 7)(\underline{x} - 2) \quad x \times x = x^2$$

    Multiply the two outside terms:

    $$(\underline{x} + 7)(x \underline{-2}) \quad x \times -2 = -2x$$

    Multiply the two inside terms:

    $$(x + \underline{7})(\underline{x} - 2) \quad 7 \times x = 7x$$

    Finally, multiply the two last terms:

    $$(x + \underline{7})(x \underline{-2}) \quad 7 \times -2 = -14$$

    Add these four products together and simplify by combining like terms:

    $$x^2 \underline{-2x + 7x} - 14 = x^2 + 5x - 14$$

2. **$-x^2 - 8x + 9$.** FOIL the parentheses. Multiply the two first terms, the two outside terms, the two inside terms, and the two last terms:

    $$(\underline{x} - 1)(\underline{-x} - 9) \quad x \times -x = -x^2$$
    $$(\underline{x} - 1)(-x \underline{-9}) \quad x \times -9 = -9x$$
    $$(x \underline{-1})(\underline{-x} - 9) \quad -1 \times -x = x$$
    $$(x \underline{-1})(-x \underline{-9}) \quad -1 \times -9 = 9$$

Add these four products together and simplify by combining like terms:

$$-x^2 \underline{-9x + x} + 9 = -x^2 - 8x + 9$$

3. **$6x^2 + 12x - 8$.** Begin by FOILing the parentheses: Multiply the first, outside, inside, and last terms:

$$\begin{aligned} (\underline{x}-2)(\underline{x}-4) & \quad x \times x^2 \\ (\underline{x}-2)(x\underline{-4}) & \quad x \times -4 = -4x \\ (x\underline{-2})(\underline{x}-4) & \quad -2 \times x = -2x \\ (x\underline{-2})(x\underline{-4}) & \quad -2 \times -4 = 8 \end{aligned}$$

Add these four products together and put the result inside one set of parentheses, replacing the two sets of parentheses that were originally there:

$$6x - (x-2)(x-4) + 7x^2 = 6x - (x^2 - 4x - 2x + 8) + 7x^2$$

The remaining set of parentheses is preceded by a minus sign, so change the sign of every term in there to its opposite and drop the parentheses:

$$= 6x - x^2 + 4x + 2x - 8 + 7x^2$$

Now simplify the expression by combining like terms and reordering your solution:

$$\begin{aligned} &= \underline{6x + 4x + 2x} - x^2 + 7x^2 - 8 \\ &= 6x^2 + 12x - 8 \end{aligned}$$

4. **$-4x^4 + 22x^3 - 4x^2 + 20x + 3$.** Begin by FOILing the parentheses, multiplying the first, outside, inside, and last terms:

$$\begin{aligned} (\underline{x^2}+1)(\underline{x}-5) & \quad x^2 \times x = x^3 \\ (\underline{x^2}+1)(x\underline{-5}) & \quad x^2 \times -5 = -5x^2 \\ (x^2+\underline{1})(\underline{x}-5) & \quad 1 \times x = x \\ (x^2+\underline{1})(x\underline{-5}) & \quad 1 \times -5 = -5 \end{aligned}$$

Add these four products together and put the result inside one set of parentheses, replacing the two sets of parentheses that were originally there:

$$3 - 4x(x^2+1)(x-5) + 2x^3 = 3 - 4x(x^3 - 5x^2 + x - 5) + 2x^3$$

The remaining set of parentheses is preceded by the term $-4x$ with no sign in between, so multiply $-4x$ by every term in there and then drop the parentheses:

$$= 3 - 4x^4 + 20x^3 - 4x^2 + 20x + 2x^3$$

Now simplify the expression by combining like terms:

$$\begin{aligned} &= -4x^4 + \underline{20x^3 + 2x^3} - 4x^2 + 20x + 3 \\ &= -4x^4 + 22x^3 - 4x^2 + 20x + 3 \end{aligned}$$

19

UNMASKING MR. X: ALGEBRAIC EQUATIONS

IN THIS CHAPTER

- Using variables (such as *x*) in equations
- Knowing some quick ways to solve for *x* in simple equations
- Understanding the *balance scale method* for solving equations
- Rearranging terms in an algebraic equation
- Isolating algebraic terms on one side of an equation
- Removing parentheses from an equation
- Cross-multiplying to remove fractions

When it comes to algebra, solving equations is the main event.

Solving an algebraic equation means finding out what number the variable (usually *x*) stands for. Not surprisingly, this process is called *solving for x,* and when you know how to do it, your confidence — not to mention your grades in your algebra class — will soar through the roof.

This chapter is all about solving for *x*. First, I show you a few informal methods to solve for *x* when an equation isn't too difficult. Then I show you how to solve more difficult equations by thinking of them as a balance scale.

The balance scale method is really the heart of algebra (yes, algebra has a heart, after all!). When you understand this simple idea, you're ready to solve more complicated equations, using all the tools I show you in Chapter 18, such as simplifying expressions and removing parentheses. You find out how to extend these skills to algebraic equations. Finally, I show you how cross-multiplying (see Chapter 8) can make solving algebraic equations with fractions a piece of cake.

By the end of this chapter, you'll have a solid grasp of a bunch of ways to solve equations for the elusive and mysterious *x*.

UNDERSTANDING ALGEBRAIC EQUATIONS

An algebraic equation is an equation that includes at least one variable — that is, a letter (such as *x*) that stands for a number. *Solving* an algebraic equation means finding out what number *x* stands for.

In this section, I show you the basics of how a variable like *x* works its way into an equation in the first place. Then I show you a few quick ways to *solve for x* when an equation isn't too difficult.

Using x in equations

As you discover in Chapter 4, an *equation* is a mathematical statement that contains an equals sign. For example, here's a perfectly good equation:

$$7 \times 9 = 63$$

At its heart, a variable (such as *x*) is nothing more than a placeholder for a number. You're probably used to equations that use other placeholders: One number is purposely left as a blank or replaced by an underline or a question mark, and you're supposed to fill it in. Usually, this number comes after the equals sign. For example:

$$8 + 2 =$$
$$12 - 3 = \underline{}$$
$$14 \div 7 = ?$$

As soon as you're comfortable with addition, subtraction, or whatever, you can switch the equation around a bit:

$$9 + \underline{} = 14$$
$$? \times 6 = 18$$

When you stop using underlines and question marks and start using variables such as *x* to stand for the part of the equation you want to figure out, bingo! You have an algebra problem:

$$4 + 1 = x$$
$$12 \div x = 3$$
$$x - 13 = 30$$

Choosing among four ways to solve algebraic equations

You don't need to call an exterminator just to kill a bug. Similarly, algebra is strong stuff, and you don't always need it to solve an algebraic equation.

Generally, you have four ways to solve algebraic equations such as the ones I introduce earlier in this chapter. In this section, I introduce them in order of difficulty.

Eyeballing easy equations

You can solve easy problems just by looking at them. This is called solving equations by *inspection*. For example:

$$5 + x = 6$$

When you look at this problem, you can see that $x = 1$. When a problem is this easy and you can see the answer, you don't need to go to any particular trouble to solve it.

Rearranging slightly harder equations

When you can't see an answer just by looking at a problem, sometimes rearranging the problem helps to turn it into one that you can solve using a Big Four operation. For example:

$$6x = 96$$

You can rearrange this problem using inverse operations, as I show you in Chapter 3, changing multiplication to division:

$$x = \frac{96}{6}$$

Now solve the problem by division (long or otherwise) to find that $x = 16$.

Guessing and checking equations

You can solve some equations by guessing an answer and then checking to see whether you're right. For example, suppose you want to solve the following equation:

$$3x + 7 = 19$$

To find out what x equals, start by guessing that $x = 2$. Now check to see whether you're right by substituting 2 for x in the equation:

$$3(2) + 7 = 13 \quad \text{WRONG! (13 is less than 19.)}$$
$$3(5) + 7 = 22 \quad \text{WRONG! (22 is greater than 19.)}$$
$$3(4) + 7 = 19 \quad \text{RIGHT!}$$

With only three guesses, you found that $x = 4$.

Applying algebra to more difficult equations

When an algebraic equation gets hard enough, you find that looking at it and rearranging it just isn't enough to solve it. For example:

$$11x - 13 = 9x + 3$$

You probably can't tell what x equals just by looking at this problem. You also can't solve it just by rearranging it, using an inverse operation. And guessing and checking would be very tedious. Here's where algebra comes into play.

Algebra is especially useful because you can follow mathematical rules to find your answer. Throughout the rest of this chapter, I show you how to use the rules of algebra to turn tough problems like this one into problems that you can solve.

Examples

Q. In the equation $x + 3 = 10$, what's the value of x?

A. $x = 7$. You can solve this problem through simple inspection. Because $7 + 3 = 10$, $x = 7$.

Q. Solve the equation $7x = 224$ for x.

A. $x = 32$. Turn the problem around using the inverse of multiplication, which is division:

$$7 \times x = 224 \text{ means } 224 \div 7 = x.$$

Now you can solve the problem easily using long division:

$$224 \div 7 = 32$$

Q. Find the value of x in the equation $8x - 20 = 108$.

A. $x = 16$. Guess what you think the answer may be. For example, perhaps it's $x = 10$:

$$8(10) - 20 = 80 - 20 = 60$$

Because 60 is less than 108, that guess was too low. Try a higher number — say, $x = 20$:

$$8(20) - 20 = 160 - 20 = 140$$

Because 140 is greater than 108, this guess was too high. But now you know that the right answer is between 10 and 20. So now try $x = 15$:

$$8(15) - 20 = 120 - 20 = 100$$

The number 100 is only a little less than 108, so this guess was only a little too low. Try $x = 16$:

$$8(16) - 20 = 128 - 20 = 108$$

The result 108 is correct, so $x = 16$.

Q. Solve for x: $8x^2 - x + x^2 + 4x - 9x^2 = 18$.

A. $x = 6$. Rearrange the expression on the left side of the equation so that all like terms are next to each other:

$$8x^2 + x^2 - 9x^2 - x + 4x = 18$$

Combine like terms:

$$3x = 18$$

Notice that the x^2 terms cancel each other out. Because $3(6) = 18$, you know that $x = 6$.

Practice Questions

1. Solve for x in each case just by looking at the equation.

   a. $x + 5 = 13$

   b. $18 - x = 12$

   c. $4x = 44$

   d. $30/x = 3$

2. Use the correct inverse operation to rewrite and solve each problem.

   a. $x + 41 = 97$

   b. $100 - x = 58$

   c. $13x = 273$

   d. $238/x = 17$

3. Find the value of x in each equation by guessing and checking.

   a. $19x + 22 = 136$

   b. $12x - 17 = 151$

   c. $19x - 8 = 600$

   d. $x^2 + 3 = 292$

4. Simplify the equation and then solve for x using any method you like:

   a. $x^5 - 16 + x + 20 - x^5 = 24$

   b. $5xy + x - 2xy + 27 - 3xy = 73$

   c. $6x - 3 + x^2 - x + 8 - 5x = 30$

   d. $-3 + x^3 + 4 + x - x^3 - 1 = 2xy + 7 - x - 2xy + x$

Practice Answers

1. **a.** $x + 5 = 13$; $x = 8$, because $8 + 5 = 13$.

   b. $18 - x = 12$; $x = 6$, because $18 - 6 = 12$

   c. $4x = 44$; $x = 11$, because $4(11) = 44$

   d. $\frac{30}{x} = 3$; $x = 10$, because $\frac{30}{10} = 3$

2. **a.** $x + 41 = 97$; $x = 56$. Change the addition to subtraction: $x + 41 = 97$ is the same as $97 - 41$, so $x = 56$.

   b. $100 - x = 58$; $x = 42$. Change the subtraction to addition: $100 - x = 58$ means the same thing as $100 - 58 = x$, so $x = 42$.

c. $3x = 273$; $\boldsymbol{x = 21}$. Change the multiplication to division: $13x = 273$ is equivalent to $\frac{273}{13} = x$, so $x = 21$.

d. $\frac{238}{x} = 17$; $\boldsymbol{x = 14}$. Switch around the division: $\frac{238}{x} = 17$ means $\frac{238}{17} = x$, so $x = 14$.

3. **a.** $19x + 22 = 136$; $\boldsymbol{x = 6}$. First, try $x = 10$:

$$19(10) + 22 = 190 + 22 = 212$$

212 is greater than 136, so this guess is too high. Try $x = 5$:

$$19(5) + 22 = 95 + 22 = 117$$

117 is only a little less than 136, so this guess is a little too low. Try $x = 6$:

$$19(6) + 22 = 114 + 22 = 136$$

136 is correct, so $x = 6$.

b. $12x - 17 = 151$; $\boldsymbol{x = 14}$. First, try $x = 10$:

$$12(10) - 17 = 120 - 17 = 103$$

103 is less than 151, so this guess is too low. Try $x = 20$:

$$12(20) - 17 = 240 - 17 = 223$$

223 is greater than 151, so this guess is too high. Therefore, x is between 10 and 20. Try $x = 15$:

$$12(15) - 17 = 180 - 17 = 163$$

163 is a little greater than 151, so this guess is a little too high. Try $x = 14$:

$$12(14) - 17 = 168 - 17 = 151$$

151 is correct, so $x = 14$.

c. $19x - 8 = 600$; $\boldsymbol{x = 32}$. First, try $x = 10$:

$$19(10) - 8 = 190 - 8 = 182$$

182 is much less than 600, so this guess is much too low. Try $x = 30$:

$$19(30) - 8 = 570 - 8 = 562$$

562 is still less than 600, so this guess is still too low. Try $x = 35$:

$$19(35) - 8 = 665 - 8 = 657$$

657 is greater than 600, so this guess is too high. Therefore, x is between 30 and 35. Try $x = 32$:

$$19(32) - 8 = 608 - 8 = 600$$

600 is correct, so $x = 32$.

d. $x^2 + 3 = 292$; **$x = 17$.** First, try $x = 10$:

$$10^2 + 3 = 100 + 3 = 103$$

103 is less than 292, so this guess is too low. Try $x = 20$:

$$20^2 + 3 = 400 + 3 = 403$$

403 is greater than 292, so this guess is too high. Therefore, x is between 10 and 20. Try $x = 15$:

$$15^2 + 3 = 225 + 3 = 228$$

228 is less than 292, so this guess is too low. Therefore, x is between 15 and 20. Try $x = 17$:

$$17^2 + 3 = 289 + 3 = 292$$

292 is correct, so $x = 17$.

4.   **a.** $x^5 - 16 + x + 20 - x^5 = 24$; **$x = 20$.** Rearrange the expression on the left side of the equation so that all like terms are next to each other:

$$\underline{x^5 - x^5} + x + \underline{20 - 16} = 24$$

Combine like terms:

$$x + 4 = 24$$

Notice that the two x^5 terms cancel each other out. Because 20 + 4 = 24, you know that $x = 20$.

b. $5xy + x - 2xy + 27 - 3xy = 73$; **$x = 46$.** Rearrange the expression on the left side of the equation:

$$\underline{5xy - 2xy - 3xy} + x + 27 = 73$$

Combine like terms:

$$x + 27 = 73$$

Notice that the three xy terms cancel each other out. Because $x + 27 = 73$ means $73 - 27 = x$, you know that $x = 46$.

c. $6x - 3 + x^2 - x + 8 - 5x = 30$; **$x = 5$.** Rearrange the expression on the left side of the equation so that all like terms are adjacent:

$$\underline{6x - x - 5x} + x^2 + \underline{8 - 3} = 30$$

Combine like terms:

$$x^2 + 5 = 30$$

Notice that the three x terms cancel each other out. Try $x = 10$:

$$10^2 + 5 = 100 + 5 = 105$$

105 is greater than 30, so this guess is too high. Therefore, x is between 0 and 10. Try $x = 5$:

$$5^2 + 5 = 25 + 5 = 30$$

This result is correct, so $x = 5$.

d. $-3 + x^3 + 4 + x - x^3 - 1 = 2xy + 7 - x - 2xy + x$; $x = 7$. Rearrange the expression on the left side of the equation:

$$\underline{-3 + 4 - 1} + \underline{x^3 - x^3} + x = 2xy + 7 - x - 2xy + x$$

Combine like terms:

$$x = 2xy + 7 - x - 2xy + x$$

Notice that the three constant terms cancel each other out, and so do the two x^3 terms. Now rearrange the expression on the right side of the equation:

$$x = 2xy - 2xy + 7 - x + x$$

Combine like terms:

$$x = 7$$

Notice that the two xy terms cancel each other out, and so do the two x terms. Therefore, $x = 7$.

THE BALANCING ACT: SOLVING FOR X

As I show you in the preceding section, some problems are too complicated to find out what the variable (usually x) equals just by eyeballing it or rearranging it. For these problems, you need a reliable method for getting the right answer. I call this method the *balance scale*.

The balance scale allows you to *solve for x* — that is, find the number that x stands for — in a step-by-step process that always works. In this section, I show you how to use the balance scale method to solve algebraic equations.

Striking a balance

Remember: The equals sign in any equation means that both sides balance. To keep that equals sign, you have to maintain that balance. In other words, whatever you do to one side of an equation, you have to do to the other.

For example, here's a balanced equation:

$$\underset{\Delta}{1 + 2 = 3}$$

If you add 1 to one side of the equation, the scale goes out of balance.

$$\frac{1+2+1 \neq 3}{\Delta}$$

But if you add 1 to *both* sides of the equation, the scale stays balanced:

$$\frac{1+2+1 = 3+1}{\Delta}$$

You can add any number to the equation, as long as you do it to both sides. And in math, *any number* means x:

$$1+2+x = 3+x$$

Remember that x is the same wherever it appears in a single equation or problem.

This idea of changing both sides of an equation equally isn't limited to addition. You can just as easily subtract an x, or even multiply or divide by x, as long as you do the same to both sides of the equation:

$$\text{Subtract: } 1+2-x = 3-x$$
$$\text{Multiply: } (1+2)x = 3x$$
$$\text{Divide: } \frac{1+2}{x} = \frac{3}{x}$$

Using the balance scale to isolate x

The simple idea of balance is at the heart of algebra, and it enables you to find out what x is in many equations. When you solve an algebraic equation, the goal is to *isolate x* — that is, to get x alone on one side of the equation and some number on the other side. In algebraic equations of middling difficulty, this is a three-step process:

1. **Get all constants (non-x terms) on one side of the equation.**
2. **Get all x-terms on the other side of the equation.**
3. **Divide to isolate x.**

For example, take a look at the following problem:

$$11x - 13 = 9x + 3$$

As you follow the steps, notice how I keep the equation balanced at each step:

1. **Get all the constants on one side of the equation by adding 13 to both sides of the equation:**

$$11x - 13 = 9x + 3$$
$$\underline{+13 = \quad +13}$$
$$11x \quad = 9x + 16$$

Because you've obeyed the rules of the balance scale, you know that this new equation is also correct. Now the only non-*x* term (16) is on the right side of the equation.

2. **Get all the *x*-terms on the other side by subtracting 9*x* from both sides of the equation:**

$$
\begin{array}{r}
11x = 9x + 16 \\
\underline{-9x - 9x } \\
2x = 16
\end{array}
$$

Again, the balance is preserved, so the new equation is correct.

3. **Divide by 2 to isolate *x*:**

$$\frac{2x}{2} = \frac{16}{2}$$
$$x = 8$$

To check this answer, you can simply substitute 8 for *x* in the original equation:

$$11(8) - 13 = 9(8) + 3$$
$$88 - 13 = 72 + 3$$
$$75 = 75 \checkmark$$

This checks out, so 8 is the correct value of *x*.

Example

Q. Use the balance scale method to find the value of *x* in the equation $5x - 6 = 3x + 8$.

A. *x* = 7. To get all constants on the right side of the equation, add 6 to both sides, which causes the –6 to cancel out on the left side of the equation:

$$
\begin{array}{r}
5x - 6 = 3x + 8 \\
\underline{+6 + 6} \\
5x = 3x + 14
\end{array}
$$

The right side of the equation still contains a 3*x*. To get all *x* terms on the left side of the equation, subtract 3*x* from both sides:

$$
\begin{array}{r}
5x = 3x + 14 \\
\underline{-3x -3x } \\
2x = 14
\end{array}
$$

Divide by 2 to isolate *x*:

$$\frac{2x}{2} = \frac{14}{2}$$
$$x = 7$$

Practice Questions

1. Use the balance scale method to find the value of x in the equation $9x - 2 = 6x + 7$.

2. Solve the equation $10x - 10 = 8x + 12$ using the balance scale method.

3. Find the value of x in $4x - 17 = x + 22$.

4. Solve for x: $15x - 40 = 11x + 4$.

Practice Answers

1. **$x = 3$.** To get all constants on the right side of the equation, add 2 to both sides:

$$\begin{array}{r} 9x - 2 = 6x + 7 \\ \underline{+2 \qquad +2} \\ 9x \quad = 6x + 9 \end{array}$$

To get all x terms on the left side, subtract $6x$ from both sides:

$$\begin{array}{r} 9x \ = 6x + 9 \\ \underline{-6x \quad -6x} \\ 3x \ = \qquad 9 \end{array}$$

Divide by 3 to isolate x:

$$\frac{3x}{3} = \frac{9}{3}$$
$$x = 3$$

2. **$x = 11$.** Move all constants on the right side of the equation by adding 10 to both sides:

$$\begin{array}{r} 10x - 10 = 8x + 12 \\ \underline{+10 \qquad +10} \\ 10x \quad = 8x + 22 \end{array}$$

To get all x terms on the left side, subtract $8x$ from both sides:

$$\begin{array}{r} 10x = \quad 8x + 22 \\ \underline{-8x \quad -8x} \\ 2x = \qquad 22 \end{array}$$

Divide by 2 to isolate x:

$$\frac{2x}{2} = \frac{22}{2}$$
$$x = 11$$

3. **$x = 13$.** Add 17 to both sides to get all constants on the right side of the equation:

$$4x - 17 = x + 22$$
$$\underline{\quad +17 \qquad +17 \quad}$$
$$4x \qquad = x + 39$$

Subtract x from both sides to get all x terms on the left side:

$$4x = \quad x + 39$$
$$\underline{-x \quad -x \quad}$$
$$3x = \qquad 39$$

Divide by 3 to isolate x:

$$\frac{3x}{3} = \frac{39}{3}$$
$$x = 13$$

4. **$x = 11$.** To get all constants on the right side of the equation, add 40 to both sides:

$$15x - 40 = 11x + 4$$
$$\underline{\quad +40 \qquad +40 \quad}$$
$$15x \qquad = 11x + 44$$

To get all x terms on the left side, subtract $11x$ from both sides:

$$15x = \quad 11x + 44$$
$$\underline{-11x \quad -11x \quad}$$
$$4x = \qquad 44$$

Divide by 4 to isolate x:

$$\frac{4x}{4} = \frac{44}{4}$$
$$x = 11$$

REARRANGING EQUATIONS AND ISOLATING X

When you understand how algebra works like a balance scale, as I show you in the preceding section, you can begin to solve more-difficult algebraic equations. The basic strategy is always the same: Changing both sides of the equation equally at every step, try to isolate x on one side of the equation.

In this section, I show you how to put your skills from Chapter 18 to work solving equations. First, I show you how rearranging the terms in an expression is similar to rearranging them in an algebraic equation. Next, I show you how removing parentheses from an equation can help you solve it. Finally, you discover how cross-multiplication is useful for solving algebraic equations with fractions.

Rearranging terms on one side of an equation

Rearranging terms becomes all-important when working with equations. For example, suppose you're working with this equation:

$$5x - 4 = 2x + 2$$

When you think about it, this equation is really two expressions connected with an equals sign. And of course, that's true of *every* equation. That's why everything you find out about expressions in Chapter 18 is useful for solving equations. For example, you can rearrange the terms on one side of an equation. So here's another way to write the same equation:

$$-4 + 5x = 2x + 2$$

And here's a third way:

$$-4 + 5x = 2 + 2x$$

This flexibility to rearrange terms comes in handy when you're solving equations.

Moving terms to the other side of the equals sign

Earlier in this chapter, I show you how an equation is similar to a balance scale. For example, take a look at Figure 19-1.

Figure 19-1: Showing how an equation is similar to a balance scale.

© John Wiley & Sons, Inc.

To keep the scale balanced, if you add or remove anything on one side, you must do the same on the other side. For example:

$$2x - 3 = 11$$
$$\underline{-2x \qquad -2x}$$
$$-3 = 11 - 2x$$

Now take a look at these two versions of this equation side by side:

$$2x - 3 = 11 \qquad -3 = 11 - 2x$$

In the first version, the term $2x$ is on the left side of the equals sign. In the second, the term $-2x$ is on the right side. This example illustrates an important rule.

Remember: When you move any term in an expression to the other side of the equals sign, change its sign (from plus to minus or from minus to plus).

As another example, suppose you're working with this equation:

$$4x - 2 = 3x + 1$$

You have x's on both sides of the equation, so say you want to move the $3x$. When you move the term $3x$ from the right side to the left side, you have to change its sign from plus to minus (technically, you're subtracting $3x$ from both sides of the equation).

$$4x - 2 - 3x = 1$$

After that, you can simplify the expression on the left side of the equation by combining like terms:

$$x - 2 = 1$$

At this point, you can probably see that $x = 3$ because $3 - 2 = 1$. But just to be sure, move the -2 term to the right side and change its sign:

$$x = 1 + 2$$
$$x = 3$$

To check this result, substitute a 3 wherever x appears in the original equation:

$$4x - 2 = 3x + 1$$
$$4(3) - 2 = 3(3) + 1$$
$$12 - 2 = 9 + 1$$
$$10 = 10 \checkmark$$

As you can see, moving terms from one side of an equation to the other can be a big help when you're solving equations.

Examples

Q. Find the value of x in the equation $7x - 6 = 4x + 9$.

A. $x = 5$. Rearrange the terms of the equation so that the x terms are on one side and the constants are on the other. I do this in two steps:

$$7x - 6 = 4x + 9$$
$$7x = 4x + 9 + 6$$
$$7x - 4x = 9 + 6$$

Combine like terms on both sides of the equation:

$$3x = 15$$

Divide by 3 to isolate x.

$$\frac{3x}{3} = \frac{15}{3}$$
$$x = 5$$

Q. Find the value of x in the equation $3 - (7x - 13) = 5(3 - x) - x$.

A. $x = 1$. Before you can begin rearranging terms, remove the parentheses on both sides of the equation. On the left side, the parentheses are preceded by a minus sign, so change the sign of every term and remove the parentheses:

$$3 - 7x + 13 = 5(3 - x) - x$$

On the right side, no sign comes between the 5 and the parentheses, so multiply every term inside the parentheses by 5 and remove the parentheses:

$$3 - 7x + 13 = 15 - 5x - x$$

Now you can solve the equation in three steps. Put the x terms on one side and the constants on the other, remembering to switch the signs as needed:

$$-7x = 15 - 5x - x - 3 - 13$$
$$-7x + 5x + x = 15 - 3 - 13$$

Combine like terms on both sides of the equation:

$$-x = -1$$

Divide by -1 to isolate x:

$$\frac{-x}{-1} = \frac{-1}{-1}$$
$$x = 1$$

Practice Questions

1. Rearrange the equation $10x + 5 = 3x + 19$ *to* solve for x.

2. Find the value of x by rearranging the equation $4 + (2x + 6) = 7(x - 5)$.

3. Solve $-[2(x + 7) + 1] = x - 12$ for x.

4. Find the value of x: $-x^3 + 2(x^2 + 2x + 1) = 4x^2 - (x^3 + 2x^2 - 18)$.

A Practice Answers

1. **$x = 2$.** Rearrange the terms of the equation so that the x terms are on one side and the constants are on the other. I do this in two steps:

$$10x + 5 = 3x + 19$$
$$10x = 3x + 19 - 5$$
$$10x - 3x = 19 - 5$$

Combine like terms on both sides:

$$7x = 14$$

Divide by 7 to isolate x:

$$\frac{7x}{7} = \frac{14}{7}$$
$$x = 2$$

2. **$x = 9$.** Before you can begin rearranging terms, remove the parentheses on both sides of the equation. On the left side, the parentheses are preceded by a plus sign, so just drop them:

$$4 + (2x + 6) = 7(x - 5)$$
$$4 + 2x + 6 = 7(x - 5)$$

On the right side, no sign comes between the number 7 and the parentheses, so multiply 7 by every term inside the parentheses and then drop the parentheses:

$$4 + 2x + 6 = 7x - 35$$

Now you can solve for x by rearranging the terms of the equation. Group the x terms on one side and the constants on the other. I do this in two steps:

$$4 + 6 = 7x - 35 - 2x$$
$$4 + 6 + 35 = 7x - 2x$$

Combine like terms on both sides:

$$45 = 5x$$

Divide by 5 to isolate x:

$$\frac{45}{5} = \frac{5x}{5}$$
$$9 = x$$

3. **$x = -1$.** Before you can begin rearranging terms, remove the parentheses on the left side of the equation. Start with the inner parentheses, multiplying 2 by every term inside that set:

$$-[2(x+7)+1] = x-12$$
$$-[2x+14+1] = x-12$$

Next, remove the remaining parentheses, switching the sign of every term within that set:

$$-2x-14-1 = x-12$$

Now you can solve for x by rearranging the terms of the equation:

$$-2x-14-1+12 = x$$
$$-14-1+12 = x+2x.$$

Combine like terms on both sides:

$$-3 = 3x$$

Divide by 3 to isolate x:

$$\frac{-3}{3} = \frac{3x}{3}$$
$$-1 = x$$

4. **$x = 4$.** Before you can begin rearranging terms, multiply the terms in the left-hand parentheses by 2 and remove the parentheses on both sides of the equation:

$$-x^3+2(x^2+2x+1) = 4x^2-(x^3+2x^2-18)$$
$$-x^3+2x^2+4x+2 = 4x^2-(x^3+2x^2-18)$$
$$-x^3+2x^2+4x+2 = 4x^2-x^3-2x^2+18$$

Rearrange the terms of the equation:

$$-x^3+2x^2+4x+2-4x^2+x^3+2x^2 = 18$$
$$-x^3+2x^2+4x-4x^2+x^3+2x^2 = 18-2$$

Combine like terms on both sides (notice that the x^3 and x^2 terms all cancel out):

$$4x = 16$$

Divide by 4 to isolate x:

$$\frac{4x}{4} = \frac{16}{4}$$
$$x = 4$$

Removing parentheses from equations

Chapter 18 gives you a treasure trove of tricks for simplifying expressions, and they come in handy when you're solving equations. One key skill from that chapter is removing parentheses from expressions. This tactic is also indispensable when you're solving equations.

For example, suppose you have the following equation:

$$5x + (6x - 15) = 30 - (x - 7) + 8$$

Your mission is to get all the x terms on one side of the equation and all the constants on the other. As the equation stands, however, x terms and constants are "locked together" inside parentheses. In other words, you can't isolate the x terms from the constants. So before you can isolate terms, you need to remove the parentheses from the equation.

Recall that an equation is really just two expressions connected by an equals sign. So you can start working with the expression on the left side. In this expression, the parentheses begin with a plus sign (+), so you can just remove them:

$$5x + \underline{6}x - 15 = 30 - (x - 7) + 8$$

Now move on to the expression on the right side. This time, the parentheses come right after a minus sign (–). To remove them, change the sign of both terms inside the parentheses: x becomes $-x$, and -7 becomes 7:

$$5x + 6x - 15 = 30 \underline{-x} + 7 + 8$$

Bravo! Now you can isolate x terms to your heart's content. Move the $-x$ from the right side to the left, changing it to x:

$$5x + 6x - 15 \underline{+x} = 30 + 7 + 8$$

Next, move -15 from the left side to the right, changing it to 15:

$$5x + 6x + x = 30 + 7 + 8 \underline{+15}$$

Now combine like terms on both sides of the equation:

$$12x = 30 + 7 + 8 + 15$$
$$12x = 60$$

Finally, get rid of the coefficient 12 by dividing:

$$\frac{12x}{12} = \frac{60}{12}$$
$$x = 5$$

As usual, you can check your answer by substituting 5 into the original equation wherever x appears:

$$5x + (6x - 15) = 30 - (x - 7) + 8$$
$$5(5) + [6(5) - 15] = 30 - (5 - 7) + 8$$
$$25 + (30 - 15) = 30 - (-2) + 8$$
$$25 + 15 = 30 + 2 + 8$$
$$40 = 40 \checkmark$$

Here's one more example:

$$11 + 3(-3x + 1) = 25 - (7x - 3) - 12$$

As in the preceding example, start out by removing both sets of parentheses. This time, however, on the left side of the equation, you have no sign between 3 and ($-3x + 1$). But again, you can put your skills from Chapter 18 to use. To remove the parentheses, multiply 3 by both terms inside the parentheses:

$$11 - 9x + 3 = 25 - (7x - 3) - 12$$

On the right side, the parentheses begin with a minus sign, so remove the parentheses by changing all the signs inside the parentheses:

$$11 - 9x + 3 = 25 - 7x + 3 - 12$$

Now you're ready to isolate the x terms. I do this in one step, but take as many steps as you want:

$$-9x + 7x = 25 + 3 - 12 - 11 - 3$$

At this point, you can combine like terms:

$$-2x = 2$$

To finish, divide both sides by –2:

$$x = -1$$

Copy this example, and work through it a few times with the book closed.

Example

Q. What is the value of x in the equation $x - (7x + 2) = -3(4x - 5) - 13 + 4x$

A. 7. To begin, remove parentheses:

$$x - (7x + 2) = -3(4x - 5) - 13 + 4x$$
$$x - 7x - 2 = -12x + 15 - 13 + 4x$$

Combine like terms on each side of the equation:

$$-6x - 2 = -8x + 12$$

Isolate x:

$$2x - 2 = 12$$
$$2x = 14$$
$$x = 7$$

Practice Questions

1. Solve for x: $5x + (-17 - 7x) = 3(x + 9) + 1$

2. What is the value of x in this equation? $-7(10 - x) + 115 = 2(x + 60) + 30$

3. Solve: $6x - 3(1 + x) + (5 - x) = 4(x + 1) + 7 - (x - 3)$

4. Find x: $4x - (3x + (x - 1)) = -(x + (6x + 31 + 5(2 - x)) - 9x)$

Practice Answers

1. **–9.** Begin by removing parentheses:

$$5x + (-17 - 7x) = 3(x + 9) + 1$$
$$5x - 17 - 7x = 3x + 27 + 1$$

Combine like terms:

$$-17 - 2x = 3x + 28$$

Isolate x:

$$-17 = 5x + 28$$
$$-45 = 5x$$
$$-9 = x$$

2. **21.** Begin by removing parentheses:

$$-7(10 - x) + 115 = 2(x + 60) + 30$$
$$-70 + 7x + 115 = 2x + 120 + 30$$

Combine like terms:

$$7x + 45 = 2x + 150$$

Isolate x:

$$5x + 45 = 150$$
$$5x = 105$$
$$x = 21$$

3. **–12.** Begin by removing parentheses:

$$6x - 3(1 + x) + (5 - x) = 4(x + 1) + 7 - (x - 3)$$
$$6x - 3 - 3x + 5 - x = 4x + 4 + 7 - x + 3$$

Combine like terms:

$$2x + 2 = 3x + 14$$

Isolate x:

$$2 = x + 14$$
$$-12 = x$$

4. **6.** Begin by removing parentheses, starting on the inside and working your way out:

$$4x - (3x + (x - 1)) = -(x + (6x + 31 + 5(2 - x)) - 9x)$$
$$4x - (3x + x - 1) = -(x + (6x + 31 + 10 - 5x) - 9x)$$
$$4x - 3x - x + 1 = -(x + 6x + 31 + 10 - 5x - 9x)$$
$$4x - 3x - x + 1 = -x - 6x - 31 - 10 + 5x + 9x$$

Combine like terms:

$$1 = 7x - 41$$

Isolate x:

$$42 = 7x$$
$$6 = x$$

Cross-multiplying

In algebra, cross-multiplication helps to simplify equations by removing unwanted fractions (and, honestly, when are fractions ever wanted?). As I discuss in Chapter 8, you can use cross-multiplication to find out whether two fractions are equal. You can use this same idea to solve algebra equations with fractions, like this one:

$$\frac{x}{2x - 2} = \frac{2x + 3}{4x}$$

This equation looks hairy. You can't do the division or cancel anything out because the fraction on the left has two terms in the denominator, and the fraction on the right has two terms in the numerator (see Chapter 18 for info on dividing algebraic terms). However, an important piece of information that you have is that the fraction equals the fraction. So if you cross-multiply these two fractions, you get two results that are also equal:

$$x(4x) = (2x + 3)(2x - 2)$$

At this point, you have something you know how to work with. The left side is easy:

$$4x^2 = (2x + 3)(2x - 2)$$

The right side requires a bit of FOILing (flip to Chapter 18 for details):

$$4x^2 = 4x^2 - 4x + 6x - 6$$

Now all the parentheses are gone, so you can isolate the x terms. Because most of these terms are already on the right side of the equation, isolate them on that side:

$$6 = 4x^2 - 4x + 6x - 4x^2$$

Combining like terms gives you a pleasant surprise:

$$6 = 2x$$

The two x^2 terms cancel each other out. You may be able to eyeball the correct answer, but here's how to finish:

$$\frac{6}{2} = \frac{2x}{2}$$
$$3 = x$$

To check your answer, substitute 3 back into the original equation:

$$\frac{x}{2x - 2} = \frac{2x + 3}{4x}$$
$$\frac{3}{2(3) - 2} = \frac{2(3) + 3}{4(3)}$$
$$\frac{3}{6 - 2} = \frac{6 + 3}{12}$$
$$\frac{3}{4} = \frac{3}{4} \checkmark$$

So the answer $x = 3$ is correct.

Examples

Q. Use cross-multiplication to solve the equation $\frac{2x}{3} = x - 3$.

A. **$x = 9$.** Cross-multiply to get rid of the fraction in this equation. To do this, turn the right side of the equation into a fraction by inserting a denominator of 1:

$$\frac{2x}{3} = \frac{x - 3}{1}$$

Now cross-multiply:

$$2x(1) = 3(x - 3)$$

Remove the parentheses from both sides (as I show you in Chapter 18):

$$2x = 3x - 9$$

At this point, you can rearrange the equation and solve for x, as I show you earlier in this chapter:

$$2x - 3x = -9$$
$$-x = -9$$
$$\frac{-x}{-1} = \frac{-9}{-1}$$
$$x = 9$$

Q. Use cross-multiplication to solve the equation $\frac{2x+1}{x+1} = \frac{6x}{3x+1}$.

A. $x = 1$. In some cases, after you cross-multiply, you may need to FOIL one or both sides of the resulting equation. First, cross-multiply to get rid of the fraction bar in this equation:

$$(2x+1)(3x+1) = 6x(x+1)$$

Now remove the parentheses on the left side of the equation by FOILing (as I show you in Chapter 18):

$$6x^2 + 2x + 3x + 1 = 6x(x+1)$$

To remove the parentheses from the right side, multiply $6x$ by every term inside the parentheses, and then drop the parentheses:

$$6x^2 + 2x + 3x + 1 = 6x^2 + 6x$$

At this point, you can rearrange the equation and solve for x:

$$1 = 6x^2 + 6x - 6x^2 - 2x - 3x$$

Notice that the two x^2 terms cancel each other out:

$$1 = 6x - 2x - 3x$$
$$1 = x$$

Practice Questions

1. Rearrange the equation $\frac{x+5}{2} = \frac{-x}{8}$ to solve for x.

2. Find the value of x by rearranging the equation $\frac{3x+5}{7} = x-1$.

3. Solve the equation $\frac{x}{2x-5} = \frac{2x+3}{4x-7}$.

4. Find the value of x in this equation: $\frac{2x+3}{4-8x} = \frac{6-x}{4x+8}$.

A Practice Answers

1. $x = -4$. Remove the fraction from the equation by cross-multiplying:

$$\frac{x+5}{2} = \frac{-x}{8}$$
$$8(x+5) = -2x$$

Multiply to remove the parentheses from the left side of the equation:

$$8x + 40 = -2x$$

At this point, you can solve for x:

$$40 = -2x - 8x$$
$$40 = -10x$$
$$\frac{40}{-10} = \frac{-10x}{-10}$$
$$-4 = x$$

2. $x = 3$. Change the right side of the equation to a fraction by attaching a denominator of 1. Remove the fraction bar from the equation by cross-multiplying:

$$\frac{3x+5}{7} = \frac{x-1}{1}$$
$$3x + 5 = 7(x-1)$$

Multiply 7 by each term inside the parentheses to remove the parentheses from the right side of the equation:

$$3x + 5 = 7x - 7$$

Now solve for x:

$$5 = 7x - 7 - 3x$$
$$5 + 7 = 7x - 3x$$
$$12 = 4x$$
$$3 = x$$

3. $x = 5$. Remove the fractions from the equation by cross-multiplying:

$$\frac{x}{2x-5} = \frac{2x+3}{4x-7}$$
$$x(4x-7) = (2x+3)(2x-5)$$

Remove the parentheses from the left side of the equation by multiplying through by x; remove parentheses from the right side of the equation by FOILing:

$$4x^2 - 7x = 4x^2 - 10x + 6x - 15$$

Rearrange the equation:

$$4x^2 - 7x - 4x^2 + 10x - 6x = -15$$

Notice that the two x^2 terms cancel each other out:

$$-7x + 10x - 6x = -15$$
$$-3x = -15$$
$$\frac{-3x}{-3} = \frac{-15}{-3}$$
$$x = 5$$

4. $x = 0$. Remove the fractions from the equation by cross-multiplying:

$$\frac{2x+3}{4-8x} = \frac{6-x}{4x+8}$$

$$(2x+3)(4x+8) = (6-x)(4-8x)$$

FOIL both sides of the equation to remove the parentheses:

$$8x^2 + 16x + 12x + 24 = 24 - 48x - 4x + 8x^2$$

At this point, rearrange terms so you can solve for x:

$$8x^2 + 16x + 12x + 24 + 48x + 4x - 8x^2 = 24$$
$$8x^2 + 16x + 12x + 48x + 4x - 8x^2 = 24 - 24$$

Notice that the x^2 terms and the constant terms drop out of the equation:

$$16x + 12x + 48x + 4x = 0$$
$$80x = 0$$
$$\frac{80x}{80} = \frac{0}{80}$$
$$x = 0$$

20

PUTTING MR. X TO WORK: ALGEBRA WORD PROBLEMS

IN THIS CHAPTER

- Solving algebra word problems in simple steps
- Choosing variables
- Using charts

Word problems that require algebra are among the toughest problems that students face — and the most common. Teachers just love algebra word problems because they bring together a lot of what you know, such as solving algebra equations (Chapter 19) and turning words into numbers (see Chapters 5 and 12). And standardized tests virtually always include these types of problems.

In this chapter, I show you a five-step method for using algebra to solve word problems. Then I give you a bunch of examples that take you through all five steps.

Along the way, I give you some important tips that can make solving word problems easier. First, I show you how to choose a variable that makes your equation as simple as possible. Next, I give you practice organizing information from the problem into a chart. By the end of this chapter, you'll have a solid understanding of how to solve a wide variety of algebra word problems.

SOLVING ALGEBRA WORD PROBLEMS IN FIVE STEPS

Everything from Chapter 19 comes into play when you use algebra to solve word problems, so if you feel a little shaky on solving algebraic equations, flip back to those chapters for some review.

Throughout this section, I use the following word problem as an example:

> In three days, Alexandra sold a total of 31 tickets to her school play. On Tuesday, she sold twice as many tickets as on Wednesday. And on Thursday, she sold exactly 7 tickets. How many tickets did Alexandra sell on each day, Tuesday through Thursday?

Organizing the information in an algebra word problem by using a chart or picture is usually helpful. Here's what I came up with:

Tuesday:	Twice as many as on Wednesday
Wednesday:	?
Thursday:	7
Total:	31

At this point, all the information is in the chart, but the answer still may not be jumping out at you. In this section, I outline a step-by-step method that enables you to solve this problem — and much harder ones as well.

Here are the five steps for solving most algebra word problems:

1. **Declare a variable.**

2. **Set up the equation.**

3. **Solve the equation.**

4. **Answer the question that the problem asks.**

5. **Check your answer.**

Declaring a variable

As you know from Chapter 18, a variable is a letter that stands for a number. Most of the time, you don't find the variable x (or any other variable, for that matter) in a word problem. That omission doesn't mean you don't need algebra to solve the problem. It just means that you're going to have to put x into the problem yourself and decide what it stands for.

Remember: When you *declare a variable,* you say what that variable means in the problem you're solving.

Here are some examples of variable declarations:

Let m = the number of dead mice that the cat dragged into the house.

Let p = the number of times Marianne's husband promised to take out the garbage.

Let c = the number of complaints Arnold received after he painted his garage door purple.

In each case, you take a variable (m, p, or c) and give it a meaning by attaching it to a number.

Notice that the earlier chart for the sample problem has a big question mark next to *Wednesday.* This question mark stands for *some number,* so you may want to declare a variable that stands for this number. Here's how you do it:

Let w = the number of tickets that Alexandra sold on Wednesday.

Tip: Whenever possible, choose a variable with the same initial as what the variable stands for. This practice makes remembering what the variable means a lot easier, which will help you later in the problem.

For the rest of the problem, every time you see the variable *w,* keep in mind that it stands for the number of tickets that Alexandra sold on Wednesday.

Setting up the equation

After you have a variable to work with, you can go through the problem again and find other ways to use this variable. For example, Alexandra sold twice as many tickets on Tuesday as on Wednesday, so she sold $2w$ tickets on Tuesday. Now you have a lot more information to fill in on the chart:

Tuesday:	Twice as many as on Wednesday	$2w$
Wednesday:	?	w
Thursday:	7	7
Total:	31	31

You know that the total number of tickets, or the sum of the tickets she sold on Tuesday, Wednesday, and Thursday, is 31. With the chart filled in like that, you're ready to set up an equation to solve the problem:

$$2w + w + 7 = 31$$

Solving the equation

After you set up an equation, you can use the tricks from Chapter 19 to solve the equation for *w.* Here's the equation one more time:

$$2w + w + 7 = 31$$

For starters, remember that $2w$ really means $w + w$. So on the left, you know you really have $w + w + w$, or $3w$; you can simplify the equation a little bit, as follows:

$$3w + 7 = 31$$

The goal at this point is to try to get all the terms with *w* on one side of the equation and all the terms without *w* on the other side. So on the left side of the equation, you want to get rid of the 7. The inverse of addition is subtraction, so subtract 7 from both sides:

$$\begin{aligned} 3w + 7 &= 31 \\ -7 \quad &-7 \\ \hline 3w \quad &= 24 \end{aligned}$$

You now want to isolate w on the left side of the equation. To do this, you have to undo the multiplication by 3, so divide both sides by 3:

$$\frac{3w}{3} = \frac{24}{3}$$
$$w = 8$$

Answering the question

You may be tempted to think that, after you've solved the equation, you're done. But you still have a bit more work to do. Look back at the problem, and you see that it asks you this question:

> How many tickets did Alexandra sell on each day, Tuesday through Thursday?

At this point, you have some information that can help you solve the problem. The problem tells you that Alexandra sold 7 tickets on Thursday. And because $w = 8$, you now know that she sold 8 tickets on Wednesday. And on Tuesday, she sold twice as many on Wednesday, so she sold 16. So Alexandra sold 16 tickets on Tuesday, 8 on Wednesday, and 7 on Thursday.

Checking your work

To check your work, compare your answer to the problem, line by line, to make sure every statement in the problem is true:

> In three days, Alexandra sold a total of 31 tickets to her school play.

That part is correct because $16 + 8 + 7 = 31$.

> On Tuesday, she sold twice as many tickets as on Wednesday.

Correct, because she sold 16 tickets on Tuesday and 8 on Wednesday.

> And on Thursday, she sold exactly 7 tickets.

Yep, that's right, too, so you're good to go.

E

Example

Q. Three sisters spent the day at the pool swimming laps. Tory swam 25 more laps than Sara, who swam twice as many laps as Petra. If together Tory and Sara swam 185 laps, how many laps did Petra swim?

   a. Let p equal the number of laps that Petra swam. How many laps did Sara and Tory swim in terms of p.

   b. Set up an equation and solve it for p.

c. Answer the question.

d. Check your work.

A. **a.** Sara = 2*p*; Tory = 2*p* + 25.

  b. *p* = 40

$$2p + 2p + 25 = 185$$
$$4p + 25 = 185$$
$$4p = 160$$
$$p = 40$$

c. Petra swam 40 laps.

d. If Petra swam 40 laps, Sarah swam 80 laps (40 × 2 = 80), and Tory swam 105 laps (80 + 25 = 105), so Sarah and Tory swam 185 laps (80 + 105 = 185).

Practice Questions

1. Annette and Benjamin are both selling books of raffle tickets for their school play. Benjamin sells four fewer than twice as many books as Annette, and together they sell 29 books. How many books did Annette sell?

   a. Let *a* equal the number of books that Annette sold. How many did Benjamin sell, in terms of *a*?

   b. Set up an equation and solve it for *a*.

   c. Answer the question.

2. A museum houses a collection of works by artist Maxine Brioche. The collection includes 8 more sculptures than mobiles and three times as many canvasses than mobiles, for a total of 38 pieces in all. How many mobiles does the collection boast?

   a. Let *m* equal the number of mobiles in the collection. How many sculptures and canvasses does the collection include, in terms of *m*?

   b. Set up an equation and solve it for *m*.

   c. Answer the question.

   d. Check your work.

3. Three friends are playing pinochle. Roy has half as many points as Keith, and Keith has 210 points fewer than Brad. If together, the three men have a total of 960 points, how many points does Brad have?

    a. Let r = the number of points that Roy has. How many points do Keith and Brad have, in terms of r?

    b. Set up an equation and solve it for r.

    c. Answer the question.

    d. Check your work.

4. Alison read a book for school in three days. On the second day, she read three times as many pages as she had read on the first day, and then on the third day she finished the book by reading 130 fewer pages than she had read on the first two days combined. If the book has a total of 358 pages, how many pages did she read of it on the third day?

    a. Let f = the number of pages that Alison read on the first day. How many pages did she read on the second and third days, in terms of f?

    b. Set up an equation and solve it for f.

    c. Answer the question.

    d. Check your work.

Practice Answers

1.   **a.** Benjamin = $2a - 4$.

    b. $a = 11$.

$$a + 2a - 4 = 29$$
$$3a - 4 = 29$$
$$3a = 33$$
$$a = 11$$

    c. Annette sold 11 books.

    d. If Annette sold 11 books, Benjamin sold 18 books ($2 \times 11 - 4 = 18$), and $11 + 18 = 29$.

2.   **a.** Sculptures = $m + 8$; canvasses = $3m$.

    b. $m = 6$.

$$m + m + 8 + 3m = 38$$
$$5m + 8 = 38$$
$$5m = 30$$
$$m = 6$$

    c. The collection includes 6 mobiles.

    d. If the collection has 6 mobiles, it has 14 sculptures (6 + 8 = 14), and it has 18 canvasses (6 × 3 = 18), so it has a total of 38 pieces (6 + 14 + 18 = 38).

3. **a.** Keith = 2r; Brad = 2r + 210.

    b. r = 150.

$$r + 2r + 2r + 210 = 960$$
$$5r + 210 = 960$$
$$5r = 750$$
$$r = 150$$

    c. Brad has 510 points (2 × 150 + 210 = 300 + 210 = 510).

    d. If Roy has 150 points, Keith has 300 points (150 × 2 = 300), and Brad has 510 points (300 + 210 = 500), so all together the three men have 960 points (150 + 300 + 510 = 960).

4. **a.** Second day = 3f; Third day = f + 3f − 130 = 4f − 130

    b. f = 61.

$$f + 3f + 4f - 130 = 358$$
$$8f - 130 = 358$$
$$8f = 488$$
$$f = 61$$

    c. Alison read 114 pages on the third day (4 × 61 − 130 = 244 − 130 = 114).

    d. If Alison read 61 pages on the first day, she read 183 pages on the second day (61 × 3 = 183), and 114 pages on the third day (61 + 183 − 130 = 114), so she read a total of 358 pages (61 + 183 + 114 = 358).

CHOOSING YOUR VARIABLE WISELY

Remember: Declaring a variable is simple, as I show you earlier in this chapter, but you can make the rest of your work a lot easier when you know how to choose your variable wisely. Whenever possible, choose a variable so that the equation you have to solve has no fractions, which are much more difficult to work with than whole numbers.

For example, suppose you're trying to solve this problem:

> Irina has three times as many clients as Toby. If they have 52 clients all together, how many clients does each person have?

The key sentence in the problem is "Irina has *three times as many* clients as Toby." It's significant because it indicates a relationship between Irina and Toby that's based on either *multiplication or division*. And to avoid fractions, you want to avoid division wherever possible.

Tip: Whenever you see a sentence that indicates you need to use either multiplication or division, choose your variable to represent the *smaller* number. In this case, Toby has fewer clients than Irina, so choosing *t* as your variable is the smart move.

Suppose you begin by declaring your variable as follows:

Let t = the number of clients that Toby has.

Then, using that variable, you can make this chart:

Irina	$3t$
Toby	t

No fraction! To solve this problem, set up this equation:

$$\text{Irina} + \text{Toby} = 52$$

Plug in the values from the chart:

$$3t + t = 52$$

Now you can solve the problem easily, using what I show you in Chapter 19:

$$4t = 52$$
$$t = 13$$

Toby has 13 clients, so Irina has 39. To check this result — which I recommend highly earlier in this chapter! — note that 13 + 39 = 52.

Now suppose that, instead, you take the opposite route and decide to declare a variable as follows:

Let i = the number of clients that Irina has.

Given that variable, you have to represent Toby's clients using the fraction $\frac{i}{3}$, which leads to the same answer but a *lot* more work.

Example

Q. Abby lives twice as many miles from school as Kaitlin, and Kaitlin lives 7 miles further from school than Giada does. If Abby lives 20 miles further from the school than Giada, how far does Kaitlin live from it?

a. Declare a variable using the initial of one of the girls, so that the resulting equation will not have fractions.

b. Represent the values of the other two girls in terms of this variable.

c. Set up the equation and solve it.

d. Answer the question.

e. Check your work.

A. **a.** Let g = the number of miles that Giada lives from school.

b. Kaitlin = $g + 7$; Abby = $2(g + 7) = 2g + 14$.

c. $g = 6$

$$2g + 14 = g + 20$$
$$g + 14 = 20$$
$$g = 6$$

d. Kaitlin lives 13 miles from school ($6 + 7 = 13$).

e. If Giada lives 6 miles from school, Kaitlin lives 13 miles from school ($6 + 7 = 13$), and Abby lives 26 miles from school ($13 \times 2 = 26$), so Abby lives 20 miles further from school than Giada does ($26 = 6 + 20$).

Practice Questions

1. Boris earned $165 less than Carlos, and Carlos earned twice as much as Tran. If Boris and Tran earned a total of $1,425, how much did Carlos earn?

   a. Declare a variable using the initial of one of the boys, so that the resulting equation will not have fractions.

   b. Represent the amounts that the other two boys earned in terms of this variable.

   c. Set up the equation and solve it.

   d. Answer the question.

   e. Check your work.

2. Darryl needs to buy a jacket, a pair of shoes, and a pair of pants for a job interview. The pair of shoes he wants costs $65 less than the jacket, and the price of the pants is one-fifth that of the jacket. If the total cost of the three items is $254, what will Darryl pay for the shoes?

   a. Declare a variable using the initial of one of the items, so that the resulting equation will not have fractions.

   b. Represent the values of the other two items in terms of this variable.

   c. Set up the equation and solve it.

   d. Answer the question.

   e. Check your work.

3. Evan, Jack, and Nate are all playing a game in which the winning score is 100. Jack's score is one more than twice Nate's score, and one-third Evan's score. If Evan has 73 more points than Nate, how many points does Jack need to win the game?

    a. Declare a variable using the initial of one of the boys, so that the resulting equation will not have fractions.

    b. Represent the other two boys' scores in terms of this variable.

    c. Set up the equation and solve it.

    d. Answer the question.

    e. Check your work.

4. Evelyn participated in a bicycle race with three parts: a flat leg, an uphill leg, and a downhill leg. The flat leg took her 20 minutes less than twice the length of the uphill leg, and the uphill leg took her 20 minutes less than twice the length of the downhill leg. If her total time was 3 hours and 34 minutes, how long did Evelyn take to complete the flat leg of the race?

    a. Declare a variable using the initial of one of the legs of the race, so that the resulting equation will not have fractions.

    b. Represent the values of the other two legs in terms of this variable.

    c. Set up the equation and solve it.

    d. Answer the question.

    e. Check your work.

Practice Answers

1.   **a.** Let t = the number of dollars that Tran earned.

    b. Carlos = $2t$; Boris = $2t - 165$.

    c. $t = 530$

$$t + 2t - 165 = 1{,}425$$
$$3t - 165 = 1{,}425$$
$$3t = 1{,}590$$
$$t = 530$$

    d. Carlos earned $1,060. ($530 \times 2 = 1{,}060$)

    e. If Tran earned 530 and Carlos earned $1,060, Boris earned $895 ($1{,}060 - 165 = 895$), so Boris and Tran earned a total of $1,425 ($895 + 530 = 1{,}425$).

2. **a.** Let p = the price of the pants.

    b. Jacket = $5p$; Shoes = $5p - 65$.

    c. $p = 29$.

    $$p + 5p + 5p - 65 = 254$$
    $$11p - 65 = 254$$
    $$11p = 319$$
    $$p = 29$$

    d. Darryl will pay $80 for the shoes. ($29 \times 5 - 65 = 145 - 65 = 80$)

    e. If Darryl pays $29 for the pants, and $80 for the shoes, he pays $145 for the jacket ($29 \times 5 = 145$), so he pays a total of $254 ($29 + 80 + 145 = 254$).

3. **a.** Let n = Nate's score.

    b. Jack = $2n + 1$; Evan = $3(2n + 1) = 6n + 3$.

    c. $n = 14$.

    $$6n + 3 = n + 73$$
    $$5n + 3 = 73$$
    $$5n = 70$$
    $$n = 14$$

    d. Jack needs 71 points to win the game. (Jack's score is $14 \times 2 + 1 = 29$, and $100 - 29 = 71$.)

    e. If Nate has 14 points and Jack has 29 points, then Evan has 87 points ($14 \times 6 + 3 = 84 + 3 = 87$), so Evan has 73 more points than Nate ($87 - 14 = 73$).

4. **a.** Let d = the length of time Evelyn took for the downhill leg.

    b. Uphill = $2d - 20$; flat = $2(2d - 20) - 20 = 4d - 40 - 20 = 4d - 60$.

    c. $d = 42$.

    $$d + 2d - 20 + 4d - 60 = 214 \text{ (3 hours and 34 minutes = 214 minutes.)}$$
    $$7d - 80 = 214$$
    $$7d = 294$$
    $$d = 42$$

    d. Evelyn took 1 hour and 48 minutes to complete the flat leg of the race. ($42 \times 4 - 60 = 160 - 68 = 108$ minutes = 1 hour and 48 minutes.)

    e. If Evelyn took 42 minutes to complete the downhill leg and 108 minutes to complete the flat leg, she took 64 minutes to complete the uphill leg ($42 \times 2 - 20 = 84 - 20 = 64$), so she took a total of 3 hours and 34 minutes to complete the race ($42 + 64 + 108 = 214$ minutes = 3 hours and 34 minutes.)

SOLVING MORE-COMPLEX ALGEBRAIC PROBLEMS

Algebra word problems become more complex when the number of people or things you need to find out increases. In this section, the complexity increases to four and then five people. When you're done, you should feel comfortable solving algebra word problems of significant difficulty.

Charting four people

As in the previous section, a chart can help you organize information so you don't get confused. Here's a problem that involves four people:

Alison, Jeremy, Liz, and Raymond participated in a canned goods drive at work. Liz donated three times as many cans as Jeremy, Alison donated twice as many as Jeremy, and Raymond donated 7 more than Liz. Together the two women donated two more cans than the two men. How many cans did the four people donate altogether?

The first step, as always, is declaring a variable. Remember that, to avoid fractions, you want to declare a variable based on the person who brought in the fewest cans. Liz donated more cans than Jeremy, and so did Alison. Furthermore, Raymond donated more cans than Liz. So because Jeremy donated the fewest cans, declare your variable as follows:

Let j = the number of cans that Jeremy donated.

Now you can set up your chart as follows:

Jeremy	J
Liz	$3j$
Alison	$2j$
Raymond	Liz $+ 7 = 3j + 7$

This setup looks good because, as expected, there are no fractional amounts in the chart. The next sentence tells you that the women donated two more cans than the men, so make a word problem, as I show you in Chapter 5:

$$\text{Liz} + \text{Alison} = \text{Jeremy} + \text{Raymond} + 2$$

You can now substitute into this equation as follows:

$$3j + 2j = j + 3j + 7 + 2$$

With your equation set up, you're ready to solve. First, isolate the algebraic terms:

$$3j + 2j - j - 3j = 7 + 2$$

Combine like terms:

$$j = 9$$

Almost without effort, you've solved the equation, so you know that Jeremy donated 9 cans. With this information, you can go back to the chart, plug in 9 for *j*, and find out how many cans the other people donated: Liz donated 27, Alison donated 18, and Raymond donated 34. Finally, you can add up these numbers to conclude that the four people donated 88 cans altogether.

To check the numbers, read through the problem and make sure they work at every point in the story. For example, together Liz and Alison donated 45 cans, and Jeremy and Raymond donated 43, so the women really did donate 2 more cans than the men.

Crossing the finish line with five people

Here's one final example, the most difficult in this chapter, in which you have five people to work with.

> Five friends are keeping track of how many miles they run. So far this month, Mina has run 12 miles, Suzanne has run 3 more miles than Jake, and Kyle has run twice as far as Victor. But tomorrow, after they all complete a 5-mile run, Jake will have run as far as Mina and Victor combined, and the whole group will have run 174 miles. How far has each person run so far?

The most important point to notice in this problem is that there are two sets of numbers: the miles that all five people have run up to *today* and their mileage including *tomorrow.* And each person's mileage tomorrow will be 5 miles greater than his or her mileage today. Here's how to set up a chart:

	Today	*Tomorrow (Today + 5)*
Jake		
Kyle		
Mina		
Suzanne		
Victor		

With this chart, you're off to a good start to solve this problem. Next, look for that statement early in the problem that connects two people by either multiplication or division. Here it is:

> Kyle has run *twice as far* as Victor.

Because Victor has run fewer miles than Kyle, declare your variable as follows:

> Let *v* = the number of miles that Victor has run up to *today.*

Notice that I added the word *today* to the declaration to be very clear that I'm talking about Victor's miles *before* the 5-mile run tomorrow.

At this point, you can begin filling in the chart:

	Today	*Tomorrow (Today + 5)*
Jake		
Kyle	$2v$	$2v + 5$
Mina	12	17
Suzanne		
Victor	v	$v + 5$

As you can see, I left out the information about Jake and Suzanne because I can't represent it using the variable v. I've also begun to fill in the *Tomorrow* column by adding 5 to my numbers in the *Today* column.

Now I can move on to the next statement in the problem:

> But tomorrow . . . Jake will have run as far as Mina and Victor combined. . . .

I can use this to fill in Jake's information:

	Today	*Tomorrow (Today + 5)*
Jake	$17 + v$	$17 + v + 5$
Kyle	$2v$	$2v + 5$
Mina	12	17
Suzanne		
Victor	v	$v + 5$

In this case, I first filled in Jake's *tomorrow* distance ($17 + v + 5$) and then subtracted 5 to find out his *today* distance. Now I can use the information that today Suzanne has run 3 more miles than Jake:

	Today	*Tomorrow (Today + 5)*
Jake	$17 + v$	$17 + v + 5$
Kyle	$2v$	$2v + 5$
Mina	12	17
Suzanne	$17 + v + 3$	$17 + v + 8$
Victor	v	$v + 5$

With the chart filled in like this, you can begin to set up your equation. First, set up a word equation, as follows:

> Jake tomorrow + Kyle tomorrow + Mina tomorrow + Suzanne tomorrow + Victor tomorrow = 174

Now just substitute information from the chart into this word equation to set up your equation:

$$17 + v + 5 + 2v + 5 + 17 + 17 + v + 8 + v + 5 = 174$$

As always, begin solving by isolating the algebraic terms:

$$v + 2v + v + v = 174 - 17 - 5 - 5 - 17 - 17 - 8 - 5$$

Next, combine like terms:

$$5v = 100$$

Finally, to get rid of the coefficient in the term $5v$, divide both sides by 5:

$$\frac{5v}{5} = \frac{100}{5}$$
$$v = 20$$

You now know that Victor's total distance up to *today* is 20 miles. With this information, you substitute 20 for v and fill in the chart, as follows:

	Today	*Tomorrow (Today + 5)*
Jake	37	42
Kyle	40	45
Mina	12	17
Suzanne	40	45
Victor	20	25

The *Today* column contains the answers to the question the problem asks. To check this solution, make sure that every statement in the problem is true. For example, tomorrow the five people will have run a total of 174 miles because

$$42 + 45 + 17 + 45 + 25 = 174$$

Copy down this problem, close the book, and work through it for practice.

Example

Q. Ray and Donna both receive $10.00 allowance each week from their parents. This week, before receiving their allowances, Ray had three times as much money as his sister. After receiving their allowances, he had $13 more than twice as much as her. How much money did they have together after receiving their allowances?

a. Declare a variable to represent the amount that one of the siblings had before receiving his or her allowance.

b. Make a table to organize the amounts the siblings had before and after receiving their allowances, and fill it in.

c. Set up an equation and solve it.

d. Answer the question.

e. Check your work.

A. a. **Let d = the amount that Donna had before receiving her allowance**

b.

	Before Allowances	*After Allowances*
Donna	d	$d + 10$
Ray	$3d$	$3d + 10$

c. $d = 23$

$$2(d+10)+13 = 3d+10$$
$$2d+20+13 = 3d+10$$
$$2d+33+3d+10$$
$$33 = d+10$$
$$23 = d$$

d. **Together, Donna and Ray had \$112 after receiving their allowances. (23 + 10 + (23 × 3) + 10 = 33 + 69 + 10 = 112)**

e. **Before receiving their allowances, Donna had \$23 and Ray had \$69 (23 × 3 = 69), so they had a total of \$92. Afterwards, they had a total of \$20 more, so they had \$112 total.**

Q

Practice Questions

1. A college class that has 73 students includes twice as many freshman as sophomores, seven fewer juniors than sophomores, three times as many juniors as seniors. How many of freshman are in the class?

a. Declare a variable to represent the number in the smallest group of students.

b. Write expressions to represent the number of students in the other three groups.

c. Set up and solve this equation.

d. Answer the question.

e. Check your work.

2. Anthony is on the football team at school, and his brother Dylan is on the wrestling team. At the start of this school year, Anthony's weight was 110% of Dylan's. But then on instructions from their coaches, Anthony gained 8 pounds and Dylan lost 5 pounds. By the end of the school year, Anthony's weight was 120% of Dylan's. What did the two boys weigh at the end of the school year?

a. Declare a variable to represent the weight of one of the boys at the start of the school year.

b. Make a table to organize the amounts the boys weighed at the start and end of the year, and fill it in.

c. Set up an equation and solve it.

d. Answer the question.

e. Check your work.

3. Three friends received bonuses at work. Artie received twice as much money as Barry, and Elaine received $150 more than Barry. They all placed their windfalls in the same mutual fund, which increased by 30% over the next year. The result was that Artie ended up with $260 more than Elaine. How much money did Barry make on his mutual fund investment?

a. Declare a variable to represent amount that one employee received as a bonus.

b. Make a table to organize the amounts the employees' bonuses and the results of their investments, and fill it in.

c. Set up an equation and solve it.

d. Answer the question.

e. Check your work.

Practice Answers

1. **Let s = the number of seniors.** There are more freshman than sophomores, more sophomores than junior, and more juniors than seniors, so the smallest group is seniors.

a. **juniors = $3s$; sophomores = $3s + 7$; freshman = $2(3s + 7)$.**

b. $s + 3s + (3s + 7) + 2(3s + 7) = 73$. The four groups added together total 73 students.

c. 4.

$$s = 3s + (3s + 7) + 2(3s + 7) = 73$$
$$s + 3s + 3s + 7 + 6s + 14 = 73$$
$$13s + 21 = 73$$
$$13s = 52$$
$$s = 4$$

d. 38.

e. If there are 4 seniors, there are 12 juniors, 19 sophomores, and 38 freshman, which equals 73 students.

2. **a.** Let d = Dylan's weight at the start of the school year.

b.

	Start of year	End of year
Dylan	d	$d - 5$
Anthony	$1.1d$	$1.1d + 8$

c. $d = 140$.

$$1.2(d - 5) = 1.1d + 8$$
$$1.2d - 6 = 1.1d + 8$$
$$0.1d - 6 = 8$$
$$0.1d = 14$$
$$d = 140$$

d. At the end of the year, Dylan weighed 135 pounds and Anthony weighed 162. ($140 - 5 = 135$; $140 \times 1.1 + 8 = 154 + 8 = 162$)

e. If Dylan weighed 140 at the start of the year, Anthony weighed 154 ($140 \times 14 = 154$). Afterwards, Dylan weighed 135 and Anthony weighed 162, so Anthony weighed 120% of Dylan's weight ($135 \times 1.2 = 162$).

3. **a.** Let b equal the amount that Barry received as a bonus.

b.

	Bonus	After investment
Artie	$2b$	$1.3(2b) = 2.6b$
Barry	b	$1.3b$
Elaine	$b + 150$	$1.3(b + 150) = 1.3b + 195$

c. $b = 350$.

$$2.6b = 1.3b + 195 + 260$$
$$2.6b = 1.3b + 455$$
$$1.3b = 455$$
$$b = 350$$

d. Barry made $105 on his investment. (30% of 350 = $105)

e. If Barry received a $350 bonus, then Artie received $700 ($350 \times 2 = 700$) and Elaine received $500 ($350 + 150 = 500$). Each earned 30% on his or her investment, so Artie earned $210, bringing his total up to $910, and Elaine earned $150, bringing her total up to $650. Thus, Artie ended up with $260 more than Elaine ($910 - 260 = 650$).

the part of

TENS

WEB EXTRA *Math is full of important concepts. To discover ten of the most interesting, go to www.dummies.com/extras/ucanbasicmathprealgebra.*

IN THIS PART . . .

- Discover tricks to help you avoid making common mathematical mistakes.

- Expand your understanding of math by learning how to distinguish between different kinds of numbers: natural numbers, integers, rational (and irrational) numbers, algebraic numbers, and more.

21

TEN LITTLE MATH DEMONS
THAT TRIP PEOPLE UP

IN THIS CHAPTER

- Knowing the multiplication table once and for all
- Understanding negative numbers
- Distinguishing factors and multiples
- Working confidently with fractions
- Seeing what algebra is really all about

The ten little math demons I cover in this chapter plague all sorts of otherwise smart, capable folks like you. The good news is that they're not as big and scary as you may think, and they can be dispelled more easily than you may have dared believe. Here, I present ten common math demons, with a short explanation to set them on a path away from you.

KNOWING THE MULTIPLICATION TABLE

A sketchy knowledge of multiplication can really hold back an otherwise good math student. Here's a quick quiz: the ten toughest problems from the multiplication table.

$$8 \times 7 = \underline{\quad} \qquad 9 \times 9 = \underline{\quad}$$
$$7 \times 9 = \underline{\quad} \qquad 6 \times 8 = \underline{\quad}$$
$$6 \times 6 = \underline{\quad} \qquad 8 \times 9 = \underline{\quad}$$
$$7 \times 7 = \underline{\quad} \qquad 9 \times 6 = \underline{\quad}$$
$$8 \times 8 = \underline{\quad} \qquad 7 \times 6 = \underline{\quad}$$

Can you do this, 10 for 10, in 20 seconds? If so, you're a multiplication whiz. If not, you may want to get a set of flash cards and practice with them a bit until you nail the multiplication table once and for all!

ADDING AND SUBTRACTING NEGATIVE NUMBERS

It's easy to get confused when adding and subtracting negative numbers. To begin, think of adding a number as moving *up* and subtracting a number as moving *down*. For example:

$2 + 1 - 6$ means *up* 2, *up* 1, *down* 6

So if you go *up* 2 steps, then *up* 1 more step, and then *down* 6 steps, you've gone a total of 3 steps *down;* therefore, 2 + 1 – 6 = –3.

Here's another example:

$$-3 + 8 - 1 \qquad \text{means } down \text{ 3, } up \text{ 8, } down \text{ 1}$$

This time, go *down* 3 steps, then *up* 8 steps, and then *down* 1 step, you've gone a total of 4 steps *up;* therefore, –3 + 8 – 1 = 4.

Tip: You can turn every problem involving negative numbers into an up-and-down example. The way to do this is by combining adjacent signs:

• Combine a plus and minus as a *minus* sign.

• Combine two minus signs as a *plus* sign.

For example:

$$-5 + (-3) - (-9)$$

In this example, you see a plus sign and a minus sign together (between the 5 and the 3), which you can combine as a minus sign. You also see two minus signs (between the 3 and the 9), which you can combine as a plus sign:

$$-5 - 3 + 9 \qquad \text{means } down \text{ 5, } down \text{ 3, } up \text{ 9}$$

This technique allows you use your up-and-down skills to solve the problem: *Down* 5 steps, then *down* 3 steps, and *up* 9 steps leaves you 1 step *up;* therefore, –5 + (–3) – (–9) = 1.

See Chapter 3 for more on adding and subtracting negative numbers.

MULTIPLYING AND DIVIDING NEGATIVE NUMBERS

When you multiply or divide a positive number by a negative number (or vice versa), the answer is always negative. For example:

$$2 \times (-4) = -8 \qquad 14 \div (-7) = -2$$
$$-3 \times 5 = -15 \qquad -20 \div 4 = -5$$

When you multiply two negative numbers, remember this simple rule: Two negatives always cancel each other out and equal a positive.

$$-8 \times (-3) = 24 \qquad -30 \div (-5) = 6$$

For more on multiplying and dividing negative numbers, see Chapter 3.

KNOWING THE DIFFERENCE BETWEEN FACTORS AND MULTIPLES

Lots of students get factors and multiples confused because they're so similar. Both are related to the concept of divisibility. When you divide one number by another and the answer has no remainder, the first number is *divisible* by the second. For example:

$$12 \div 3 = 4 \;\rightarrow\; 12 \text{ is } \textit{divisible} \text{ by } 3$$

When you know that 12 is divisible by 3, you know two other things as well:

3 is a *factor* of 12   and   12 is a *multiple* of 3

In the positive numbers, the factor is always the *smaller* of the two numbers and the multiple is always the *larger*.

For more on factors and multiples, see Chapter 7.

REDUCING FRACTIONS TO LOWEST TERMS

Math teachers usually request (or force!) their students to use the smallest-possible version of a fraction — that is, to reduce fractions to lowest terms.

To reduce a fraction, divide the *numerator* (top number) and *denominator* (bottom number) by a *common factor,* a number that they're both divisible by. For example, 50 and 100 are both divisible by 10, so

$$\frac{50}{100} = \frac{50 \div 10}{100 \div 10} = \frac{5}{10}$$

The resulting fraction, $\frac{5}{10}$, can still be further reduced, because both 5 and 10 are divisible by 5:

$$\frac{5}{10} = \frac{5 \div 5}{10 \div 5} = \frac{1}{2}$$

When you can no longer make the numerator and denominator smaller by dividing by a common factor, the result is a fraction that's reduced to lowest terms.

See Chapter 8 for more on reducing fractions.

ADDING AND SUBTRACTING FRACTIONS

Adding and subtracting fractions that have the same denominator is pretty simple: Perform the operation (adding or subtracting) on the two numerators and keep the denominators the same.

$$\frac{2}{7} + \frac{3}{7} = \frac{5}{7} \qquad \frac{8}{9} - \frac{7}{9} = \frac{1}{9}$$

When two fractions have different denominators, you can add or subtract them without finding a common denominator by using cross-multiplication, as shown here:

$$\text{To add: } \frac{3}{5} + \frac{1}{4} = \frac{(3 \times 4) + (5 \times 1)}{5 \times 4} = \frac{17}{20}$$

$$\text{To subtract: } \frac{2}{3} - \frac{1}{5} = \frac{(2 \times 5) - (3 \times 1)}{3 \times 5} = \frac{7}{15}$$

For more on adding and subtracting fractions, see Chapter 9.

MULTIPLYING AND DIVIDING FRACTIONS

To multiply fractions, multiply their two numerators to get the numerator of the answer, and multiply their two denominators to get the denominator. For example:

$$\frac{3}{10} \times \frac{7}{8} = \frac{21}{80}$$

To divide two fractions, turn the problem into multiplication by taking the *reciprocal* of the second fraction — that is, by flipping it upside-down. For example:

$$\frac{2}{7} \div \frac{5}{6} = \frac{2}{7} \times \frac{6}{5}$$

Now multiply the two resulting fractions:

$$\frac{2}{7} \times \frac{6}{5} = \frac{12}{35}$$

For more on multiplying and dividing fractions, see Chapter 9.

IDENTIFYING ALGEBRA'S MAIN GOAL: FIND X

Everything in algebra is, ultimately, for one purpose: Find x (or whatever the variable happens to be). Algebra is really just a bunch of tools to help you do that. In Chapter 18, I give you these tools. Chapter 19 focuses on the goal of finding x. And in Chapter 20, you use algebra to solve word problems that would be much more difficult without algebra to help.

KNOWING ALGEBRA'S MAIN RULE: KEEP THE EQUATION IN BALANCE

The main idea of algebra is simply that an equation is like a balance scale: Provided that you do the same thing to both sides, the equation stays balanced. For example, consider the following equation:

$$8x - 12 = 5x + 9$$

To find x, you can do anything to this equation as long as you do it equally to both sides. For example:

Add 2: $8x - 12 = 5x + 9$ becomes $8x - 10 = 5x + 11$

Subtract $5x$: $8x - 12 = 5x + 9$ becomes $3x - 12 = 9$

Multiply by 10: $8x - 12 = 5x + 9$ becomes $80x - 120 = 50x + 90$

Each of these steps is valid. One, however, is more helpful than the others, as you see in the next section.

For more on algebra, see Chapters 18 through 20.

SEEING ALGEBRA'S MAIN STRATEGY: ISOLATE X

The best way to find x is to *isolate it* — that is, get x on one side of the equation with a number on the other side. To do this while keeping the equation balanced requires great cunning and finesse. Here's an example, using the equation from the preceding section:

Original problem	$8x - 12 = 5x + 9$
Subtract $5x$	$3x - 12 = 9$
Add 12	$3x = 21$
Divide by 3	$x = 7$

As you can see, the final step isolates x, giving you the solution: $x = 7$.

For more on algebra, see Chapters 18 through 20.

22

TEN IMPORTANT NUMBER SETS TO KNOW

IN THIS CHAPTER

- Identifying counting numbers, integers, rational numbers, and real numbers
- Discovering imaginary and complex numbers
- Looking at how transfinite numbers represent higher levels of infinity

The more you find out about numbers, the stranger they become. When you're working with just the counting numbers and a few simple operations, numbers seem to develop a landscape all their own. The terrain of this landscape starts out uneventful, but as you introduce other sets, it soon turns surprising, shocking, and even mind blowing. In this chapter, I take you on a mind-expanding tour of ten sets of numbers.

I start with the familiar and comfy counting numbers. I continue with the integers (positive and negative counting numbers and 0), the rational numbers (integers and fractions), and real numbers (all numbers on the number line). I also take you on a few side routes along the way. The tour ends with the bizarre and almost unbelievable transfinite numbers. And in a way, the transfinite numbers bring you back to where you started: the counting numbers.

Each of these sets of numbers serves a different purpose, some familiar (such as accounting and carpentry), some scientific (such as electronics and physics), and a few purely mathematical. Enjoy the ride!

COUNTING ON COUNTING (OR NATURAL) NUMBERS

The *counting numbers* — also called the *natural numbers* — are probably the first numbers you ever encountered. They start with 1 and go up from there:

$$\{1, 2, 3, 4, 5, 6, 7, 8, 9, 10, 11, 12, \ldots\}$$

Remember: The three dots (or ellipsis) at the end tell you that the sequence of numbers goes on forever — in other words, it's infinite.

The counting numbers are useful for keeping track of tangible objects: stones, chickens, cars, cell phones — anything that you can touch and that you don't plan to cut into pieces.

The set of counting numbers is *closed* under both addition and multiplication. In other words, if you add or multiply any two counting numbers, the result is also a counting number. But the set isn't closed under subtraction or division. For example, if you subtract $2 - 3$, you get -1, which is a negative number, not a counting number. And if you divide $2 \div 3$, you get $\frac{2}{3}$, which is a fraction.

If you place 0 in the set of counting numbers, you get the set of *whole numbers*.

IDENTIFYING INTEGERS

The set of *integers* includes the counting numbers (see the preceding section), the negative counting numbers, and 0:

$$\{\ldots, -6, -5, -4, -3, -2, -1, 0, 1, 2, 3, 4, 5, 6, \ldots\}$$

The dots, or ellipses, at the beginning and the end of the set tell you that the integers are infinite in both the positive and negative directions.

Because the integers include the negative numbers, you can use them to keep track of anything that can potentially involve debt. In today's culture, it's usually money. For example, if you have $100 in your checking account and you write a check for $120, you find that your new balance drops to –$20 (not counting any fees that the bank charges!).

The set of integers is *closed* under addition, subtraction, and multiplication. In other words, if you add, subtract, or multiply any two integers, the result is also an integer. But the set isn't closed under division. For example, if you divide the integer 2 by the integer 5, you get the fraction $\frac{2}{5}$, which isn't an integer.

KNOWING THE RATIONALE BEHIND RATIONAL NUMBERS

The *rational numbers* include the integers (see the preceding section) and all the fractions between the integers. Here, I list only the rational numbers from –1 to 1 whose denominators (bottom numbers) are positive numbers less than 5:

$$-1 \ldots -\frac{3}{4} \ldots -\frac{2}{3} \ldots -\frac{1}{2} \ldots -\frac{1}{3} \ldots -\frac{1}{4} \ldots 0 \ldots \frac{1}{4} \ldots \frac{1}{3} \ldots \frac{1}{2} \ldots \frac{2}{3} \ldots \frac{3}{4} \ldots 1$$

The ellipses tell you that between any pair of rational numbers are an infinite number of other rational numbers — a quality called the *infinite density* of rational numbers.

Rational numbers are commonly used for measurement in which precision is important. For example, a ruler wouldn't be much good if it measured length only to the nearest inch. Most rulers measure length to the nearest $\frac{1}{16}$ of an inch, which is close enough for most purposes. Similarly, measuring cups, scales, precision clocks, and thermometers that allow you to make measurements to a fraction of a unit also use rational numbers. (See Chapter 14 for more on units of measurement.)

The set of rational numbers is closed under the Big Four operations. In other words, if you add, subtract, multiply, or divide any two rational numbers, the result is always another rational number.

MAKING SENSE OF IRRATIONAL NUMBERS

In a sense, the irrational numbers are a sort of catchall; every number on the number line that isn't rational is irrational.

By definition, no *irrational number* can be represented as a fraction, nor can an irrational number be represented as either a terminating decimal or a repeating decimal (see Chapter 10 for more about these types of decimals). Instead, an irrational number can be approximated only as a *nonterminating, nonrepeating decimal:* The string of numbers after the decimal point goes on forever without creating a pattern.

The most famous example of an irrational number is π, which represents the circumference of a circle with a diameter of 1 unit. Another common irrational number is $\sqrt{2}$, which represents the diagonal distance across a square with a side of 1 unit. In fact, all square roots of nonsquare numbers (such as $\sqrt{3}$, $\sqrt{5}$, and so forth) are irrational numbers.

Irrational numbers fill out the spaces in the real number line. (The *real number line* is just the number line you're used to, but it's continuous; it has no gaps, so every point is paired with a number.) These numbers are used in many cases where you need not just a high level of precision, as with the rational numbers, but the *exact* value of a number that you can't represent as a fraction.

Irrational numbers come in two varieties: *algebraic numbers* and *transcendental numbers.* I discuss both types of numbers in the sections that follow.

ABSORBING ALGEBRAIC NUMBERS

To understand *algebraic numbers,* you need a little information about polynomial equations. A *polynomial equation* is an algebraic equation that meets the following conditions:

- Its operations are limited to addition, subtraction, and multiplication. In other words, you don't have to divide by a variable.

- Its variables are raised only to positive, whole-number exponents.

You can find out more about polynomials in *Algebra For Dummies,* by Mary Jane Sterling (Wiley). Here are some polynomial equations:

$$2x + 14 = (x+3)^2$$
$$2x^2 - 9x - 5 = 0$$

Every algebraic number shows up as the solution of at least one polynomial equation. For example, suppose you have the following equation:

$$x^2 = 2$$

You can solve this equation as $x = \sqrt{2}$. Thus, $\sqrt{2}$ is an algebraic number whose approximate value is 1.4142135623. . . (see Chapter 3 for more information on square roots).

MOVING THROUGH TRANSCENDENTAL NUMBERS

A *transcendental number,* in contrast to an algebraic number (see the preceding section), is *never* the solution of a polynomial equation. Like the irrational numbers, transcendental numbers are a sort of catchall: Every number on the number line that isn't algebraic is transcendental.

The best-known transcendental number is π, whose approximate value is 3.1415926535. . . . Its uses begin in geometry but extend to virtually all areas of mathematics. (See Chapter 15 for more on π.)

Other important transcendental numbers come about when you study *trigonometry,* the math of right triangles. The values of trigonometric functions — such as sines, cosines, and tangents — are often transcendental numbers.

Another important transcendental number is *e,* whose approximate value is 2.718281828459. . . . The number *e* is the base of the natural logarithm, which you probably won't use until you get to pre-calculus or calculus. People use *e* to do problems on compound interest, population growth, radioactive decay, and the like.

GETTING GROUNDED IN REAL NUMBERS

The set of *real numbers* is the set of all rational and irrational numbers (see the earlier sections). The real numbers comprise every point on the number line.

Like the rational numbers (see "Knowing the Rationale behind Rational Numbers," earlier in this chapter), the set of real numbers is closed under the Big Four operations. In other words, if you add, subtract, multiply, or divide any two real numbers, the result is always another real number.

TRYING TO IMAGINE IMAGINARY NUMBERS

An *imaginary number* is any real number multiplied by $\sqrt{-1}$.

To understand what's so strange about imaginary numbers, it helps to know a bit about square roots. The *square root* of a number is any value that, when multiplied by itself, gives you that number. For example, the square root of 9 is 3 because $3 \times 3 = 9$. And the square root of 9 is also –3 because $-3 \times -3 = 9$. (See Chapter 3 for more on square roots and multiplying negative numbers.)

The problem with finding $\sqrt{-1}$ is that it isn't on the real number line (because $\sqrt{-1}$ isn't in the set of real numbers). If it were on the real number line, it would be a positive number, a negative number, or 0. But when you multiply any positive number by itself, you get a positive number. And when you multiply any negative number by itself, you also get a positive number. Finally, when you multiply 0 by itself, you get 0.

If $\sqrt{-1}$ isn't on the real number line, where is it? That's a good question. For thousands of years, mathematicians believed that the square root of a negative number was simply meaningless. They banished it to the mathematical nonplace called *undefined,* which is the same place they kept fractions with a denominator of 0. In the 19th century, however, mathematicians began to find these numbers useful and found a way to incorporate them into the rest of math.

Mathematicians designated $\sqrt{-1}$ with the symbol *i.* Because it didn't fit onto the real number line, *i* got its own number line, which looks a lot like the real number line. Figure 22-1 shows some numbers that form the imaginary number line.

Figure 22-1:
Numbers on the imaginary number line.

$$-3i \quad -2i \quad -i \quad 0 \quad i \quad 2i \quad 3i$$

© *John Wiley & Sons, Inc.*

Even though these numbers are called imaginary, mathematicians today consider them no less real than the real numbers. And the scientific application of imaginary numbers to electronics and physics has verified that these numbers are more than just figments of someone's imagination.

GRASPING THE COMPLEXITY OF COMPLEX NUMBERS

A *complex number* is any real number (see "Getting Grounded in Real Numbers," earlier in this chapter) plus or minus an imaginary number (see the preceding section). Consider some examples:

$$1+i \quad 5-2i \quad -100+10i$$

You can turn any real number into a complex number by just adding $0i$ (which equals 0):

$$3 = 3+0i \qquad -12 = -12+0i \quad 3.14 = 3.14+0i$$

These examples show you that the real numbers are just a part of the larger set of complex numbers.

Like the rational numbers and real numbers (check out the sections earlier in this chapter), the set of complex numbers is closed under the Big Four operations. In other words, if you add, subtract, multiply, or divide any two complex numbers, the result is always another complex number.

GOING BEYOND THE INFINITE WITH TRANSFINITE NUMBERS

The *transfinite numbers* are a set of numbers representing different levels of infinity. Consider this for a moment: The counting numbers (1, 2, 3, . . .) go on forever, so they're infinite. But there are *more* real numbers than counting numbers.

In fact, the real numbers are *infinitely more infinite* than the counting numbers. Mathematician Georg Cantor proved this fact. He also proved that, for every level of infinity, you can find another level that's even higher. He called these ever-increasing levels of infinity *transfinite,* because they transcend, or go beyond, what you think of as infinite.

The lowest transfinite number is $\aleph_0$ (aleph null), which equals the number of elements in the set of counting numbers ($\{1, 2, 3, 4, 5, . . .\}$). Because the counting numbers are infinite, the familiar symbol for infinity (∞) and $\aleph_0$ mean the same thing.

The next transfinite number is $\aleph_1$ (aleph one), which equals the number of elements in the set of real numbers. This is a higher order of infinity than ∞.

The sets of integers, rational, and algebraic numbers all have $\aleph_0$ elements. And the sets of irrational, transcendental, imaginary, and complex numbers all have $\aleph_1$ elements.

Higher levels of infinity exist, too. Here's the set of transfinite numbers:

$$\{\aleph_0, \aleph_1, \aleph_2, \aleph_3, ...\}$$

The ellipsis tells you that the sequence of transfinite numbers goes on forever — in other words, that it's infinite. As you can see, on the surface, the transfinite numbers look similar to the counting numbers (in the first section of this chapter). That is, the set of transfinite numbers has $\aleph_0$ elements.

23

TEN CURIOUS TYPES OF NUMBERS

Numbers seem to have personalities all their own. For example, even numbers are go-along numbers that break in half so you can carry them more conveniently. Odd numbers are more stubborn and don't break apart so easily. Powers of ten are big friendly numbers that are easy to add and multiply, whereas most other numbers are prickly and require special attention. In this chapter, I introduce you to some interesting types of numbers, with properties that other numbers don't share.

SQUARE NUMBERS

When you multiply any number by itself, the result is a *square number*. For example,

$$1^2 = 1 \times 1 = 1$$
$$2^2 = 2 \times 2 = 4$$
$$3^2 = 3 \times 3 = 9$$
$$4^2 = 4 \times 4 = 16$$
$$5^2 = 5 \times 5 = 25$$

Therefore, the sequence of square numbers begins as follows:

1, 4, 9, 16, 25, ...

To see why they're called square numbers, look at the arrangement of coins in squares in Figure 23-1.

Figure 23-1: The first five square numbers.

$1^2 = 1$ $2^2 = 4$ $3^2 = 9$ $4^2 = 16$ $5^2 = 25$

What's really cool about the list of square numbers is that you can get it by adding the odd numbers (3, 5, 7, 9, 11, 13, . . .), beginning with 3, to each preceding number in the list:

Square Number	Preceding Number + Odd Number	Sum
$2^2 = 4$	$1^2 + 3$	$1 + 3 = 4$
$3^2 = 9$	$2^2 + 5$	$4 + 5 = 9$
$4^2 = 16$	$3^2 + 7$	$9 + 7 = 16$
$5^2 = 25$	$4^2 + 9$	$16 + 9 = 25$
$6^2 = 36$	$5^2 + 11$	$25 + 11 = 36$
$7^2 = 49$	$6^2 + 13$	$36 + 13 = 49$

TRIANGULAR NUMBERS

When you add up any sequence of consecutive positive numbers starting with 1, the result is a *triangular number.* For example,

$$1 = 1$$
$$1 + 2 = 3$$
$$1 + 2 + 3 = 6$$
$$1 + 2 + 3 + 4 = 10$$
$$1 + 2 + 3 + 4 + 5 = 15$$

So, the sequence of triangular numbers begins as follows:

$$1, 3, 6, 10, 15, \ldots$$

Triangular numbers' shapely name makes sense when you begin arranging coins in triangles. Check out Figure 23-2.

Figure 23-2:
The first five triangular numbers.

1   3   6   10   15

CUBIC NUMBERS

If you're feeling that the square and triangular numbers are too flat, add a dimension and begin playing with the *cubic numbers.* You can generate a cubic number by multiplying any number by itself three times:

$$1^3 = 1 \times 1 \times 1 = 1$$
$$2^3 = 2 \times 2 \times 2 = 8$$
$$3^3 = 3 \times 3 \times 3 = 27$$
$$4^3 = 4 \times 4 \times 4 = 64$$
$$5^3 = 5 \times 5 \times 5 = 125$$

The sequence of cubic numbers begins as follows:

1, 8, 27, 64, 125, …

Cubic numbers live up to their name. Look at the cubes in Figure 23-3.

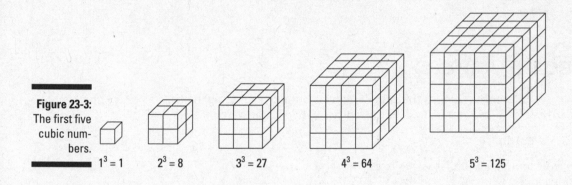

Figure 23-3: The first five cubic numbers.

$1^3 = 1$ $2^3 = 8$ $3^3 = 27$ $4^3 = 64$ $5^3 = 125$

FACTORIAL NUMBERS

In math, the exclamation point (!) means *factorial,* so you read 1! as *one factorial.* You get a factorial number when you multiply any sequence of consecutive positive numbers, starting with the number itself and counting down to 1. For example,

$$1! = 1$$
$$2! = 2 \times 1 = 2$$
$$3! = 3 \times 2 \times 1 = 6$$
$$4! = 4 \times 3 \times 2 \times 1 = 24$$
$$5! = 5 \times 4 \times 3 \times 2 \times 1 = 120$$

Thus, the sequence of factorial numbers begins as follows:

1, 2, 6, 24, 120, …

Factorial numbers are very useful in *probability,* which is the mathematics of how likely an event is to occur. With probability problems, you can figure out how likely you are to win the lottery or estimate your chances of guessing your friend's locker combination within the first few tries.

POWERS OF TWO

Multiplying the number 2 by itself repeatedly gives you the *powers of two*. For example,

$$2^1 = 2$$
$$2^2 = 2 \times 2 = 4$$
$$2^3 = 2 \times 2 \times 2 = 8$$
$$2^4 = 2 \times 2 \times 2 \times 2 = 16$$
$$2^5 = 2 \times 2 \times 2 \times 2 \times 2 = 32$$

Powers of two are the basis of binary numbers, which are important in computer applications. They're also useful for understanding Fermat numbers, which I discuss later in this chapter.

PERFECT NUMBERS

Any number that equals the sum of its own factors (excluding itself) is a *perfect number*. To see how this works, find all the factors of 6 (as I show you in Chapter 7):

6: 1, 2, 3, 6

Now add up all these factors except 6:

$$1 + 2 + 3 = 6$$

These factors add up to the number you started with, so 6 is a perfect number. The next perfect number is 28. First, find all the factors of 28:

28: 1, 2, 4, 7, 14, 28

Now add up all these factors except 28:

$$1 + 2 + 4 + 7 + 14 = 28$$

Again, these factors add up to the number you started with, so 28 is a perfect number. Perfect numbers are few and far between. The sequence of perfect numbers begins with the following five numbers:

6; 28; 496; 8,128; 33,550,336; ...

You can use the same method I outline to check 496 and beyond by yourself.

AMICABLE NUMBERS

Amicable numbers are similar to perfect numbers, except they come in pairs. The sum of the factors of one number (excluding the number itself) is equal to the second number, and vice versa. For example, one amicable pair is 220 and 284. To see why, first find all the factors of each number:

220: 1, 2, 4, 5, 10, 11, 20, 22, 44, 55, 110, 220
284: 1, 2, 4, 71, 142, 284

For each number, add up all the factors except the number itself:

$$1+2+4+5+10+11+20+22+44+55+110 = 284$$
$$1+2+4+71+142 = 220$$

Notice that the factors of 220 add up to 284, and the factors of 284 add up to 220. That's what makes this pair of numbers amicable.

The next-lowest pair of amicable numbers is 1,184 and 1,210. You can either trust me on this one or do the calculation yourself.

PRIME NUMBERS

Any number that has exactly two factors — 1 and itself — is called a *prime number*. For example, here are the first few prime numbers:

2, 3, 5, 7, 11, 13, 17, 19, …

There are infinitely many prime numbers — that is, they go on forever. See Chapter 6 for more on prime numbers.

MERSENNE PRIMES

Any number that's 1 less than a power of two (which I discuss earlier in this chapter) is called a *Mersenne number* (named for French mathematician Marin Mersenne). Therefore, every Mersenne number is of the following form:

$2^n - 1$ (where n is a nonnegative integer)

When a Mersenne number is also a prime number (see the preceding section), it's called a *Mersenne prime.* For example,

$$2^2 - 1 = 4 - 1 = 3$$
$$2^3 - 1 = 8 - 1 = 7$$
$$2^5 - 1 = 32 - 1 = 31$$
$$2^7 - 1 = 128 - 1 = 127$$
$$2^{13} - 1 = 8,192 - 1 = 8,191$$

Mersenne primes are of interest to mathematicians because they possess properties that ordinary prime numbers don't have. One of these properties is that they tend to be easier to find than other prime numbers. For this reason, the search for the largest known prime number is usually a search for a Mersenne prime.

FERMAT PRIMES

A *Fermat number* (named for mathematician Pierre de Fermat) is of the following form:

$2^{2^{\wedge}n} + 1$ (where n is a nonnegative integer)

The $^\wedge$ symbol means that you're finding a power, so with this formula, you first find 2^n; then you use that answer as an exponent on 2. For example, here are the first five Fermat numbers:

$2^{2^{\wedge}0} + 1 = 2^1 + 1 = 3$

$2^{2^{\wedge}1} + 1 = 2^2 + 1 = 5$

$2^{2^{\wedge}2} + 1 = 2^4 + 1 = 16 + 1 = 17$

$2^{2^{\wedge}3} + 1 = 2^8 + 1 = 256 + 1 = 257$

$2^{2^{\wedge}4} + 1 = 2^{16} + 1 = 65,536 + 1 = 65,537$

As you can see, Fermat numbers grow very quickly. When a Fermat number is also a prime number (see earlier in this chapter), it's called a *Fermat prime*. As it happens, the first five Fermat numbers are also Fermat primes (testing this is fairly simple for the first four numbers above and *much* harder for the fifth).

24

TEN ALTERNATIVE NUMERAL AND NUMBER SYSTEMS

The distinction between numbers and numerals is subtle but important. A number is an idea that expresses how much or how many. A numeral is a written symbol that expresses a number.

In this chapter, I show you ten ways to represent numbers that differ from the Hindu-Arabic (decimal) system. Some of these systems use entirely different symbols from those you're used to; others use the symbols that you know in different ways. A few of these systems have useful applications, and the others are just curiosities. (If you like, you can always use them for sending secret messages!) In any case, you may find it fun and interesting to see how many different ways people have found to represent the numbers that you're accustomed to.

TALLY MARKS

Numbers are abstractions that stand for real things. The first known numbers came into being with the rise of trading and commerce — people needed to keep track of commodities such as animals, harvested crops, or tools. At first, traders used clay or stone tokens to help simplify the job of counting. The first numbers were probably an attempt to simplify this recording system. Over time, tally marks scratched either in bone or on clay took the place of tokens.

When you think about it, the use of tally marks over tokens indicates an increase in sophistication. Previously, one real object (a token) had represented another real object (for example, a sheep or ear of corn). After that, an *abstraction* (a scratch) represented a real object.

BUNDLED TALLY MARKS

As early humans grew more comfortable letting tally marks stand for real-world objects, the next development in numbers was probably tally marks scratched in *bundles* of 5 (fingers on one hand), 10 (fingers on both hands), or 20 (fingers and toes). Bundling provided a simple way to count larger numbers more easily.

Of course, this system is much easier to read than non-bundled scratches — you can easily multiply or count by fives to get the total. Even today, people keep track of points in games using bundles such as these.

EGYPTIAN NUMERALS

Ancient Egyptian numerals are among the oldest number systems still in use today. Egyptian numerals use seven symbols, explained in Table 24-1.

Table 24-1   Egyptian Numerals

Number	Symbol
1	Stroke
10	Yoke
100	Coil of rope
1,000	Lotus
10,000	Finger
100,000	Frog
1,000,000	Man with raised hands

Numbers are formed by accumulating enough of the symbols that you need. For example,

> 7 = 7 strokes
>
> 24 = 2 yokes, 4 strokes
>
> 1,536 = 1 lotus, 5 coils of rope, 3 yokes, 6 strokes

In Egyptian numbers, the symbol for 1,000,000 also stands for infinity (∞).

BABYLONIAN NUMERALS

Babylonian numerals, which came into being about 4,000 years ago, use two symbols:

> 1 = Y   10 = <

For numbers less than 60, numbers are formed by accumulating enough of the symbols you need. For example,

> 6 = YYYYYY
>
> 34 = <<< YYYY
>
> 59 = <<<<< YYYYYYYYY

For numbers 60 and beyond, Babylonian numerals use place value based on the number 60. For example,

61 = Y Y (one 60 and one 1)

124 = YY YYYY (two 60s and four 1s)

611 = < < Y(ten 60s and eleven 1s)

Unlike the decimal system that you're used to, Babylonian numbers had no symbol for zero to serve as a placeholder, which causes some ambiguity. For example, the symbol for 60 is the same as the symbol for 1.

ANCIENT GREEK NUMERALS

Ancient Greek numerals were based on the Greek letters. The numbers from 1 to 999 were formed using the symbols in Table 24-2.

Table 24-2 Numerals Based on the Greek Alphabet

Ones	Tens	Hundreds
1 = α (alpha)	10 = Ι (iota)	100 = ρ (rho)
2 = β (beta)	20 = κ (kappa)	200 = σ (sigma)
3 = γ (gamma)	30 = λ (lambda)	300 = τ (tau)
4 = δ (delta)	40 = μ (mu)	400 = υ (upsilon)
5 = ε (epsilon)	50 = ν (nu)	500 = φ (phi)
6 = ° (digamma)	60 = ξ (xi)	600 = χ (chi)
7 = ζ (zeta)	70 = ο (omicron)	700 = ψ (psi)
8 = η (eta)	80 = π (pi)	800 = ω (omega)
9 = θ (theta)	90 = ϟ (koppa)	900 = ϡ (sampi)

ROMAN NUMERALS

Although Roman numerals are over 2,000 years old, people still use them today, either decoratively (for example, on clocks, cornerstones, and Super Bowl memorabilia) or when numerals distinct from decimal numbers are needed (for example, in outlines). Roman numerals use seven symbols, all of which are capital letters in the Latin alphabet (which pretty much happens to be the English alphabet as well):

I = 1   V = 5   X = 10   L = 50
C = 100   D = 500   M = 1,000

Most numbers are formed by accumulating enough of the symbol that you need. Generally, you list the symbols in order, from highest to lowest. Here are a few examples:

3 = III	8 = VIII	20 = XX	70 = LXX
300 = CCC	600 = DC	2,000 = MM	

Numbers that would contain 4s or 9s in the decimal system are formed by transposing two numbers to indicate subtraction. When you see a smaller symbol come before a larger one, you have to subtract the smaller value from the number that comes after it:

4 = IV	9 = IX	40 = XL
90 = XC	400 = CD	900 = CM

These two methods of forming numbers are sufficient to represent all decimal numbers up to 3,999:

37 = XXXVII	664 = DCLXIV
1,776 = MDCCLXXVI	1,999 = MCMXCIX

Higher numbers are less frequent, but you form them by putting a bar over a symbol, beginning with a bar over V for 5,000 and ending with a bar over M for 1,000,000. The bar means that you need to multiply by 1,000.

MAYAN NUMERALS

Mayan numerals developed in South America during roughly the same period that Roman numerals developed in Europe. Mayan numerals use two symbols: dots and horizontal bars. A bar is equal to 5, and a dot is equal to 1. Numbers from 1 to 19 are formed by accumulating dots and bars. For example,

3 = 3 dots

7 = 2 dots over 1 bar

19 = 4 dots over 3 bars

Numbers from 20 to 399 are formed using these same combinations, but raised up to indicate place value. For example,

21 = raised 1 dot, 1 dot (one 20 + one 1)

86 = raised 4 dots, 1 dot over 1 bar (four 20s + one 5 + one 1)

399 = raised 4 dots over 3 bars, 4 dots over 3 bars (nineteen 20s + three 5s + four 1s)

As you can see, Mayan place value is based on the number 20 rather than the number 10 that we use. Numbers from 400 to 7,999 are formed similarly, with an additional place — the 400s place.

Because Mayan numerals use place value, there's no limit to the magnitude of numbers that you can express. This fact makes Mayan numerals more mathematically advanced than either Egyptian or Roman numerals. For example, you could potentially use Mayan numerals to represent astronomically large numbers — such as the number of stars in the known universe or the number of atoms in your body — without changing the basic rules of the system. This sort of representation would be impossible with Egyptian or Roman numerals.

BASE-2 (BINARY) NUMBERS

Binary numbers use only two symbols: 0 and 1. This simplicity makes binary numbers useful as the number system that computers use for data storage and computation.

Like the decimal system you're most familiar with, binary numbers use place value (see Chapter 2 for more on place value). Unlike the decimal system, binary place value is based not on powers of ten (1, 10, 100, 1,000, and so forth) but on powers of two (2^0, 2^1, 2^2, 2^3, 2^4, 2^5, 2^6, 2^7, 2^8, 2^9, and so on), as seen in Table 24-3 (see Chapter 3 for more on powers).

Table 24-3   Binary Place Values

512s	256s	128s	64s	32s	16s	8s	4s	2s	1s

Notice that each number in the table is exactly twice the value of the number to its immediate right. Note also that the base-2 number system is based on the *base* of a bunch of exponents (see Chapter 3, which covers powers). You can use this chart to find out the decimal value of a binary number. For example, suppose you want to represent the binary number 1101101 as a decimal number. First, place the number in the binary chart, as in Table 24-4.

Table 24-4   Breaking Down a Binary Number

512s	256s	128s	64s	32s	16s	8s	4s	2s	1s
			1	1	0	1	1	0	1

The table tells you that this number consists of one 64, one 32, no 16s, one 8, one 4, no 2s, and one 1. Add these numbers up, and you find that the binary number 1101101 equals the decimal number 109:

$$64 + 32 + 8 + 4 + 1 = 109$$

To translate a decimal number into its binary equivalent, use whole-number division to get a quotient and a remainder (as I explain in Chapter 3). Start by dividing the number you're translating into the next-highest power of 2. Keep dividing powers of 2 into the remainder of the result. For example, here's how to find out how to represent the decimal number 83 as a binary number:

$$83 \div 64 = \mathbf{1} \text{ r } 19$$
$$19 \div 32 = \mathbf{0} \text{ r } 19$$
$$19 \div 16 = \mathbf{1} \text{ r } 3$$
$$3 \div 8 = \mathbf{0} \text{ r } 3$$
$$3 \div 4 = \mathbf{0} \text{ r } 3$$
$$3 \div 2 = \mathbf{1} \text{ r } 1$$
$$1 \div 1 = \mathbf{1} \text{ r } 0$$

So the decimal number 83 equals the binary number 1010011 because 64 + 16 + 2 + 1 = 83.

BASE-16 (HEXADECIMAL) NUMBERS

The computer's first language is binary numbers. But in practice, humans find binary numbers of any significant length virtually undecipherable. Hexadecimal numbers, however, are readable to humans and still easily translated into binary numbers, so computer programmers use hexadecimal numbers as a sort of common language when interfacing with computers at the deepest level, the level of hardware and software design.

The hexadecimal number system uses all ten digits 0 through 9 from the decimal system. Additionally, it uses six more symbols:

A = 10	B = 11	C = 12
D = 13	E = 14	F = 15

Hexadecimal is a place-value system based on powers of 16, as shown in Table 24-5.

Table 24-5   Hexadecimal Place Values

1,048,576s	65,536s	4,096s	256s	16s	1s

As you can see, each number in the table is exactly 16 times the number to its immediate right.

Here are a few examples of hexadecimal numbers and their equivalent representations in decimal notation:

$$3B = (3 \times 16) + 11 = 59$$
$$289 = (2 \times 256) + (8 \times 16) + 9 = 649$$
$$ABBA = (10 \times 4,096) + (11 \times 256) + (11 \times 16) + 10 = 43,962$$
$$B00B00 = (11 \times 1,048,576) + (11 \times 256) = 11,537,152$$

PRIME-BASED NUMBERS

One wacky way to represent numbers unlike any of the others in this chapter is prime-based numbers. Prime-based numbers are similar to decimal, binary, and hexadecimal numbers (which I describe earlier) in that they use place value to determine the value of digits. But unlike these other number systems, prime-based numbers are based not on addition but on multiplication. Table 24-6 shows a place value chart for prime-based numbers.

Table 24-6   Prime-Based Place Values

31s	29s	23s	19s	17s	13s	11s	7s	5s	3s	2s

You can use the table to find the decimal value of a prime-based number. For example, suppose you want to represent the prime-based number 1,204 as a decimal number. First, place the number in the table, as shown in Table 24-7.

Table 24-7   Breaking Down a Prime-Based Number

31s	29s	23s	19s	17s	13s	11s	7s	5s	3s	2s
							1	2	0	4

As you may have guessed, the table tells you that this number consists of one 7, two 5s, no 3s, and four 2s. But instead of adding these numbers together, you *multiply* them:

$$7 \times 5 \times 5 \times 2 \times 2 \times 2 \times 2 = 2,800$$

To translate a decimal number into its prime-based equivalent, factor the number and place its factors into the chart. For example, suppose you want to represent the decimal number 60 as a prime-based number. First, decompose 60 into its prime factors (as I show you in Chapter 7):

$$60 = 2 \times 2 \times 3 \times 5$$

Now count the number of twos, threes, and fives and place these in Table 24-6. The result should look like Table 24-8.

Table 24-8   Finding the Prime-Based Equivalent of 60

31s	29s	23s	19s	17s	13s	11s	7s	5s	3s	2s
								1	1	2

So, 60 in prime-based notation is 112.

Interestingly, multiplication with prime-based numbers looks like addition with decimal numbers. For example, in decimal numbers, 9 × 10 = 90. The prime-based equivalents of the factors and product — 9, 10, and 90 — are 20, 101, and 121. So here's how to do the same multiplication in prime-based numbers:

$$20 \times 101 = 121$$

As you can see, this multiplication looks more like addition. Even weirder is that 1 in decimal notation is represented as 0 in prime-based notation. This makes sense when you think about it, because multiplying by 1 is very similar to adding 0.

INDEX

B

C

Q

Z

About the Author

MARK ZEGARELLI is a teacher, tutor, and co-founder of Simple Step Learning (www.simplestep.co), a web-based educational resource. He is the author of numerous *For Dummies* books, including *SAT Math For Dummies, ACT Math For Dummies,* and *Calculus II For Dummies*. He holds degrees in both English and math from Rutgers University and lives in Long Branch, New Jersey.

Dedication

This book is for my nephew, Jacob Thomas Cianflone.

Author's Acknowledgments

This is my tenth *For Dummies* book and, as always, the ongoing support of my editors has made it a wonderful journey. Many thanks to Lindsay Lefevere, Elizabeth Kuball, Becky Moening, and Chris Mark.

And once again, I pour out endless gratitude to my awesome assistants, Scott Rosenstein and Nick Petruncio, for their constant support and unfailing good cheer.

Publisher's Acknowledgments

EXECUTIVE EDITOR: Lindsay Lefevere

PROJECT EDITOR: Elizabeth Kuball

COPY EDITOR: Elizabeth Kuball

TECHNICAL EDITOR: Becky Moening

PRODUCTION EDITOR: Suresh Srinivasan

COVER IMAGE: Beardwood/Wiley